# Cultural Communications Between China and The Outside World Throughout History

Fuwei Shen

# Cultural Communications Between China and The Outside World Throughout History

上海人民出版社

palgrave macmillan

Fuwei Shen
Suzhou, China

*Translated by*
Rui Lyu
Tongji University
Shanghai, China

ISBN 978-981-97-4695-8 ISBN 978-981-97-4696-5 (eBook)
https://doi.org/10.1007/978-981-97-4696-5

Jointly published with Shanghai People's Publishing House
The print edition is not for sale in China (Mainland). Customers from China (Mainland) please order the print book from: Shanghai People's Publishing House.
ISBN of the Co-Publisher's edition: 978-7-208-14766-9

The publication of the translation work is under the subsidy of Chinese Fund for the Humanities and Social Sciences



This Palgrave Macmillan imprint is published by the registered company Springer Nature Singapore Pte Ltd.
The registered company address is: 152 Beach Road, #21-01/04 Gateway East, Singapore 189721, Singapore

Fuwei Shen has died before the publication of this book.

# Contents

CHAPTER 1

# Cultural Intercourse Between China and the Western Regions Before the Qin Dynasty

## 1.1 Cultural Flow in the Neolithic Age

During the primitive social and matrilineal communal periods, humans formed different groups and engaged in activities in different regions, thus creating various civilizations. Due to the limitation of activity ability, people from different groups have very few opportunities to contact each other, and the information of cultural transmission is fragile, so they have to be in the Paleolithic era, which used coarse stone tools for millions of years. In the Neolithic Age, with the improvement of production capacity, contact between people from different regions was expanding daily, and thus, the pulse of cultural dissemination was also strengthened. Although the creation of civilizations in various areas has its characteristics, ranging from large to small, the product of civilizations is a fundamental activity commonly carried out by people living in different regions with shared cultural traits. Any foreign culture can only catalyze the birth of a new culture.

The Yellow River Basin has entered the ranks of only a few advanced cultures since prehistoric times. The middle and lower reaches of the Yellow River are also places where the ancestors of the Chinese nation have lived, multiplied, and worked for generations. The Book of *Shangshu · Yugong* divided the country into nine states and referred to the middle and lower reaches of the Yellow River, where the political, economic, and cultural centers of the Huaxia ethnic group were located, as China,



F. Shen, *Cultural Communications Between China and The Outside World Throughout History*,
https://doi.org/10.1007/978-981-97-4696-5_1

also known as the Central Plains. It is equivalent to the northern part of Henan, the southern part of Shanxi and Hebei, the eastern part of Shaanxi, and the western part of Shandong, with an area of nearly 500,000 square kilometers. This land covers a size larger than the oldest centers of civilization in the world and has also spread its culture to its vast neighboring areas.

The ancient civilizations of Mesopotamia and the Indus River Valley were relatively close to China's western region. They had long had contact with China's western part, although this contact could only be carried out indirectly and through many intermediate ethnic groups. Scholars from Western countries have written this connection, but it has become the basis for the Western interpretation of Chinese culture. They initially attributed the origin of Chinese culture to the remains of a developed ethnic group from the West. Still, they later believed that the developed culture of Western nations played a decisive role in the emergence and development of Chinese culture. The study of the history of cultural exchange between China and the West has been heavily overshadowed for two to three hundred years due to the theory of Western origin disseminated by Western scholars on the origin of Chinese culture.

In 1758, Joseph de Guignes, a French orientalist, arbitrarily believed that Chinese originated from Egyptian hieroglyphs and was the creation of Egyptian immigrants. In the first half of the nineteenth century, the French sinologist Piot claimed that the oldest ethnic groups in China were only barbaric hunters, and it was not until after 300 BC that ethnic groups from the northwest brought the agricultural economy and pottery-making technology to the local Chinese people. The most typical example is the Englishman Terrien de Lacouperie's "Bak migration theory". In 1894, in his book *Western Origin of the Early Chinese Civilization from 2,300 B.C. to 200 A.D.*, he made the legends in ancient Chinese history a reproduction of the ancient history of Mesopotamia. The chief of the Yellow River Basin ethnic group, Emperor Huang, was transformed into King Nakhunte (Nai Hwang ti), and the Han ethnic group was assumed to be the Bak ethnic group who migrated from West Asia to Xinjiang and Gansu in 2300 BC.[1] In 1913, Ball from Oxford

[1] Terrien de Lacouperie, *Western Origin of the Early Chinese Civilization from 2,300 B.C. to 200 A.D.* London: ASHER&Co., 1894.

University further developed this theory and wrote it in his book *Chinese and the Sumerians*.[2]

Since the twentieth century, the viewpoint of communication theory has been all the rage. The school represented by Elliott Smith regards the ancient civilization of Egypt as the only source of all civilizations in the world. The communication school created a monism of the origin of civilization, took Egypt or other Mediterranean civilizations as the authentic world civilization, and described the European and American cultures that have flourished in the past 500 years as the direct successors of Mediterranean civilizations, casting a paradigm of European civilization-centrism, and artificially elevating the historical status of European civilization. The communication school regards ancient Chinese culture as an extension of Egyptian and Greek cultures, and the proposed transmission route is from the southern part of the Caspian Sea to Shaanxi via Central Asia. Chinese culture is usually included in the cultural circles of Greece, South Russia, and Central Asia. Now, no one believes that Chinese culture originated from Egypt or Babylon, but the "determinism of foreign influence" created by the communication school still casts a shadow over the issue of the origin of ancient Chinese culture. This fallacy started with "Chinese culture from the West" in Western European countries and later in the "foreign information determinism" spread by Soviet academic circles. Both theories make a big fuss about the origin of the painted pottery culture that appeared in the late Neolithic Age and the bronze culture of the Shang Dynasty, trying to place the occurrence of ancient Chinese culture under the wing of a particular Western culture.

"Painted pottery culture from the West" is a hypothesis by Anderson of Sweden in 1923. According to this speculation, the Yangshao culture in the Yellow River Basin is the culture of the Europa people who immigrated from Central Asia and South Russia. In 1921, a late Neolithic site was discovered in Yangshao Village, Mianchi County, Henan Province. The unearthed objects include stone tools, bone tools, and pottery, as well as stone spinning wheels and painted ceramics. The surface of painted pottery is red, polished inside and outside, and there are various painted patterns, such as geometric patterns and Jomon patterns, on the vessel's surface, so it is called painted pottery culture. The excavator Anderson estimated that the Yangshao culture was around 2500

[2] C. J. Ball, *Chinese and Sumerian*. London: Oxford University Press, 1913 (Translator's note).

BC and identified this culture as coming from Anno and Tripoli in Central Asia.[3]Later, Yangshao cultural sites were discovered successively in Shaguotun, Liaoning, and various places in Gansu. The excavator continued to publicize his speculation and compared this painted pottery with the painted pottery culture unearthed in northern Greece, Galicia, Ukraine, Trypillia, and Anau sites, enclosing it in the European Neolithic cultural zone.[4]

The Anau-painted pottery culture relates to the Jeitun culture discovered in South Turkmenistan. The painted pottery unearthed from the Chopan site of the Jeitun culture is earlier than the I-a period of the Anau culture, corresponding to about the end of the 5th millennium BC to the beginning of the 4th millennium BC.[5] The painted pottery of Central Asia is connected with the painted pottery of Iran and South Russia, and the ages are also very different. According to the current carbon determination, the Trypillian culture is slightly earlier than Gansu's late Neolithic period. It appeared in Ukraine in the AD. It entered its prosperous period from the fifth century BC to the fourth century BC.[6] According to the chronological sequence, in the area north of the Caspian Sea, the direction of this spread is not from west to east.

Did Yangshao-painted pottery culture originate from Anau and Trip-illia? Andersen's main argument is that the background color of painted pottery is similar to simple geometric patterns. Even Schmidt, the excavator of Anau culture, thinks it is not easy to convince. There is a big difference in the cultural outlook between the two. The stone and pottery wares produced by scholars in Yangshao each have their characteristics, and the pottery wares are incredibly unique in the shape of the Central Plains culture. The central motifs are not the same in painted pottery decoration, even with some similarities. The age of Yangshao culture

[3] J. G. Anderson, An Early Chinese Culture. *Bulletin of the Geological Survey of China*, 5, 1923, p. 23.

[4] J. G. Anderson, *Children of the Yellow Earth*. London: Kegan Paul, Trench, Trubner & Co., Ltd, 1934; "Researches into the Prehistory of the Chinese", *Bulletin of the Museum of Far Eastern Antiquities* 15, 1943, p. 287.

[5] M. E. Masson and B. M. Masson, The Archaeological Culture of Central Asia's Gold-Stone Age and the Bronze Age, *Journal of World History*, 5(1), 1959. Cited from Chinese version in *Archaeology,* No. 3, 1960.

[6] Hans Quitta, The C14 Chronology of the Central and South-East European Neolithic, *Antiquity*, XII, 1967, pp. 263–270.

has been scientifically identified. The Yangshao Culture site discovered in Banpo Village, Xi'an, in 1954, was determined by radiocarbon from 4115 ± 110 BC to 3635 ± 105 BC. 2 Therefore, Yangshao Culture was not only earlier than Anau Culture but also earlier than Tripillian culture by more than 1000 years. Therefore, the Yangshao culture is by no means from the West, and there is no such thing as "the residents of the plateau brought high-level culture to China". The Central Plains painted pottery culture was produced due to Central Asia's "migration of painted pottery culture residents" (Vasiliev L.S.).

Although "striking similarities" exist between the painted themes of Miaodigou and Tripillian pottery (especially the C1 stage), all Western artifacts similar to the Yangshao culture can only be due to natural formation rather than coincidental. Because the change of ornamentation has its development process, it does not have to be produced through human transmission.[7]

Painted pottery culture is still a hot topic in Western archaeological circles because painted pottery culture generally exists in Central Asia, West Asia, and Europe, and there is a gap between China's painted pottery culture and the Paleolithic Age that needs to be filled. The gap caused the late Neolithic culture to suddenly come to the middle and lower reaches of the Yellow River, so the painted pottery culture has become a kind of coincidence. Now, archaeological discoveries have answered. The Neolithic site discovered in Xianren Cave, Wannian County, Jiangxi Province, in 1962 was dated to 800 B.C., 10,000 years ago, according to carbon dating. The Hemudu culture discovered in Yuyao, Zhejiang Province, in 1975, carbon dated as 5000 to 4750 BC. Mesolithic ruins have been found in Xiachuan, Qinshui, Shanxi, and Lingjing, Xuchang, Henan. Neolithic cultures have also successively found their distribution sites. Extensive excavations and investigations of the Yangshao culture layer by archaeologists in Shaanxi, Gansu, Shandong, Henan, and the lower reaches of the Yangtze River[8] have confirmed that the Shanxi-Shaanxi Loess Plateau is the hometown of the Yangshao culture, so the Yangshao culture is a systematic and distinct painted pottery culture native to the Central Plains.

[7] C. B. Kiselev, The Neolithic and Bronze Ages in China, *Soviet Archeology*, 4, 1960, p. 246.

[8] SU Bingqi, Some Questions About Yangshao Culture, *Journal of Archeology*, 1, 1965, pp. 55–56.

The Yangshao and Longshan cultures developed after they grew and prospered in the Yellow River Basin and its adjacent areas. After archaeological excavation, the ancient cultural relics found in Shaanxi, Gansu, and Xinjiang are generally earlier in the east than in the west, revealing the process that the Yangshao Culture centered in Henan and Shaanxi and the Longshan Culture in Henan (the second phase of Miaodigou in Shanxian County) spreading from east to west.[9] In 1957, at the Majiayao-Wajiaping site in Lintao, Gansu. From 1962 to 1965, in Weiyuan, Lintao, Pingliang, Tianshui, Wushan, and other places, strata accumulation where the Majiayao culture was superimposed on the Yangshao culture was discovered.[10] The Longshan Culture in Henan Province, which was at the same time as the Majiayao Culture, had its early Miaodigou Phase II, its carbon date was 2310–95 BC, and its late Luoyang Wangwan Phase III, its late Luoyang Wangwan Phase III, its carbon dating date was 2000 BC 95 years. The second stage of Miaodigou Culture is between the Yangshao Culture and the Majiayao Culture and is in the transition period of the Yangshao Culture. Its discovery proves that the Shandong Longshan Culture (Black Pottery Culture) is the eastern branch of the Yangshao Culture in the Central Plains, and the Majiayao Culture is the western branch of the Yangshao Culture in the Central Plains. In Gansu, following the Majiayao culture (including the three types of Majiayao, Banshan, and Machang), the Qijia culture coexisted with the Longshan culture. The painted pottery culture was distributed in Gansu and Qinghai after the Yangshao culture and before the Shang and Zhou cultures. Significantly influenced by the Central Plains painted pottery culture.

It is clear that the painted pottery culture of the Central Plains, starting from the Yangshao Culture, spread widely in China. The Shan, Shaya, and Yixing river basins are the terminals of the westward spread of the Central Plains painted pottery culture as we know it now. Archaeological data have firmly pointed out that the Central Plains region in the middle and lower reaches of the Yellow River is the birthplace of ancient Chinese culture. After the birth of the Central Plains culture, it spread to the

[9] An Zhimin, Brief Discussion on the Chronology of Neolithic Culture in China, *Archaeology*, 6, 1972.

[10] A brief report on the archaeological survey of Lintao and Linxia counties in Gansu, *Archaeological Newsletter*, 9, 1958; Briefing on the Clearing-up of Majiayao and Machang Type Tombs in Lanzhou, *Cultural Relics*, 6, 1975.

surrounding areas. It showed a transmission direction from east to west in the Western Region.

Clearly, the painted pottery culture of the Central Plains, which started from the Yangshao Culture, spread widely in China. Hetian, Pishan, Shaya, and Yili River basins in the west of Xinjiang are the terminals of the westward transmission of the painted pottery culture of the Central Plains as we know it now. Archaeological data have eloquently pointed out that the Central Plains region in the middle and lower reaches of the Yellow River is the birthplace of ancient Chinese culture. After its birth, the Central Plains culture spread to the surrounding areas. It showed a transmission direction from east to west in the Western Region.

In recent years, in the cemetery of Tianshan North Road found in the Hami area, we saw the "transitional type" pottery in western Gansu, which fell in the Siba culture between 1700 and 1300 BC. This period was when the ancient European tribes of the Afanasievo culture in southern Siberia and its successor Andronovo culture, migrated southward across the Altai Mountains and Tianshan Mountains to Central Asia. Due to the eastward migration of these Indo-European ethnic groups and the short-headed Eastern Mediterranean ethnic groups of Central Asia, there was the first collision and fusion between the East and the West in China's Xinjiang region. This period coincided with the Shang and Zhou dynasties between 1300 and 1000 BC.

## 1.2 The Earliest Contact Between Hua Xia Culture and the Eurasian Steppe Culture in the Shang and Zhou Dynasties

### *Shang Culture and Karasuk Culture*

The earliest use of bronze in China was a semicircular brass piece around 4500 BC, found in the Jiangzhai site in Lintong, Shaanxi, in 1973. The Majiayao culture in the upper reaches of the Yellow River is one of the earliest areas where copper smelting technology was developed. In 1978, a copper knife containing 6–10% tin was unearthed from the copper smelting site in Linjia, Dongxiang, eastern Gansu. It is a tin bronze ware with an absolute age of 2740 BC, comparable to when the world's earliest bronzeware appeared in the Near East.

In later research, it was clarified that during approximately the same period in the middle and lower reaches of the Yellow River, copper smelting processes mainly consisting of tin bronze and brass were formed, while in the upper reaches of the Yellow River, copper smelting processes ranging from red copper, bell copper to tin bronze were experienced. After the Xia and Shang dynasties, China's copper smelting technology gradually matured, promoting the Central Plains region first to enter the era of civilization.

Shang culture is a bronze culture characterized by the peak of bronze casting, once centered around the Central Plains region and spreading to neighboring regions. The commercial activities of the Shang Dynasty were more active than before, with areas ranging from Hexigten Banner in Inner Mongolia and Karaqin in Liaoning in the North to Ningxiang in Hunan, Changning, and Qingjiang in Jiangxi in the South, from Chenggu in Shaanxi in the west to Haiyang in Shandong in the east.

At present, essential sites of the early Shang Dynasty bronze culture that have been discovered include Erlitou in Yanshi, Henan, Luodamiao in Zhengzhou, Erligang in Zhengzhou, and the Erligang type site in Panlongcheng, Huangpi, Hubei. The bronze casting site discovered in Erligang, Zhengzhou, contains many copper slag and pottery models, with cluster, spear, and knife models being the most common. The manufacturing of bronze tools and weapons has already taken a prominent position. The bronze casting site discovered in Erligang, Zhengzhou, contains many copper slag and pottery molds, with cluster molds, wok molds, and knife molds being the most common. The manufacturing of bronze tools and weapons has taken a prominent position. The green copper crucibles unearthed include the large-mouth container style (clay pottery) and the large-mouth jar style (sand pottery). The objects melted and cast include common vessels such as tripods, cooking pots, wine vessels, ampullas, and large bronze weapons such as knives, dagger-axes, spears, and battle axes.

The patterns on copperware include Taotie designs, dragon designs, cloud designs, dot designs, triangle designs, and vortex designs. The cast objects are mostly magnificent in shape and gorgeous in design. The production of bronze ware in the late Shang Dynasty, whether in terms of technology or art, has entered a period of maturity and reached a stage of perfection. The superb casting technology and the grand scale of smelting reached a technological height that surpassed Egypt and Babylon, which had already entered the Bronze Age. Moreover, the

Shang Dynasty bronze ware casting used a piece-mold casting method,[11] utterly different from the Near East, West Asia, Central Asia, and Siberian copper techniques that used the lost-wax casting method. It formed its system in the history of bronze technology development.

At the same time as the Xia Shang bronze culture, there are two bronze cultural centers in the northern region of Asia. One is the Afanasievo culture (about 2000–1500 BC), the subsequent Andronovo culture (1500–1200 BC), and the Karasuk culture (1200–700 BC) developed in the Minusinsk Basin in the upper reaches of the Yenisey River in southern Siberia; The other is the Namazga V period and VI period culture (2000–1000 BC) with a long history centered around south Turkmenistan. So far, no Bronze Age sites have been found in the vast area south of 50 degrees north latitude and west of southern Turkmenistan.[12] Southern Turkmenistan is known for its red copper culture,[13] so the academic community recognizes that the early Bronze Age in Central Asia and Siberia began in the fifteenth to thirteenth centuries BC,[14] roughly parallel to the Early Shang culture. In the southern region of Central Asia, from the east of the Caspian Sea to the lower reaches of the Zeravshan River, until the late Bronze Age in the second half of the second millennium BC, red bronze artifacts were still smelted, with daggers, knives, spears, pins, cones, and triangular arrowheads being widespread. This region is connected and interacts with the Andronovo culture in Siberia, the southeastern European culture in the Caucasus, and the lower reaches of the Volga River through the grassland bronze culture in northern Central Asia. They may have had some contact with Tangshan in Hebei,

[11] Noel Barnard, *Bronze Casting and Bronze Alloys in Ancient China*, Routledge, 1961, Ch. 5.

[12] А. Монгант, *Археология СССР*, Институт археологии РАН, 1955, 387.

[13] Southern Turkmenistan is the earliest and most prosperous metallurgical center in Central Asia. The Chalcolithic period began in 4000 BC and is very similar in culture to northern Iran. Kuftin is divided into the first to sixth periods of the Namazga culture, with each period mainly featuring bronze artifacts. In 3000 BC, metal artifacts began to spread in Central Asia. The characteristic of the cultures of South Turkmenistan and North Iran is that the majority are made of red bronze, while bronze is much less common. Even during the development period of the Bronze Age, the fourth and fifth periods of the Namazga culture were still dominated by red copper, with brass and bronze only occasionally appearing.

[14] A. P. Okladnikov, *Ancient Population of Siberia and Its Cultures,* Cambridge, MA: Peabody Museum, 1959, pp. 22–23.

Ejin Horo Banner in Inner Mongolia, and the Hexi Corridor after 200 BC, where red copper culture was discovered. From the end of 200 BC to the beginning of 100 BC, the Chust culture was forming in Fergana. The remains of the Naryn Pass indicate a certain degree of connection between this culture and Chinese culture.[15] Small pieces of bronze found in the ruins of Balikun, Turpan, Lopnur, Yiwu, and other places have suggested that Xinjiang residents have played a role as cultural communicators between grassland nomad and nomadic settled peoples through the grasslands and desert roads that run across the north and south of Tianshan Mountains.

The discovery of the Erlitou culture and the excavation of the Erligang Early Shang site in Zhengzhou provide a positive answer to the origin of Chinese bronze culture in the Yellow River basin. From a comparative chronological perspective, it is possible to distinguish the nature of its connection with the bronze culture of Siberia.

The attempt to describe the bronze culture in the Yellow River Basin as originating from the input of Siberian metallurgy[16] or to argue that the high achievement of bronze culture in the late Shang Dynasty was due to the influence of the Seima culture on the shape changes of certain artifacts during the Shang Dynasty, which led to significant development and improvement in the quality of production of Anyang type artifacts.[17] Although this argument does not believe that the bronze culture of the Shang Dynasty originated from foreign ethnic groups, it also exaggerates the influence of European civilization.

The metallurgical technology of the Afanasievo or Andronovo cultures in Siberia is minimal and cannot have any noteworthy impact on the metallurgical technology of the distant Yellow River basin. The bronze cultural center of Minusinsk in Siberia was influenced by the Saima culture of the Volga River and Og River in Europe after 1600 BC. Seima culture was discovered in 1914–1922 and appeared between 1600 and 1300 BC. Some factors in Seima culture spread eastward to Baikal Lake through

[15] В. М. Массон (Ленинград). Изучение энеолита и бронзового века Средней Азии, *Советская археология*, СА, 1957, № 4, 53.

[16] Max Loehr, *Chinese Bronze Age Weapons: The Werner Jannings Collection in the Chinese National Palace Museum, Peking*, Ann Arbor: University of Michigan Press, 1956, pp. 103–105.

[17] Kiselev, The Relationship between Bronze Culture in the Soviet Union and Chinese Merchant Culture, *Archaeology*, (2), 1960.

Minusinsk and southward to southern Inner Mongolia, where they met with the Shang culture. The influence of the Seima culture extends from west to east and can be seen in the Andronovo culture of West Siberia, the Yenisei River basin, and the Altai region. The Karasuk culture found in the Yenisei River basin in 1929 completed the southern spread of Saima culture from Outer Baikal. Karasuk culture and the culture of the late Shang Dynasty in Anyang have a lot in common. The early Karasuk culture introduced white jade rings and curved knives from the Saima culture into Anyang, but the differences are only secondary. The decoration on the bronze hilt end of the Saima culture is a wild horse, while the decoration on the bronze hilt end of the Shang dynasty is a horse head. The Seima culture hollow axe has an inverted triangular pattern, while the Shang Dynasty copper axe has a Taotie pattern.

The influence exerted by Shang culture on Karasuk culture, with its superb pottery and copper casting technology, far exceeds that of the northern culture it absorbed. Among the relics of the Karasuk culture found in the Transbaikal region are more than 30 pieces of pottery tripods and pottery stews, and the artifacts unearthed at the same time include typical specimens of the Karasuk culture—bronze scimitars and semicircular ornaments. This semicircular ornament was discovered in Ordos, and its transitional form appeared in Mongolia. It spread northward from the Hetao area. The more decadent remains of the Karasuk culture in the Yenisei River Basin are due to the direct influence of the same type of artifacts from the Shang culture. The bronze knife and the bow with two curved ends and a straight middle are very similar to those found in Anyang. The crank sword that appeared in the late Shang Dynasty (unearthed in Linzheyu, Baode, Shanxi Province in 1971) was made in northern China, and its distribution area reaches as far as Mongolia and the Transbaikal region. Still, it is not seen in Liaoning and Minusinsk regions. The objects from the Outer Baikal region and Mongolia belong to the late Karasuk culture. The concave daggers of the Karasuk culture, the beast-shaped daggers, and the curved daggers of the Tagar culture (800–100 B.C.) are similar to the willow-leaf swords widespread in the Yellow River Basin and northern China during the Western Zhou Dynasty.[18] The dissemination of commercial culture in the north reached the upper reaches of the Yenisei River and the Altai

[18] Wuen, About Bronze Short Swords in Northern China, *Archaeology*, 5, 1978.

region. In the Waibaikal Lake region, local bronze and pottery were affected. At the same time, in the Yenisei River basin, this influence was limited to the form of bronze ware.[19] The representative artifacts in the Minusinsk funerary objects are believed to have been brought by the southeast Chinese northern ethnic groups when they migrated, especially the swords and daggers, as well as the short swords and some ornaments, which were brought from China by these northern ethnic groups.

### *King Mu of Zhou's Western Tour and Horse Riding Culture*

At the beginning of the Zhou Dynasty, there were a total of nine countries living west of the Huaxia ethnic group. Through these ethnic settlements, Zhou had contact with the ethnic groups west of Congling. During the reign of King Cheng of Zhou (1042–1021 BC), the rebellion of the enslavers in the Yin Dynasty was quelled, and neighboring ethnic groups came to pay tribute. The influence of Zhou extended to the northwest region, making the pottery li, bronze age, and arrowhead unearthed in the Hexi Corridor and Huangshui Basin during this period similar to similar artifacts in the Central Plains of the Western Zhou Dynasty. Zhou maintained a particular connection with the northern and southern Tianshan Mountains in Xinjiang, and Han immigrants also arrived east of Congling. Since then, China has learned to raise and ride horses from the horse riding ethnic group in Congling. The domestication of horses and the resulting equestrian culture were initially infused into the civilization center of Mesopotamia by the Kassite people who rode horses. The Kassite people and the Babylonian dynasty were at the same time, and after the fall of the Babylonian dynasty, the Kassite people continued to rule the region for 400 years. Later, the Hyksos, Hittites, Egyptians, Assyrian people, and Persians all domesticated horses and linked chariots with horses. The horse riding culture spread eastward. The nomads living in Tianshan Mountain and Hexi Corridor learned to ride horses first in China.

However, the connection between Zhou and the northwest region was always constrained by the Rongdi tribe. During the reign of King Mu of Zhou (976 BC–922 BC), the power of Quanrong expanded and hindered the communication between Zhou and the tribes of the northwest region.

[19] С.В. Киселёв, Древняя история Южной Сибири. М.: Изд-во Академии Наук СССР, 1951. 145.

Therefore, King Mu marched westward toward Quanrong and opened up the path to the grasslands of the northwest. King Mu of Zhou's Western Expedition was recorded in the "*Bamboo Annals*" and "*Records of the Grand Historian*". These records reflect the strong connection between the Yellow River Basin and the upper reaches of the Sil River in Central Asia after the tenth century BC. The road from Guanzhong to the west, once passing through the Hexi Corridor in the north of the Qilian Mountains and once passing through the Qaidam Basin in the south of the Qilian Mountains, has established connections with Xinjiang. The grassland road on Tianshan North Road and the oasis road on Tianshan South Road have become the main transportation routes between China and the West. The opening of this channel relied entirely on the nomadic tribes of the grasslands that traveled through this area. Through these nomadic Qiang people, Sai people, and later Yue Zhi, the people of various ethnic groups in Central Asia and the people of different ethnic groups in northwest China established friendly trade relations, and they had closer economic and cultural ties with each other than before. The Biography of Emperor Mu recorded that King Mu of Zhou presented silk, bronze ware, and shell coins to various tribal chieftains every time he went, and chieftains from different regions also gave him a large number of horses, cows, sheep, and wine. The mass eastward transportation of jade from Xinjiang and the westward transmission of silk and bronze from the Central Plains region became essential transport between China and the West during this period. These goods' westward transportation terminals crossed the Congling Mountains and extended to the Urals, Polo grasslands, and the Iranian Plateau.

## 1.3 The Emerging Silk Road in the Spring and Autumn Period and the Warring States Period

### *Migration of Scythians*

During the Spring and Autumn Period (770–476 BC), the state of Qin became stronger and stronger. Duke Mu of Qin (659–621 BC) conquered the eight kingdoms of Xirong and expanded the territory to the northwest to develop trade relations. Qin traded silk and metal objects for livestock, fur, and jade from nomadic tribes. During the Warring States period (475–21 BC), the Eight Kingdoms of Xirong were successively annexed by Qin, which allowed for smooth transportation between Qin and the Hexi Corridor.

To the west of Xirong was a nomadic tribe, Rong, surnamed Yun, who lived near Dunhuang. They were initially Scythians and lived in Guazhou for generations. During the Spring and Autumn Period, the Yue ethnic group became more muscular and drove the Scythians westward. At this time, the Scythians moved westward along the Tianshan Mountains and scattered in the north of the Tianshan Mountains, including the vast grasslands from Altai Mountain to the Balkhash Lake. Some farmers and herders settle in the oasis around the Talimu Basin south of the Tianshan Mountains. Most of them use East Iranian. The Yuezhi clan, who lives east of Xinjiang and the Tianshan Mountains, also speaks Eastern Iranian and is of the same lineage as the Scythians who speak Northern Iranian in northern Central Asia. In the eighth century BC, some Scythians who lived north of Central Asia migrated from the northwest of Central Asia to the northwest of the Black Sea under pressure from the Sarmatians, who also spoke Northern Iranian. They established frequent trade relations with the Greek colonial city-states in the Black Sea in the sixth century BC. The road from the northern foothills of the Tianshan Mountains to the Central Asia and Kuban grasslands appears unusually smooth due to the use of these northern Iranian-speaking grassland herders as intermediaries. The Scythian tribes, through their nomadic methods, served as the oldest silk traders between China and distant Greek city-states, and the Kyrgyz and Ross grasslands they roamed became the earliest places along the Silk Road.

### *The Westward Transportation of Silk*

China was the first country in the world to raise silkworms and weave silk, and until the sixth century AD, it was the only country to raise silkworms (Bombyx mori) and weave silk tapestry. In 1958, silk sheets, ribbons, and silk threads woven from silk were discovered on the lower floor of the Neolithic site in Qianshan, Wuxing, Zhejiang Province. This site is more than 4700 years old, around 2800 BC.

The Shang Dynasty made significant progress in sericulture and silk reeling, becoming a handicraft industry with substantial achievements. The silk-weaving industry made considerable progress during the Spring and Autumn period and the Warring States period. In addition to using flat or oblique horizontal looms to weave various jacquard fabrics, there were also flat brocade looms that used different colored silk threads to weave colorful brocades. Colorful embroidery was also applied to silk fabrics, resulting in various brocades. These unique and exquisite clothes attracted the nomads in the northwest and became an essential exchange of equal values. They played the role of currency among the people of all ethnic groups and, through them, flowed to all parts of the Eurasian grassland.

Since the Shang Dynasty, silk fabrics have been sold to the outside world in batches. In the late Spring and Autumn Period, the northern nomads Yiqu, Linhu, and Loufan, who came to the frontier fortress, and the Huns who rose in Mobei during the Warring States Period were accustomed to using horses. They had close economic and cultural relations with the Yuezhi and the Scythians in the west. Many skilled technicians and artisans proficient in astronomy and silk weaving have migrated in batches to the northwest ethnic minority areas. As a result, astronomy, calendars, and silk-weaving techniques in the mainland spread further to the West through the northwest region. A batch of Chinese-made silk fabrics was unearthed from the noble stone tombs discovered in the Katun River, Burial River, Ursul River, and Ulakhan River basins in the Altai region. The time of these tombs roughly belongs to the fifth century BC, equivalent to the Spring and Autumn period and the Warring States period. Qin-style mirrors were also found in tombs dating back to the fourth century BC. A unique four-wheeled carriage has been excavated from the 5$^{th}$ tomb of Pazyryk, which is believed to come from China based on its shape and the silk fabric used for burial. Silk fabrics and lacquerware from the Spring, Autumn, and Warring States periods were also discovered in 1977 at the eastern pass of Ala Gou (Toksun West) in Xinjiang. The diamond-patterned chain-style silk unearthed was a silk-weaving treasure that had just been found on the mainland. Due to export, it has been transported along the Silk Road to the foothills of the Tianshan Mountains. The owner of the vertical tomb at the east entrance of Ala Gou is a Scythian.

Along the Eurasian Grassland Silk Road, early Chinese silk artifacts have been unearthed in the cities of Sapalli Tepe in southern Uzbekistan and Palmyra in Syria. Among the 138 tombs in Sapalli Tepe, 25 have found genuine silk fragments dating back to 1700–150 BC. Since 1925, French archaeologist Pfister, R., who excavated at the Palmyra site, has discovered 40 to 50 pieces of Chinese silk fabrics, all of which belong to the products of the Han and Jin dynasties.

At the western end of the Silk Road in Greece, due to the discovery of thin and transparent clothing worn in sculptures and painted pottery portraits, some speculate that in the fifth century BC, Chinese silk had become the favorite clothing among Greek upper-class figures. These statues include the "Fate Goddess" of the Parthenon Temple (38–431 BC) and the caryatids of erechtheion, all dressed in transparent and soft silk chitons.[20] Similar silk fabrics are also found in Greek paintings. Fragile fabrics were already found on the batch-produced red pottery pots in Athens in the fifth century BC.[21] In the tomb of the wealthy Alcibiades family in the northwest work district of Athens, six pieces of silk fabrics and a bundle of thread that can be divided into three strands were unearthed. After identification, these silk fabrics were woven from Chinese silk and belonged from 430 to 400 BC.[22]

The earliest Chinese silk unearthed in Europe so far is a German noble tomb in the mid-sixth century BC. A locally made woolen sweater with Chinese silk mixed in its wool and decorative patterns was discovered in Tomb 6, excavated in Hohmichele, Baden Wurttemburg, southwestern Germany. Batches of Greek and Mediterranean artifacts were also unearthed from the tomb.[23] These Chinese silk may have been transported into Germany through the Black Sea region, as silk wool blended fabrics were also unearthed from an ancient tomb in the late sixth century BC near Hochdorf-Eberdingen in Stuttgart. These historical facts vividly delineated that the Scythians who galloped on the Eurasian grasslands served as the largest intermediary and trafficker of Chinese silk from the sixth century BC to the third century BC.

[20] G. M. Richter, Silk in Greece. *American Journal of Archaeology*, 33, 1929, pp. 27–33.

[21] *Complete Collection of World Art*, Volume 5, Tokyo, 1950.

[22] E. J. W. Barber, *Prehistoric Textiles: The Development of Cloth in the Neolithic and Bronze Ages with Special Reference to the Aegean*, Princeton University Press, 1992, p. 32.

[23] J. Briard, *The Bronze Age in Barbarian Europe*. Transl. M. Turton. London: Routledge. 1979, p. 213. E. J. W. Barber, *Prehistoric Textiles: The Development of Cloth in the Neolithic and Bronze Ages with Special Reference to the Aegean*, Princeton University Press, 1992, pp. 203–204.

### *Technical and Artistic Exchanges Between China and Central Asia, West Asia, and Europe*

The widespread distribution and interaction of the Scythian tribes on the Eurasian grasslands further developed the connection between China and Europe during the Bronze Age, manifested in similarities in various types of artifacts. After 1000 BC, South Turkmenistan in Central Asia began to enter the Iron Age. The handicraft industry was separated from agriculture. Animal husbandry had new developments, and horses gradually became the main animal husbandry object. At the same time (about the tenth century BC to the fifth century BC), Europe began the first phase of the early Iron Age Hallstatt culture. The bronze double-edged sword with horn handle (also known as a double ring) unearthed in Denmark during the Hallstatt era and the bronze ring-shaped handle sword of Russian scholars are both similar to the same type of bronze willow-leaf sword unearthed at the Chengdu Twelve Bridges Bashu Cultural Site in China in the late Shang Dynasty. The bronze swords of the Hallstatt culture have double-ring handles, half-moon-shaped guards, circular handles, and oval guards, which are different from traditional local products and are influenced by the popular sword style in the Central Plains and northern regions of China. The ring-shaped sword with a head was used throughout the Han Dynasty and profoundly impacted the grassland ethnic group.

Chinese bronzes were introduced into South Russia and Europe, as well as copper or jade ornaments with symmetrical shapes and rich patterns attached to the scabbard belt, bronze axes, and buckled collars. The artifacts introduced from Europe or the Scythians to the East and West to China include animal-shaped pot handles, animal-shaped gold ornaments, gold and silver goblets, bronze spearheads, and saddle buckles.

Horse gear, weapons, and animal motifs constitute the three main characteristics of the peoples of the Eurasian grasslands. Horses were introduced and borrowed by China from the plains of Central Asia. Duke Mu of Qin was the first to organize cavalry, and by the time of King Wuling of Zhao (325 BC–299 BC), he vigorously promoted horsemanship and archery. The Iranian-style iron and ring lock armor replaced bulky rhinoceros skin armor. The weapons of China and the Eurasian grassland ethnic groups have mutual influence, and in some aspects, such as bronze swords, spears, knives, and axes, there is a trend of diffusion from China to the north and west. As for animal patterns, they are also widely popular art in grassland areas and are prevalent in northern China, Mongolia, southern Siberia, Kazakhstan, and even along the Black Sea coast. The origin of animal patterns in the grasslands of north China is exceptionally early, dating back to the late Shang Dynasty. Animal head knives and short swords from the Shang and Zhou dynasties, rectangular decorative plates with group animal patterns from the Spring and Autumn and Warring States periods, tiger-shaped decorative plates curled into rings, and rectangular openwork bands popular during the Han dynasties have been found in the Mongolian, Outer Baikal, and Minusinsk basin. These animal pattern artifacts can be found in their ancestral forms within China. The rich and colorful circular animal designs and various other animal motifs that have been popular since the Spring and Autumn Period and the Warring States period all have the basic themes and artistic styles unique to the northern ethnic groups in China and have no inheritance relationship with the Scythia–Siberian beast patterns.[24]

In the late Spring and Autumn Period and the Warring States Period, the Scythia-Siberian beast patterns and Altai art were absorbed by the plastic arts of animal patterns in northern China, showing the repetition and imitation of specific themes, but they appeared later. The grassland ethnic groups in the north of China, especially after the rise of the Xiongnu during the Warring States period, strengthened their reference to the art of Scythia. Therefore, the Scythia–Siberian animal pattern only spread to the grassland areas in northern China in the late Warring States period (fourth century BC). Since the Neolithic Age, Chinese animal patterns have formed unique themes and artistic styles.

## 1.4 The Westward Migration of "Jichuo" and the Name of China in Western Ancient Books

### *Jade Trade Between Western Zhou State and Xinjiang*

In the pre-Qin era, China emerged as a great power in East Asia. Through the intermediary of northern nomads, it has gradually crossed the Tianshan Mountains and Kunlun Mountains. It is recognized by the surrounding nations west of the Congling Mountains and outside the

[24] Wu'en, Ancient Animal Patterns in Northern China, *Journal of Archaeology*, 1, 1981.

Qinghai Tibet Plateau. Many grassland herders in northwest China are the most critical intermediary ethnic group for conveying this information.

Herodotus (484 BC–425 BC) transcribed a movement of ethnic migration on the Eurasian grasslands after the mid-seventh century BC, based on the long poem *Arimaspea* by the Greek poet Aristeas, which recounted his travels. The surge of grassland ethnic groups also unfolded in the vast pastoral areas in the middle and upper reaches of the Yellow River, posing a threat to the Central Plains dynasty from the north and west. As a result, Duke Huan of Qi (685–643 BC) from the east launched a westward expedition, while Duke Wen of Jin (636–628 BC) and Duke Mu of Qin (659–621 BC) fought against Beidi and Xirong in the Hetao region. As a result of a series of wars, the Guifang people who had invaded Hetao during the Wuding period of the Shang Dynasty gradually migrated westward and finally scattered in the Junggar Basin to the west of Mount Taishan in the fourth century BC.

At the same time as Herodotus, the Yan people in northern China left behind the *Book of Mountains and Seas*. The "Hainei Xijing" describes the ethnic distribution from the Hexi area to the Tarim Basin. The "Wuzang Shanjing" records the places where jade is produced in Xinjiang, and the most famous and large-scale high-quality jade is almost all produced near Hotan. Hotan Jade and Yeerqiang Jade have become the source of jade widely used by the royal family in the Shang Dynasty.

Since the Zhou Dynasty, Hotan Jade has been the western treasure pursued by the royal family. The Zhou people's base is located between the upper reaches of the Wei River and the Jing River, and they attach great importance to horse farming. They rely on this emerging industry to expand their connections with distant areas. Before the establishment of the country, Jichang had a high prestige among the surrounding Qiang and Rong ethnic groups, so they were able to migrate along the jade trade route to the Kunlun Hill at the northern foot of the Kunlun Mountains, where they established a country called Chiwu.

This country of the Chiwu clan was called the "State of Western Zhou" in the "Da Huang Xi Jing" of the *Book of Mountains and Seas* written by the Qi people at the end of the sixth century BC. This Western Zhou state disappeared two hundred years later when the Chu people sorted out the *Haiwan Jing*. Instead, the "Xuanyuan Kingdom" appeared in the *Da Huang Xi Jing*. The Dahuang Xijing and the Haiwai Xijing unanimously agree that the Xuanyuan State has been in power for at least 800 years for a long time. This sentence confirms that since the migration of Jichuo to the West in the twelfth century BC, the Later Shang of the Zhou Dynasty had lived on the edge of the poor mountains (Congling) for so long. It needs to be clarified whether Xuanyuan State was the early Western Zhou State. Still, immigrants from the Central Plains region west of Mount Mildai established more than one country.

### *China in the Eyes of Ancient Persians*

Xuanzang arrived in Tashkurgan in 643, then known as Khabandha. The king claimed to be the son of the Han people and the sun god. Xuanzang recorded the local legend of the country's founding in Volume 1–2 of the *Buddhist Records of the Western World in the Tang Dynasty*. "The Persian King sent an envoy to Han for marriage. When the envoy returned here, the road between east and west was cut off due to wars. Therefore, the Han princess was placed on a solitary peak, and the surrounding area was guarded. However," Every day in the middle of the day, a man rides a horse from the sun to come here. Three months later, the war subsided, and the Persian king sent envoys to greet his bride. However, the Han princess was already pregnant. The courtiers decided to stand by on the spot and build a palace city on the stone peak, with "the princess as the queen". Later, she gave birth to a boy with outstanding talent. He built a close relationship with the neighboring countries. "Based on the delivery of Khabandha ancestors, his mother was a native of Han, and his father was the sun, so he called himself the Han Ri Tian Zhong. The royal family of Khabandha, with the same appearance as China, has a square crown of jewelry and is dressed in Hu clothing". Although the founding legend of Tashkurgan is unknown, combined with the materials of Xuanyuan Kingdom, it can be said that the Chinese royal daughter had already believed in the sun god Mithra in the Persian legend of the during the establishment of Khabandha. It can be traced back to Darius(521 BC–485 BC), who established the Persian Empire, which spanned Asia, Europe, and Africa, with its eastern border near the Congling Mountains and the upper reaches of the Indus River. Therefore, the earliest mention of China as an East Asian country was in Persia.

In the fifth century BC, the word Cini appeared in the Psalms of Fervadin, which had the same meaning as Cinistan and Cinastan used in ancient Persian to refer to China. Xuanzang translated the word into "Zhina(支那)" or "Han(汉)," which was claimed by the descendants of the Zhou Dynasty who migrated to Hetian, Yeerqiang, Tashkurgan, and other places in Xinjiang. Cini, Cinistan, or Cinastan are all the correspondence sounds of the "Zhou" or the "Ji" state. Let's consider this "Ji" state to be the "Qin" state and assume that before the unification of the six states by Emperor Qin Shihuang, people from the Western Regions referred to Chinese people as Qin people. It is entirely unreasonable to compare the historical development of Zhou people's migration in the eastern region of Congling because this system, which has been developed through the Chiwu, Western Zhou, and Xuanyuan states since Jichuo, has nothing to do with the Qin people in the Jingwei River basin. It will also be mentioned below that ancient Indian literature supports that the Ji state refers to the Western Zhou state, not the Qin state.

The "Ji" state is also mentioned in Firdausi's *Shah Namah*. This "Ji" state is located in the Moyu or Shache area west of Hetian. The residents here speak a relatively ancient Tocharian language, which belongs to the western branch of the Indo-European language family, Centum. The Tocharian people initially migrated from West Asia to Xinjiang, China, and then to Afghanistan in the second century BC, dominating Bactria. They and the Scythians who spoke Northern Iranian were the most active ethnic groups in conveying Eastern and Western cultures on the Eurasian grasslands during China's Spring and Autumn and Warring States periods. At least in the fifth century BC, the Persians had already established relations with the Kingdom of Turan, where the Tocharian people lived, and with the Ji state near Congling and Hetian, where the Qiang people lived. The languages of the Zhou and Qiang people belong to the Tibeto-Burmese languages. Still, the residents here did not need to rely on intermediary ethnic groups to deal with the Persians because the Bactrians (Great Xia) and Indians were already subjects of the Cyrus Empire.[25] According to Herodotus, India was the 20th state of the Persian Empire, and Bactria was the 12th state, all adjacent to Xinjiang, China. Therefore, Persians even came into contact with Chinese people earlier than Indians because the eastern side of Congling was the territory of the Zhou Dynasty.

25 Xenophanes, Cyropaedia, I, pp. 1–4.

### *China in Early Indian Literature*

India was also one of the earliest countries to learn about China. In the fifth century BC, after the Akhmani Dynasty occupied Sot, Bactria, and Punjab, it sent caravans to Xinjiang, including Indians. In the two major epics, *Mahabharata* and *Ramashina*, there is the China Kingdom. The *Mahabharata* had been formed at least in the early third century BC. There is China listed in the book, which can be determined that it is located in the Pishan and Shache areas of Xinjiang, China, north of Kashmir.

Written at the end of the second century BC, the *Ramashina* also mentions China. In the *Kishkindhya Kandam*, Sugriva, the Lord of the Monkeys, in giving direction to monkey leaders for the quest of Sita, says the land of the Kosakaras, which is generally interpreted as no other country than China.[26]

Kautiliya, who served in the palace of Chandra Gupta Maurya (324–300 BC), wrote a book called *Arthasastra*, which includes the term "Cinapattasca". This word can be broken down into "Cina" and "patta". Together, it means "bundled silk produced in China". This word was used in the late fourth century BC. This can be considered one of the ancient Indian documents that records China with an exact date. Rather than referring to the Qin state that extended its influence to the south-west region, the term Cinas suggests that as early as *Dahuang Xijing*, immigrants and merchants from southern China entered the northeast part of India, thus passing on the title of China. So Cinas must have been the Zhou Dynasty, not the Qin state that expanded its territory westward during the reign of Duke Mu of Qin (659–621 BC).

Moreover, the modern historical community must recognize that Duke Mu of Qin had a hegemonic career in the Central Plains. The banner that Duke Mu of Qin used then could only be to uphold the dignity of the Zhou family. On the east side of Congling, there were indeed the Chiwu Kingdom and the Western Zhou Kingdom with the Zhou flag, and there was also the Xuanyuan Kingdom to the west, adjacent to India.

[26] Radhakumud Mookerji, *Indian Shipping: A History of the Sea-borne Trade and Maritime Activity of the Indians from the Earliest Times*, Longmans, Green and Company, 1912, pp. 55–56.

In the early translations of Buddhist scriptures, the Sanskrit words related to "China" were translated as "zhina" (支那). However, the translation of "Dafang Guangzhuang Zhuangyan Jing" in the Eastern Han Dynasty has restored Zhina to "Qin". Qin "is often translated as" Han. The translations of Buddhist scriptures during the Six Dynasties were often translated into "Jin" or "Sui" according to the change of dynasties. In the Tang Dynasty, from Xuanzang onwards, "China" in the Sanskrit was usually translated into "Han". In short, the changes in translation since the Qin and Han dynasties aim to adapt to a certain historical era without exploring the origin of this word. Since Martino Martini (1614–1661), an Italian Jesuit scholar in the seventeenth century, modern scholars have translated China back into "Qin" because all Jesuits are familiar with Latin. It is inevitable that they used the later Latin pronunciation of Sinae or Thinae to draw farfetched analogies and gave strained interpretations of the true origin of the name "China" in Persian and Indian Sanskrit. The French sinologist Pothier Paul Pelliot has explained the origin of this term because the power of the Qin Kindom has extended to the Western Regions since the Spring and Autumn period and the Warring States period. However, compared to the history of Chinese development in Xinjiang since the Zhou Dynasty, there are more reasons to believe that the ancient translation of the term "China" comes from "Zhou" or "Ji" and should be longer than the theory of "Qin" or Sinae.

CHAPTER 2

# The Flourishing of Cultural Communication Between China and the Outer World: Han to Tang Dynast

## 2.1 Transportation Between the Han Dynasty and the Western Regions

### *The Western Regions During the Han Dynasty*

After the rise of the Xiongnu during the Warring States period, there were Eastern Hu in the east and the Yue clan in the west, and the Xiongnu were subject to two sides and the Xiongnu even gave their sons to the Yue clan as hostages. The Xiongnu quickly became powerful after accessing the throne in 209 BC. After 1 7 6 BC, the 26 northern states of Wusun and Loulan were under the control of the Xiongnu. The growing strength of the Xiongnu posed a significant danger to the Han Dynasty, and the national power had not yet been restored in the early 60 years of the Han Dynasty, so the Xiongnu could only adopt a policy of compromise, giving Shan Yu a large number of tillers every year, and intermarrying with the Xiongnu. However, the Xiongnu still boasted of being assertive. It possessed millions of elite soldiers who could ride and shoot well. Every autumn ripe, they would ride south to invade, carrying with them the Han's border people and wealth, forcing all parts of the Western Regions to pay taxes, give hostages to the Xiongnu nobles, and prevent commercial exchanges between the Han and Central Asian countries. After Emperor Wu of Han, Liu Che (140–87 BC), came to the throne, he contacted the Western Region countries to obtain their



F. Shen, *Cultural Communications Between China and The Outside World Throughout History*,
https://doi.org/10.1007/978-981-97-4696-5_2

support in launching a counterattack against the Xiongnu. At the same time, to get horses produced in Central Asia, improve military equipment and transportation tools, and export silk that was becoming more and more abundant, it was also urgent to develop relations with Central Asian countries. In 138 BC, Liu Che sent Zhang Qian as an envoy to the Dayuezhi to persuade the king of the Dayuezhi and the Han to unite to defeat the Xiongnu.

Zhang Qian and a posse of a few nationalities, Father Gan, led more than 100 people, set out from Chang'an, were detained by the Xiongnu while passing through Longxi, and lived there for 11 years. Later, he fled by plane and his entourage, and after a long journey of several dozen days, he reached Dawan, Kangju, and Dayuezhi. At that time, the Dayuezhi had established a new king, annexed Daxia, and was already living and working in peace; plus, it was too far away from the Central Plains and had no intention of attacking the Xiongnu. Zhang Qian lived for more than a year from the Yue clan to Baark, the capital of the Great Xia Dynasty, and did not have to return with the essentials. In 128 BC, Zhang Qian returned to China via Nanshan via Qiangzhong (present-day Qinghai) and was detained by the Xiongnu on the way. A year later, the Xiongnu died alone, and civil unrest broke out; Zhang Qian took the opportunity to escape and returned to Chang'an in 126 BC, which lasted for thirteen years and returned with only two people.

Although Zhang Qian did not achieve the purpose of concluding an alliance with the Dayuezhi, he gained knowledge of the geography, property, and military affairs of various countries in the Western Region and understood many Xiongnu internal affairs. Emperor Wu of Han thus knew that dealing with countries in Central Asia and West Asia was not only extremely meaningful in military affairs but also had a lot of benefits for the Han Dynasty economically.

In 119 BC, Wei Qing and Huo Qubing again led a large army to attack the Xiongnu and further expel the Xiongnu to northern Mobei. However, the Han army suffered heavy losses, with a severe shortage of troops and horses. Liu Che, therefore, decided to send Zhang Qian to the Western Regions again in an attempt to take advantage of the contradiction that Wusun originally lived in the Dunhuang area and later had a sharp statement with the Xiongnu and asked them to move back to their hometown by marrying Wusun. On this occasion, Zhang Qian brought with him many deputy envoys, led an entourage of 300 people, each with two horses, so that he could be sent to neighboring countries at any time, and brought tens of thousands of cattle and sheep and tens of millions of gold coins to actively develop friendly relations among the countries of the Western Regions. However, this time, Zhang Qian failed to achieve his expected goal again. The King of Wusun, Kunmo, and the nobles feared the Xiongnu, and the country was divided into three because of the dispute over the throne. Zhang Qian then sent his deputy envoys to Dawan, Kangju, Dayuezhi, Daxia, Anxia, Shendu, Yutian (Hetian), and neighboring countries. When Zhang Qian returned to China, Wusun sent an interpreter to send him and sent an envoy to Chang'an, where he saw that the Han Dynasty was rich and decadent and returned to report truthfully, greatly enhancing the prestige of the Han Dynasty. When Zhang Qian's deputy envoy returned to China, he also brought back many envoys from the countries he visited, and many countries in the northwest had formal foreign exchanges with the Han Dynasty.

After Zhang Qian's envoy to Wusun, the Han envoys mostly took the southern border of Wusun to bypass Lake Issyk-Kul and went to Central Asian countries such as Dawan and Dayuezhi. The marriage of Han and Wusun in 110 B.C. was a blow to the Xiongnu and a significant strategy for developing transportation in the Western Regions. Since then, the envoys of the Han Dynasty have gone even further to rest, Amangcai, Lixuan, Tiaozhi, and Sindhu.

In 60 BC, a coup d'état occurred in the Xiongnu. The rizhu beg Xianxianchan, who was in the western part of the Xiongnu, had to seek asylum with Han. Zheng Ji was appointed as the protector of the Western Regions. The capital was located in Wuleicheng, with a position slightly lower than that of the inland areas. From then on, the Western Region, including northern Xinjiang and the vast area east and south of Lake Balkashi, was formally included in the territory of the Han Dynasty, and Dawan and Wusun to the west and north of Pashui were all under the jurisdiction of Duhu.

The traffic between Han and Central Asian countries enters and exits the Tianshan South Road from Chang'an to Dunhuang through Wuwei, Zhangye, Jiuquan, and Anxi in the Hexi Corridor, and Yumen and Yangguan in Longle County, Dunhuang County, and from there are two north–south roads to the west.

The south road goes out from Yangguan (today's South Lake) to Yixun (today's Milan), Qiemo (today's Qiemo), Jingjue (today's Minfeng),

Juami (today's Cele), Yuyue (today's Hotan), Pishan (today's Pishan), and Shache (today's Shache) in the southwest along the Altun Mountain, from which Mingtiegai Mountain passes through Puli (today's Tashkurgan), and west along the upper reaches of the Panchi River at the northern foot of the Hindu Kush Mountain to Daryueshi (Balk) and Anxi (Hecantomfilus). After going out of Yumen Pass (today's Little Fangpan City) in the north and crossing Bailongdui (Luobu Alkali Beach) in the west to Loulan in the northwest of Luobunor, you must detour the former kingdom of Cheshi (Turpan Siyal Lake) in the north, and take Weixu (Wushaketar in the north of Bosten, Hubei), Yanzhe (today's Yanzhe), Guifa (today's Kuche), Wensu (today's Wushi), Gumo (today's Aksu), Weitou (today's Aheqi) in the southwest to Shule (today's Kashi), By donating drugs (now Ucha), they crossed the Pamir Plateau to Dawan (Fergana Basin) and Kangju (Samar Army and its vicinity). To the northwest of Kangju, one can pass through Yanban (northeast of the Aral Sea); to the south, one can reach the Dayuezhi; and to the southwest, one can get to Arsak.

### *Cultural Exchange Between the Han Dynasty and Western Regions*

The culture and scientific technology of the Han people were advanced in many aspects, and the techniques of making iron weapons and drilling wells were introduced to Central Asia during this period. During the Western Han Dynasty, the length of the iron sword ranged from 80 to 118 centimeters, and the blade of the steel sword was hardened and made solid and sharp. Moreover, there have been ring handle long knives with only one edge, which are sharp and powerful weapons that Central Asian countries lack. The escaped soldiers from the Han Dynasty directly taught iron casting technology to the ironworkers in Dawan, Kangju, and Anxi. In the second century BC, the Fergana people in present-day Uzbekistan learned to cast iron technology from the Chinese and later spread it westward to Russia. The pacifist dynasty also imported steel weapons from China, and the wooden deer was the distribution center of Chinese steel. The weapons used by the pacifist cavalry were imported from here, so Roman historian Plutarch referred to the pacifist cavalry weapons as wooden deer weapons, and the swords used were forged from Chinese steel, known for their sharpness. These Chinese steel once again flowed into Rome through Anxi.

### *The Introduction of Buddhism*

Indian Buddhism was founded in the sixth century BC. Not long after its establishment, information had already spread to China. In the first century BC, Buddhism was introduced to Khotan from Kashmir. Before this, Buddhists had already visited Xianyang. According to Zhu Shixing's "Records of the Sutras" and "Records of the White Horse Temple" in Volume 35 of the "Records of the Buddha", it is mentioned that in 242 BC, 18 people, including Shamen Lifang from the Western Regions brought Sanskrit scriptures to Xianyang.[1] Still, they did not spread them in China. Not long after Buddhism was introduced to Khotan, the Sanskrit texts were copied using unique Chinese written materials but had yet to be translated.

The upper ruling group of the Han Dynasty attached great importance to disseminating Buddhist scriptures. During Emperor Huan's reign in Yanxi (158–166 AD), he worshipped the Buddha, placing equal importance on him as Huang Lao. Since Buddhism has been advocated, translating Buddhist scriptures is particularly needed. Eminent monks from various countries in the Western Regions, such as India, the Moon clan, Anxi, and Kangju, have successively visited the East and collaborated with Han Shamen and Qing believers in Luoyang to engage in the translation of Sanskrit scriptures. It has become ancient China's most significant translation industry from the Han dynasty to the Tang and Song dynasties.

Chinese scholars began engaging in Buddhist research at the Eastern Han Dynasty's end. During the reign of Emperor Xian, the Confucian scholar Mou Rong fled from the Central Plains to Jiaozhi, where he passed on Buddhist scriptures. However, he was criticized and wrote 37 articles called "The Theory of Reason and Confusion", which was the first Buddhist treatise in China and became a precursor to the popularity of Buddhism throughout the country during the Wei, Jin, Southern, and Northern Dynasties.

[1] Please refer to Volume 1 of the "Records of the Three Treasures of the Past Dynasties" and Volume 1 of the "Golden Soup Chapter of Buddha's Dharma".

## 2.2 The Integration of Western Region Culture and Chinese Culture During the Wei, Jin, Northern, and Southern Dynasties

### *The Introduction of Sericulture and Silk-Weaving Techniques to Persia and Byzantium*

The origin of Chinese silk was transported to the West, which not only made silk fabrics a coveted material and decoration for Asian and European countries but also, with the increasing demand, silk became a daily commodity for some wealthy and powerful countries in Western Asia, such as Persia and Byzantium. This inevitably led to the westward spread of sericulture and silk-weaving techniques.

The Central Plains' sericulture and silk-weaving techniques were first promoted in Xinjiang, east of Congling. There is a story in the Tibetan language, "Yu A Guo Shou Ji", about a Chinese silk princess who introduced the method of raising silkworms and reeling silk to Khotan. The king who married the silk princess, YuChi Sheye, was the tenth king of Yu Tian. Some speculate it was around 20 years ago, but it may have been as early as the first century AD.

The technology of sericulture, silk reeling, and brocade production in Xinjiang was introduced to Fergana and Persia in Central Asia. At least in the fifth century, Persia already had its silk-weaving industry. Chinese silk fabrics have always been a model for the Persian silk industry. After becoming a silk-weaving country, a new stimulus was generated for Byzantium, which had been producing a large amount of silk. Therefore, Byzantium also tried to learn how to raise silkworms and silk reeling to solve the problem of raw material sources. At this time, India also learned to raise silkworms and make silk from China, changing the technique of using wild silk to make Qiusheya clothes. Xinjiang has become an outpost for the spread of silk to the West.

Joseph Needham believed that Western jacquard machines were passed down from China, so they were four centuries later than China. According to Forbes research, simple jacquard machines were already used in Persia, Byzantium, Syria, and Egypt in the third century but reached perfection in the late twelfth century.[2] Therefore, jacquard technology in Western Asia and Egypt initially grew under the influence

[2] Forbes: *Research on Ancient Technology*, London, 1956, vol. 4, p. 215.

of Chinese silk-weaving technology, and this influence lasted for several centuries between the second, third, seventh, and eighth centuries.

### *The Eastern Spread of Roman, Indian, and Persian Art*

During the Han and Jin dynasties, with the introduction of Buddhism, Buddhist art centered around Gandhara also began its Eastern Expedition. The grotto temples excavated in western China from the second century AD are relics that reflect the Eastern transmission of Buddhist art in the style of Gandhara and Gupta in Central India. The grotto temples were first excavated from west to east in the southern part of Xinjiang, and the Mogao Caves were built in Dunhuang in the fourth century. Later, Buddhist stone chamber temples were extensively excavated in the northern mainland of China, reaching their peak in the sixh and seventh centuries. Thus, they preserved many Buddhist themes, paintings, and sculptures reflecting contemporary customs, becoming a major art treasure trove in medieval China. The art of Gandhara, which was introduced to China during the Han and Jin dynasties, remained vibrant in Xinjiang until the seventh and eighth centuries. They received a fatal blow only after Tubo occupied various parts of Tianshan South Road. During the Northern Dynasty, from the fifth and sixth centuries, the art of Sassanian Persia became a popular art style in Xinjiang, especially in various parts of Tianshan South Road. It was in the eighth century that it was replaced by painting and sculpture in mainland China. During a long period equivalent to the Han and Tang dynasties, Xinjiang formed two cultural and economic centers, represented by Yuyue and Gaochang, along the ancient north–south transportation routes between China and the West. For nearly a thousand years, these two centers have become outposts for the mutual contact, infiltration, and integration of Chinese and Greek (Roman), Indian, and Persian cultures.

## 2.3 The Exchange of Chinese and Western Culture and Science and Technology in the Tang Dynasty

### *Tang's Communication with South Asia*

China and India had close relations in the early Tang Dynasty over the past 100 years. Not only were Buddhist monks constantly in contact, but official envoys also had frequent exchanges. In the Tang Dynasty, as many as 52 Chinese monks traveled westward to seek enlightenment, while only 16 Indian monks came to China to preach. Due to the increasing number of Chinese monks proficient in Sanskrit, Buddhist scriptures have been directly translated from Sanskrit by Chinese believers, and the interpretation of Buddhist theories has gradually deviated from Indian idioms and begun to have unique insights. In the history of cultural exchange between China and India, Xuanzang, who studied in India for 15 years and returned home with honor, is the most famous one dedicated to translating Indian classics. His name has become a symbol of cultural cooperation between China and India.

Xuanzang (596–664), initially named Chen Hui, was from the Gou family (Yanshi) in Henan province. Before going abroad, he visited famous teachers throughout the country, such as Chang'an and Chengdu, and became familiar with the teachings of Mahayana and Hinayana Buddhism. To further explore the classics of the yoga system and understand the essence of Buddhism, he decided to venture out of Yumen Pass in August 627, risking the risk of illegal immigration. He embarked on a ten thousand-mile-long march toward the Nalanda Temple in Central India.

Xuanzang visited the six sacred sites of Buddhism in Central India. He studied with many famous Buddhist masters, especially at the renowned center of Buddhist learning at Nālanda University. Xuanzang studied logic, grammar, Sanskrit, and the Yogacara school of Buddhism during his time at Nalanda.

In the spring of 643, Xuanzang returned with honors and brought back many scriptures and Buddha statues. He requested permission to return to China from Emperor Taizong of Tang, and Emperor Taizong sent people to welcome him.

After returning to his homeland, Xuanzang devoted himself to translating Buddhist scriptures and organized the 520 packages of 657 Indian

Buddhist scriptures, as well as the theories of logic and grammar that he brought back.

In July 646, according to Li Shimin's instructions, Xuanzang completed 12 volumes of his "Journey to the West in the Great Tang Dynasty" in Central Asia and India. This masterpiece categorizes and narrates the 110 countries and 28 countries that Xuanzang experienced in his over ten years of tourism life, including historical evolution, geographical divisions, ethnic origins, product customs, religious beliefs, language, and writing. The content is extensive, clear, and accurate, making it a research topic in Central Asia, Afghanistan, and other regions—the most essential firsthand information on Pakistan and India's ancient history and geography.

### *Introduction of Chinese Alchemy to Arabia and Western Europe*

Alchemy emerged in China in the second century BC. Alchemy is a discipline that arises from the combination of alchemy and the theory of immortality. Alchemy contributed to experimental chemistry and became the precursor of modern chemistry due to the preparation of elixirs of immortality and the use of drugs with the effect of "turning iron into gold." At the end of the Eastern Han Dynasty, Wei Boyang's "Zhou Yi Can Tong Qi" already knew that mercury could produce lead and lead alloys and mastered the method of mercury smelting. The Book of Changes can also obtain artificial red mercury sulfide, called silver vermilion or spirit sand. Nitrate was first applied in solution, as seen in the "Taiqing Jinye Shendan Classic" in the first century. In the "Jinye", if there is a nitrate, it can dissolve copper in impurities and make sponge-like gold. The Han Dynasty also knew the role of iron and copper salts in metal replacement; in the earliest alchemy work, "Huainan Wan Bi Shu", there is already a saying that "when zengqing gets iron, it turns into copper". During the Eastern Han Dynasty, Gan Shi and his master Han Ziya had a legend of casting gold into the sea. This chemical knowledge was a significant contribution made by alchemists to understand material changes. During the second and third centuries, during frequent trade and maritime transportation between South China, Egypt, and the Red Sea, they were introduced to Alexandria, Egypt.

Although alchemy originated early, it was known as a specialized discipline in Europe after being introduced and passed down by the Arabs in the Middle Ages. Chinese alchemy was highly prosperous during the

Tang and Song dynasties and had close contact with Arab countries. Many Persians in China specialize in selling medicine, gold, and incense, and even some are alchemists. They have played a role in promoting the spread of alchemy to the West. In Yuanhe (806–820), there was a shop set up by alchemist Wang Silang in Chang'an West City, specializing in producing medicinal gold, specifically for Arabian merchants.[3] Li Xun, the younger brother of the literary writer Li Xun, was originally a Persian expatriate who specialized in selling incense and gold elixirs.[4] This knowledge of using chemical methods to make elixirs of immortality and pursuing the "turning iron into gold" potion was introduced to Arabia, which led to the growth of Arab alchemy and its westward spread to Europe.

The founder of Arab alchemy, Jabir ibn Hayyan, served as a teacher for the Crown Prince of Caliph Khalid and had prominent characteristics as a Chinese alchemist. Starting from him, Arab alchemists have pursued the elixir "a l i k s i r" (a l i k s i r), which is the same thing as what Gehong referred to as the elixir essence or divine elixir Chabir's knowledge about mercury comes from China. Chabir's theory on metal composition is based on Chinese alchemists' understanding of the two most critical medicinal materials, mercury and sulfur. His theory on the six metals and the era of their generation is even more consistent with the statement in *Huainanzi Terrain Training*. Later, the teachings of Rhazes (865–925), the great alchemist and physician of Chabir, came down in one continuous line with Chabir. He had met Chinese scholars and passed on the great works of the Greek physician Galen in Arabic to this Chinese scholar who was visiting Baghdad. He possesses a lot of knowledge in alchemy and is directly related to this international exchange. The alchemy of Europe was introduced from Arab countries. Most of the alchemy knowledge of European alchemists Roger Bacon (1214–1292) and Albertus Magnus (1193–1280) was derived from Arab alchemist Lazi and medical expert Avicenna. Lazi's "Secret Code" benefited from two books: *The Book of Changes, Tongqi* and *Baopuzi*. The views of the German great Arzhelt on alchemy can be found in the *Book of Changes, Shen Tongqi* and *Baopuzi*.

Modern chemistry has a continuous relationship with alchemy. Terms such as elixir and zinc oxide are used; It originated from Arabic, which

[3] *Taiping Guangji*, Volume 35.

[4] *Maotingkehua*, Volume 2.

in turn originated from Chinese. Chemistry originally meant alchemy in Arabic, which originated from Arabic *al-kimiya* and was introduced to Spain in the twelfth century. It was translated as *al-Chimia* in Latin and became French *Chimie*, German Chemnie, and English *Chemistry* by the mid-sixteenth century, when the Arabian article was removed. It became modern chemistry in the eighteenth century. But this word originally originated from the ancient Chinese pronunciation of "gold" (kim), borrowed from "gold liquid" (chin-i),[5] which originally meant alchemy. This is similar to the universal solvent *al-kehest* pursued by Arab alchemists, and the "gold liquid" that ancient Chinese alchemists sought to swallow, both of which came from China, the birthplace of alchemy.

Before the classic works of ancient Chinese alchemy were introduced to the international academic interface in the 1930s, Wu Luqiang translated Wei Boyang's "Zhou Yi Can Tong Qi", Ge Hong's "Jindan Pian", and "Huangbai Pian" into English. In 1940, the Ciba Symposia recognized the history of Chinese alchemy spreading to the West. "The basic idea of Chinese alchemy was spread throughout Europe due to India, Persia, Arabia, and Islamic Spain advancing westward. Ge Hong's theories and methods, and even the terminology he used, were widely adopted by alchemists in these countries in the following centuries. If we acknowledge alchemy as a precursor to modern chemistry, then the original theories of Chinese alchemy can be regarded as the earliest norms of pharmaceutical chemistry."[6] This is a fair evaluation by the European chemical history community of the world's role played by China's ancient alchemy.

### *Export of Tang Dynasty Ceramics*

The Tang and Five Dynasties saw significant progress in porcelain-making techniques, with a significant increase in the number of porcelain kilns. Ceramic production was in a period of great development. Porcelain has been included in the bulk goods for export since the mid to late Tang Dynasty, transported from the international trade ports of Yangzhou, Mingzhou, and Guangzhou along the northwest land route and southeast coast to various parts of West Asia and the Persian Gulf, the Gulf of Aden, and the Red Sea.

[5] S. Mahdihassan, *Journal of University Bombay*, 1951, vol. 20, pp. 107–131.

[6] Alchemy in China, *Ciba Symposia*, vol. 2, No. 7, 1940.

Since the ninth century, the export of Chinese porcelain has been seen in Arab works. Ibn Guo Dabei listed famous coastal ports in China in his "Provincial Highway Chronicles" completed around 846, and mentioned the production of porcelain, water, and wrought iron from Bijing in his exports. The knowledgeable and prolific Arab scholar al Jahiz (approximately 779–869) once worked in Basra, and his book "The Way to Make Money" also mentioned an exchange agreement, one of which was to import porcelain from China. Porcelain has become an indispensable item in China's exports. Geographer Ibn Faki listed Chinese silk, Chinese porcelain, and Chinese lanterns as the three major famous brands in his "Geography" (903).

Chinese porcelain has become a popular commodity along the Persian Gulf and the Arabian Peninsula coasts. Among the gifts given by the caliph of Baghdad to a local official in 2001 were 300 pieces of Chinese porcelain. Chinese porcelain is a highly respected and precious item in Arabia. In the ninth century, Tanukhi highly appreciated over 30 Chinese porcelain pots, as they were used to store spices and maintain their fragrance for a long time.

Starting from the end of the eighth century, Chinese porcelain opened up foreign markets, pioneering the large-scale export of Song Dynasty porcelain. Whether it is Central Asia, West Asia, and Egypt that pass through the land Silk Road, or various parts of the Indian Ocean where Chinese and Arab ships travel by sea, there are fragments of Tang porcelain unearthed. Japanese ceramic scholar Mikami Tsugio called the sea route to West Asia the "ceramic road." The ceramic road and the Maritime Silk Road were both new arteries of communication between China and the West during the Tang Dynasty. The export of porcelain directly promoted the development of the maritime industry, bringing it to a new climax.

CHAPTER 3

# Transportation and Cultural Communication between China and the West During the Song and Yuan Dynasties

## 3.1 The Western Expedition of Mongolia and the Intersection of Eastern and Western Cultures

In the twelfth century, Mongolian nomadic tribes emerged in northern China, and world history underwent tremendous changes. In 1206, Temuchin (1162–1227) of the Borjigin clan unified the Mongolian grasslands and was elected as the Mongolian Great Khan, respectfully known as Genghis Khan, meaning "Heavenly Gifted Khan."[1]

In 1218–1223, Genghis Khan led a large army to carry out the first Western Expedition of the Mongol Khanate, launching the wars of annihilating the Huazhizimo (east of the Caspian Sea, south of the Sil River), suppressing the Chincha (west of the Caspian Sea, north of the Black Sea) and Ganros (west of the Volga River in Moscow and Kiev), and conquering the Kangri (north of the Caspian Sea and Aral Sea).

After the death of Genghis Khan, a 250,000-strong army led by the second sons of Shu Chi, Chincha Khan Batu and Subutai, launched their second Western Expedition in 1235–1244, officially fulfilling their plan to conquer Europe. Conquering Chincha, pacifying Russia, defeating

[1] Ala Ad Din Ata-Malik Juvaini, *The History of the World Conqueror*, Harvard University Press, 1958.



F. Shen, *Cultural Communications Between China and The Outside World Throughout History*,
https://doi.org/10.1007/978-981-97-4696-5_3

Poland, and attacking Hungary were all on the second Western Expedition. During the reign of Emperor Möngke, he sent Hülegü to lead the army and carry out the third Western Expedition in 1253–1260. This march enabled the Mongols to occupy the Caspian Sea and Mesopotamia and achieve their military goals of breaking the wood, destroying Baghdad, and seizing Syria. Since then, the Mongol Empire has confronted the Mamluk Dynasty of Muslims on the Euphrates River.

After Ögedei succeeded to the throne in 1229, Haci Helin (near today's Ulaanbaatar, Mongolia) was chosen as the capital. Helin has become a bustling international city. In Helin City, there are not only Uyghur people, Hui people, and Persians but also Hungarians, Freemans, Russians, and even the British and French. After Kublai Khan established his capital in Dadu (Beijing), he built the greatest city in the world at that time. The metropolitan area is home to nobles, officials, guardians, missionaries, astronomers, yin-yang masters, architects, doctors, engineering technicians, musicians, artists, and dancers from all over Asia and Europe.

The greatest traveler in Europe, Marco Polo (1254–1324), also followed his father and uncle Nikolai brothers to Shangdu in the Yuan Dynasty through Hormuz, Bardockshan, Heyue, and Heicheng, and completed his trans-Eurasian journey in May 1275, where he met the Great Khan Kublai Khan. Because of his intelligence, Marco Polo was deeply favored by Kublai Khan and lived in court for 17 years. His footprints in China reached as far as Yunnan and the golden teeth in western Sichuan, and he traveled south of the Dayun River to Fuzhou and Quanzhou. At the beginning of 1291, Marco Polo returned to his native Venice from Quanzhou to Hormuz by sea. A Dehua-made blue and white-glazed vial he brought is still on display in the Italian Museum.

Marco Polo's travelogues shook Europe, introducing China's politics, banknote law, printing, navigation, and shipbuilding for the first time, describing the diversity of an Eastern power, and recording many novelties he saw in Asia. His travelogues attracted a lot of criticism and absurd accusations. Still, more often, it made many Europeans fall for the prosperity and cultural prosperity of the East, pushing Columbus to sail west in search of the Japanese and Chinese cities he was looking for.

Mongolia's westward expedition, the opening of east–west traffic, and the unprecedented activity of international relations have made many foreign nationalities, people from Chincha, Oros, Azul, Kangri, Persia, Syria, Arabia, and European Christian countries emigrate to China. At the same time, Chinese and Mongolians have also migrated to India and the Arab world in batches. The migration and exchanges of these nationalities have strongly promoted the integration and absorption of Chinese and Western cultures. It ushered in a colorful and prosperous period in the cultural history of humanity.

## 3.2 Silk and Porcelain Sold Overseas During the Song and Yuan Dynasties

In the thirteenth and fourteenth centuries, Chinese sailing ships completely controlled the maritime traffic between South India and China. There are three types of Chinese ships that Ibn Battutah saw in Quelon: the large ones are called junks, the medium ones are called chai, and the third one is kakam. The Yuan Dynasty stipulated that each large ship could carry one firewood water boat and one eight-rowed boat, Proof of Inspection and Sailing to Sea with Ships (Volume 104 of the Criminal Law Annals of the Yuan Dynasty). These are the three types of Chinese ships that Ibn Battutah saw. Chinese sailboats have three to twelve sails, each carrying 1000 people, 600 sailors, 400 soldiers, and three small boats. The ship is equipped with weapons to defend pirates in the Malacca Strait and Malaba Sea, including archers, shieldmen, and rocket launchers. This giant Chinese ship has four decks. Marco Polo said that the primary material of the ship was pine, and the keel of the Song ship, which was excavated from the Houzhu area of Quanzhou Bay, was constructed with Masson pine. Ibn Battutah listed as many as twenty oars used by Chinese ships, each requiring ten to thirty people, similar in length to the mast. Marco Polo recorded that Chinese ships often had fifty or sixty cabins, and Jordanians even said they had over a hundred cabins. Ibn Battutah described that these Chinese ships were all built in Quanzhou and Guangzhou. Both domestic and foreign merchants take Chinese ships to and from India and China, and foreign merchants prefer to rent or purchase Chinese seagoing vessels. Therefore, the Yuan government stipulated that "those who buy and sell official ships for their various types of passenger trips shall be selected according to regulations" (Volume 94 of the "History of the Yuan Dynasty: Food and Goods Annals," Market Ships). In the fourteenth century, it was these Chinese sailing ships that became the dominant maritime envoys in the Indian Ocean, engaging in frequent maritime trade with the Islamic world along the coasts of India, southern Arabia, and East Africa, conveying the latest information between the East and the West.

In the early Song Dynasty, China's exports were mainly made of gold, silver, rouleau coins, lead, tin, variegated silk, and porcelain, with silk and porcelain being the two central pillars of exports during the Song and Yuan dynasties.

During the Song and Yuan dynasties, silk was exported from Guangzhou, Quanzhou, and Mingzhou to Japan, Korea, Southeast Asia, South Asia, and Arab countries. The Song Dynasty government encouraged silk production to support the army and the nation. The export volume of silk has grown rapidly. The silk production area known to Idrisi is in Quanzhou (Su Sab), where "the silk produced is exquisite and magnificent, unparalleled in the world." There is also Hangzhou (Janku), which has a glass manufacturing industry and is good at weaving silk. The Zhejiang and Fujian regions are the main bases for the export of Chinese silk fabrics.

Silk is a special kind of trading object, which enjoys a high reputation abroad and plays the role of hard currency. In 1082, the captain of the Srivijaya state presented Borneol and cloth to Sun Hui, who was in charge of Guangzhou's ship trade. Sun Hui reported to the imperial court and ordered it to be purchased into the official treasury at a price. "Upon receiving the goods, we will repay them with preferential silk treatment." In 1219, due to the excessive imports of fragrant herbs and the outflow of gold and silver, silk, brocade, and enamel were used for exchange, promoting the export of various silk products to different overseas regions.

Various exquisite silk fabrics were also gifts used by the Song Dynasty to give to foreign envoys and win the admiration of distant visitors. In 993, Arab captain Li Yawu (Ethiopia, Emirate of Afar) brought Captain Bhumi's letter to Kaifeng to pay homage to Emperor Taizong. Emperor Taizong gifted him with a robe, purple silk wrapped around his head, a pair of gold and silver phoenix bottles, and 20 pieces of silk. Brocade robes and silk bundles were the most popular gifts.

During the Yuan Dynasty, silk was mainly exported through Quanzhou. The variety of styles is unprecedented in history. Spun silk includes flower spun silk, red-green spun silk, finely spun silk, red spun silk, five-color spun silk, red spun silk, dog trace spun silk, mountain red spun silk, and fine-spun silk. There are various colors of damask silk and water damask silk; Satins include dragon satin, grass gold satin, green satin, five-color satin, brocade satin, silk satin, and Suzhou-Hangzhou five-color satin; Brocade includes Jianning Brocade, Danshan Brocade, and Flower Brocade; Silk includes silk cloth, north–south silk, blue silk cloth, Western silk cloth, white silk, and red silk cloth. The colored satin produced in Quanzhou is very famous. The Persians call it zeituni, the Castilians call it setuni, and the Italians call it zetani. Later, it was converted into French *satin* and German *seide*, both from the Latin root *seta* and Italian *zeta*.

These silk fabrics are popular around the world, according to the "Records of the Islanders", starting from Luzon in the Philippines in the east, Maluku, Banda, and Timor in Indonesia, Kozekot in India in the west, Sindh in Pakistan, Cais Island in the Persian Gulf, Basra in Iraq, Mecca in Saudi Arabia in the west, Dushuyat in Egypt on the south coast of the Mediterranean, Tangier in Morocco on the Atlantic coast, and Mogadishu in Mali as far as the south Kilva, Tanzania. The silk and Suzhou-Hangzhou five-color satin produced in Quanzhou were sold as far as Kilva Kisiwani in Tanzania. The Maritime Silk Road extends westward to the Mediterranean Sea, northwestward to Tangui on the Atlantic coast, and southwestward to Kirwa Island, the monsoon trading terminal north of Cape Delgado in East Africa.

Porcelain is also exported from the three major ports of coastal Guangzhou, Quanzhou, and Mingzhou. Since the Tang Dynasty, especially in the late Yuan Dynasty, ceramics have been transported to countries worldwide by the "ceramic road" on the sea. Song Dynasty Chinese porcelain has not only spread throughout Asia and the east coast of Africa but even entered the Mediterranean Sea and Europe. Song porcelain fragments were once published in Italy. Sultan Saladin (1172–1193), Sultan of Ayyubid of Egypt, was famous for his rich celadon. Europeans first learned about Chinese porcelain through the legendary sultan, who often dealt with the Crusaders, so celadon is called Saradon in Europe. The Song and Yuan dynasties were the most prosperous period for China's porcelain export. Porcelain export thrived from the Southern Song Dynasty to the middle of the Ming Dynasty, lasting four centuries from the twelfth to the fifteenth centuries.

Since the Song Dynasty, in all parts of Southeast Asia, especially in the Philippines, Vietnam, Cambodia, Malaysia, and Indonesia, Chinese porcelain has gradually become a daily life tool for local people. Porcelain is also a precious ceremonial and decorative item. The Philippines cherished Chinese porcelain throughout the ages, usually buried in the ground, only to be taken out for use during festivals and then buried deep into the ground after service, so many excavations are complete. In the thirteenth century, Chinese porcelain was commonly used as gifts or decorations throughout Egypt and East Africa, and Chinese porcelain plates, plates, and bowls were often inlaid in the lintels of mosques and the alcoves of the rich living rooms, becoming the most fashionable and luxurious decorative patterns. Since the fourteenth century, with the dumping of large quantities of Chinese porcelain, porcelain has penetrated deeper into every household. The coastal residents of East Africa have switched to porcelain [17], and Arabic faience and Persian glazed pottery have gradually been replaced by Chinese porcelain. This trend was more evident at the end of the Yuan and early Ming dynasties.

The export porcelain of the Song Dynasty was mainly the Yue kiln and Longquan kiln celadon. Yueyao celadon has a long history and began to be exported after the ninth century and ended at the end of the eleventh century. In its place was Longquan celadon. The earliest products started around the eleventh century, and four consecutive centuries from the twelfth to the fifteenth century were bestsellers. In the Song and Yuan Dynasties, the second largest porcelain export to Longquan celadon was blue and white porcelain and white porcelain made in Jingdezhen, Fujian, and Guangdong kilns. The emerging blue and white ware of the Yuan Dynasty was widely used for export and rare in China. During the Song and Yuan dynasties, Quanzhou kilns and Dehua kilns were also widely distributed to countries and regions, including Indochina, the Philippines, Indonesia, the Malay Peninsula, India, Pakistan, the Arabian Peninsula, and the coast of the Persian Gulf.

The celadon and white porcelain, white porcelain, blue and white porcelain, blue and white porcelain, and colored porcelain produced by various kilns in China are widely sold to all countries in the world. Asian countries such as Japan, North Korea, the Philippines, Malaysia, Thailand, Brunei, Indonesia, India, Pakistan, Iran, Iraq, Syria, Saudi Arabia, and Yemen; African countries such as Egypt, Sudan, Ethiopia, Sumali, Kenya, Tanzania, and even as far as Zimbabwe and South Africa have produced Chinese ceramics in varying quantities. Japan and the Philippines are the most found among them, and Malaysia, Iran, and Egypt are also considerable as far as Kilwa, Tanzania; a large number of Chinese porcelain from the Shishi Shi include celadon from Longquan kiln, blue and white porcelain and blue and white porcelain from Jingdezhen kiln, white-glazed ochre porcelain from Cizhou kiln and Dehua kiln from Cizhou kiln.

The Forstat ruins in the southern suburbs of Cairo, Egypt, are an underground ceramics museum; an estimated 6.7 million ceramic fragments are found. The Japanese ceramist Fujio Koyama classified it so that 12,000 Chinese ceramic pieces can be re-recognized. These fragments include Tang to early Song dynasty Yue kiln celadon, Tang to understand porcelain, Song Yuan Longquan celadon, Yuan blue and white porcelain, Ming and Qing dynasty blue and white porcelain, and Tang Sancai. Sarawak, East Malaysia: since the twentieth century, excavations around Santubang and its nearby Sungei Buah, Sungei Jaong, and Great Caveat Niah in the Sarawak River Delta have yielded a lot of Song, Yuan, and Ming porcelain and more than 1 million pieces of porcelain have been excavated in the Sarawak Museum alone for more than a decade. Song porcelain has blue and white porcelain from Fujian Dehua, Quanzhou, Guangdong Chao'an, Guangzhou West Village, celadon are Fujian, Zhejiang coastal porcelain kiln products, black porcelain to build kilns most, Cizhou kiln-carved porcelain, and Fujian, Guangdong porcelain products are similar. In the Yuan Dynasty, Dehua kilns also produced a lot of materials.

The quantity of Chinese porcelain unearthed in the Philippines is considerable, and the kiln mouth is complex. Some products are even rare in China. Unearthed porcelain belonging to the Song Dynasty is relatively rare, including Yue kiln-carved celadon, Longquan kiln-carved five tube and plum vases, Fujian Quanzhou and Dehua carved blue-white porcelain, blue-white brown spot porcelain, and Fujian black porcelain. The Yuan Dynasty had the most significant number of porcelain scholars. Mainly produced by Jingdezhen, there are varieties such as blue and white, blue and white, and underglaze red. Blue and white porcelain, adorned with brown spots, is a small vessel that may be exclusively for export to the Philippines. No specimens in the Jingdezhen ancient kiln site have been found and are rarely seen in China and other countries. Small pieces also dominate Zhejiang Longquan celadon. Fujian Dehua Kiln products include printing bottles, pots, plates, washing boxes, and other utensils. Many low-temperature green lead glaze printed flower plates and military holding utensils from Quanzhou Kiln in Fujian exist. Military holding utensils are customized fired porcelain, and there are also black glaze carved flower vases, jars, furnaces, and other utensils fired with imitation red and black carving techniques.[2] The porcelain collected by

[2] Feng Xianming, Investigation on the Sales of My Porcelain to Asia before the Yuan Dynasty, *Cultural Relics* (6), 1981.

Filipino scholars and collectors is quite complete, ranking second to none among Southeast Asian countries.

Two museums in West Asia are renowned for their exquisite and rich collection of Chinese porcelain. One is the Ardebil Shrine in northern Iran, and the other is the Serario Palace southeast of Istanbul, Turkey, now known as the Topkapu Sarayi Muzesi Museum.

During the Yuan Dynasty, porcelain flowed into Europe through Hormuxian, Levant, Egypt, and Morocco, becoming a daily tool that swept across Asia, Africa, and Europe, replacing pottery, bronze, and glass. The blue and white-glazed small vase unearthed from the Jiachunling Kiln in the third class of Dehua, Fujian, was sold to Japan and the Philippines and brought back to the country by Italian traveler Marco Polo. There is still one piece stored in an Italian museum to this day. The export of Chinese porcelain ranges from bowls, dishes, cups, plates, and cover holders to bottles, bowls, basins, washing, pots, jars, stoves, lamps, and boxes, with various designs. Four series jars, five tube jars, double ear jars, plum jars, military holding jars, round plates, and holding pots are popular in various countries overseas. The patterns and decorative techniques of Huaci have become masterpieces in the world of arts and crafts. The designs of lotus petals, scorched leaves, tangled flowers, peonies, chrysanthemums, pine and bamboo, plum blossoms, crabapple, and camellia are popular artistic images in Asia and Africa. Since the ninth century in Egypt and the eleventh century in Iran, they have followed the example of Chinese porcelain, but their products are still difficult to match those of their origin.

The influence of Chinese porcelain goes beyond this. In East Africa, the excavation of Huaci led to a re-understanding of the history of coastal areas that had long been forgotten. The excavation of Chinese porcelain allows the history of the Lamu Islands to be traced back to the ninth century. Chinese porcelain is also the key to opening the gate of East Africa in the past. Many pieces of Chinese porcelain unearthed in 56 sites in Tanganyika, especially 400 pieces of Chinese porcelain found in 30 of the ruins, make British archaeologist Moche Weller happy about the past of Tanganyika, which has been buried underground for a long time since the tenth century, and can rely on Chinese porcelain to form a series of articles. Chinese porcelain has also exposed many ancient cities, islands, and harbors in East Africa, from their slumber to their glorious past. The ancient port of Gedi in Kenya, the ancient site of Kilindni in

Mombasa, Wenguja Kuu on Zanzibar Island in Tanzania, and Mafia Kisiwani on Mafia Island, as well as the early history of Kilva Island, are all closely related to the unearthed Chinese porcelain. This shows that the East African coast, which seems to have been outside the world's major economic regions for a long time, has already had multiple exchanges with the mainland and islands in the eastern Indian Ocean.

## 3.3 Arabs Are the Bridge for the Spread of Chinese Culture to the West

China's four great inventions (papermaking, printing, gunpowder, compass) were introduced to Europe through the Arab world in the Middle Ages, exerting immeasurable power in the world and promoting the process of civilization history. Similarly, Muslim craftsmen's borrowing of Chinese silk and imitating Chinese porcelain also provided examples for future European development in this regard. The westward spread of Chinese culture was greatly accelerated by the Mongols' westward expeditions and the establishment of unprecedentedly large khanates in Asia and Europe.

Both land and sea transportation underwent significant improvements and development in the thirteenth century. Between Qara-Qorum, Dadu, and Tabriz, a regular team of merchants returns every year. Along the way, residential buildings are scattered in bustling cities. At both ends of the road, Turpan and Kaburnham were once outposts for commercial and cultural exchanges between China and the Arab world. During this period, Tabriz replaced Baghdad and became the main commercial center in Western Asia. Many Chinese people inhabit Tabriz, Rashddin once described this international city as, under the eyes of the Padishah of Islam, "philosophers, astronomers, scholars, historians, of all religions, of all sects; people of Cathay, of Máchín, of India, of Kashmir, of Tibet, of the Uighúr and other Turkish nations, Arabs and Franks."[3] The Chinese who resided in the Ilkhanate directly contributed to the Western transmission of Chinese culture.

[3] Marco Polo and da Pisa Rusticiano, translated by Henry Yule, *The Travels of Marco Polo*, 1, 76.

### *Calendar*

After Genghis Khan conquered Samarkand, the Chinese calendar became well-known to Central Asian astronomers because the famous scholar Yelü Chucai and Hui scholars exchanged knowledge of astronomical calendars there. At that time, Chinese astronomers were advanced in predicting solar and lunar eclipses and stellar observations, relying on many traditional methods to obtain reasonable data.

The Chinese calendar holds an essential position in Central Asia. The descendant of Gumuer in the fifteenth century, Wuselbo (1393–1449), was very enthusiastic about astronomical observation before ascending to the throne in 1448. The famous astronomer and mathematician Al Kashi (?–1436) was an Arab scholar proficient in the Chinese calendar. He once presided over the Samarkand Observatory and compiled the famous "Ulubert Catalogue" into four volumes in 1437. The first volume discussed the principle of intercalation in the Chinese calendar, which had been widespread throughout Asia and Europe and was widely circulated.

### *Arithmetic*

Chinese arithmetic has a particular influence on India; since the sixth century, Indians have used the digit system to calculate in the sandy disk. The four arithmetic methods of this soil plate arithmetic are roughly similar to the Chinese chip algorithm, and the representation and four operations of fractions are also the same as the Chinese fraction algorithm. The arithmetic propositions in the writings of Brahmans Gupta in the early seventh century and Maha Veda in the mid-ninth century also have a congruence problem borrowed from the Chinese *Sun Zi's Mathematical Manual* that "things do not know numbers." There is an indefinite equation problem in the Maha Vedas, similar to the "Hundred Chicken Problem" in the fifth century *Zhang Qiu Jian's Mathematical Manual*.

The decimal notation in Indian arithmetic, as well as the resulting four arithmetic operations, fractional notation, and its four arithmetic operations, the three rate method (proportional algorithm), and the "hundred chicken problem", have all been introduced to Islamic countries, and are actually derived from Chinese mathematical knowledge. In the early tenth century, Arab mathematician Al Batani began to apply tangent and cotangent in addition to the sine and cosine introduced in India, which was also closely related to China's "double difference technique."

The "surplus and deficiency" issue was discussed in the works of the ninth-century Arab mathematician Al Huachizimo. The concept of " surplus and deficiency" appeared in the "Nine Chapters of Arithmetic" compiled in the first century AD, which proposed calculating profit and loss categories. This algorithm spread from northern China to Islamic countries and was circulated in Arabic mathematical writings. It was referred to as the al-Khitan algorithm in Al-Qassi's *The Key to Arithmetic* (1427), and in the early thirteenth century, it became a topic in the Italian mathematician Fibonacci's "Book of Arithmetic" (1202). His "Book of Arithmetic" also contains data consistent with the "Mathematical Classic of Sun Tzu", such as the "Number of Things Not Known Problem" and the "Hundred Chicken Problem", which were the first mathematicians to introduce the Eastern mathematical system to Europe.

Al Kassi was a renowned figure well-versed in Chinese mathematics in the fifteenth century. His book "The Key to Arithmetic" includes four operations, square root, cubic root, "Khitan Algorithm", and "hundred chicken problem", which directly and indirectly transmitted from China. The extraction method of higher root introduced by him is completely the same as the multiplication method of Jia Xian and Qin Jiushao. The coefficient table of the binomial theorem he put forward in Chapter 5 of Volume 1 of the book is Jia Xian's "The Original Diagram of the Method of Generating a Square", which first appeared in China in the eleventh century.

There was already Sudoku in Chinese mathematics, which can be traced back to their origins using Hetu and Luoshu. Sudoku is a three-line vertical and horizontal chart invented in the Han Dynasty, which later developed into an n-line vertical and horizontal chart with eight numbers more significant than three. The vertical and horizontal Sudoku mentioned in Zhen Luan's annotations on "Records of Numeric Techniques," and Wang Ximing's *Tai Yi Golden Mirror Scripture* of the Tang Dynasty are all about Sudoku. There are also many vertical and horizontal illustrations in Wang Wensu's *Treasure of Arithmetic* and Cheng Dawei's *Calculation Orthodox*. Muslim mathematicians once again utilized these vertical and horizontal diagrams, adding new content to Chinese mathematics. Al Buni of Algeria listed various magical squares in his works, such as *The Light of Wisdom* and *Magics*. Later, it developed into the "grid computing" of Islamic countries. In 1450, there was an official record in the *Nine Chapters of Algorithm Comparison* compiled by Wu Jing, known as "written calculation." This algorithm records the multiplicand in the right row according to the grid, records the multiplier in the upstream, multiplies each multiplier digit by the multiplicand in sequence, and records the data in the corresponding grid. Finally, it is added diagonally to obtain the desired number. This algorithm is the same as in Al Cassie's "The Key to Arithmetic." Cheng Dawei referred to the grid calculation as "paving the ground brocade" in his *Calculation Orthodox* (1592).

### *Cartography*

Since Pei Xiu (224–271) used the term "Zhun Wang", Chinese cartography established the scientific cartography of rectangular grids, which was used until the Yuan and Ming dynasties. But during the same period, the mapping of Europe returned from quantitative mapping to the "world map" method under the control of religious cosmology, and the entire world was drawn as a disk with completely abandoned coordinates. Between the eighth and eleventh centuries, some wheel-shaped maps also appeared in Arabia, which made the drawing of maps geometric.

Chinese grid cartography had two effects on Western countries in the late Middle Ages: one was the return of Arabic cartography to gridding, and the other was that it played a catalytic role in the emergence of practical nautical charts in European countries. The famous Chinese grid drawing method for maps in Iran has long attracted the attention of Italians. Marino Sanuto of Italy also used grid drawing on the map of Palestine, which he created for one of his Crusader geography works in 1306. This is the only grid map Latin cartographers drew in the early days of practical nautical charts.

The Yuan Dynasty's grid maps influenced the quantitative cartography used by Islamic countries and directly prompted Europe to draw practical nautical charts. At the end of the thirteenth century, compasses were used to draw nautical charts in the Mediterranean. The earliest available type of nautical chart was the Viscante practical nautical chart from 1311, which was annotated in Italian, Spanish, Portuguese, or Catalan without any Byzantine or Arabic markings.

The scale on this practical nautical chart is an intertwined compass azimuth line or oblique line. The oblique line is a compass azimuth line extending from the center of a compass wind instrument at any selected location to all directions. Due to the intersection of compass azimuth lines or oblique lines, a rectangular grid drawing method naturally appears. Traces of this rectangular grid can be seen on the navigation map of Angelino Dusetto in 1339. The widespread emergence of practical nautical charts allowed European sailors to navigate safely over a wider sea area. According to Jane Rochen Diep's *The Book of Navigation Charts* in 1542, the method of using this practical nautical chart was for helmspersons to select a diagonal line that was as parallel as possible to the straight line between two ports or the starting point of a day's voyage, to read the sailing direction. This method is not only inseparable from the use of a compass but also closely related to the depiction of the coastal map along the route, and the compass azimuth line is similar to the needle route used by China in the thirteenth century. The extensive use of practical navigation maps in Europe in the fourteenth century was inspired by the grid maps, navigation maps, and needle routes of the Yuan Dynasty. This was the knowledge that many missionaries, envoys, navigators, and merchants who traveled between China and Europe by sea and land at that time had actually experienced and widely disseminated. It is not an exaggeration to say that quantitative cartography spread more or less westward from China, along with the study of Toremian cartography by the Arabs, promoting new developments in cartography in Europe in the fourteenth and fifteenth centuries.

### *Medical Science*

Medicine in China and Iran was exchanged with each other during the Yuan Dynasty. Many doctors were invited during the Western Expedition of Hülegü, and Chinese doctors often accompanied the army for treatment. Chinese medical literature has become an essential reference for Iranian medicine. Sun Simiao's *Qian Jin Yao Fang* was translated into Persian during the Yuan Dynasty. In 1313, Rashidedin, the author of the *Collection of Histories*, edited an encyclopedia of Chinese medicine entitled *Tanksuq namah-i Ilkhan dar funun-i ulum-i Khitai*, which covered various subjects such as pulse, anatomy, embryology, gynecology, and pharmacology. Due to the importance of Chinese pulse studies, this medical dictionary also mentions the famous physician of the Jin Dynasty, Wang Shuhe (265–317), and his *Pulse Classic*. There are three Chinese-style medical pictures attached to the book. The first picture depicts the Eight Trigrams, divided into twenty-four equal parts and matched with day and night to indicate the rise and fall of the patient's body temperature. The second is an anatomical diagram of the internal organs, depicting the heart, diaphragm, liver, and kidneys. The third picture shows the palm and wrist. The attached drawings are wholly taken from Chinese medical books. This great work in Persian has been circulating to this day, and it was translated into Turkish by Suheyl Ünver of Istanbul University and published in 1939.[4]

## *Art*

For a long time, Chinese porcelain has been a handicraft imitated by Islamic art. From the late ninth century to the early establishment of the Mamluk dynasty in the mid-thirteenth century, imitation of Chinese porcelain became a trend in Egypt. Persian imitation of Chinese porcelain has a long history and has flourished since the ninth and tenth centuries. Starting from the ninth century, white porcelain was imitated in Lai, Kashan, and Samara, and it was not until the twelfth century that white pottery bodies could be fired. During the same period, many glazed pottery imitating celadon was also produced in Yemen. Persian pottery in the twelfth and thirteenth centuries often imitated the northern Ding kiln and the southern Yingqing in glaze color. Also, it absorbed the lotus, wave, cloud, and dark flower patterns of Chinese porcelain.

During the Ilkhan period, Iranian art was deeply influenced by China, especially during the reign of the Halbanda Khan (1304–1316). Ceramics, fabrics, metalwork, and painting were all willing to accept Chinese customs. Iranian silk was then exported to Europe and Africa, and silk was once unearthed from the ancient tomb of Falluna in Italy, printed with Chinese flowers and Arabic.

During the Mongol rule, Persian painters began to imitate Chinese ink painting, and the style of their images resembled that of China. After the fourteenth century, Muslim painters in the Middle East and Near East began to imitate the flowers and plants in Chinese paintings, and lotus, duckweed, reeds, and peonies became the themes of their images. Chinese-style dragons, phoenixes, and kylins have become popular decorative paintings in Iran, and dragon statues were painted on the door of

[4] Suheyl Ünver *Tanksuk namei Ilhan der Fünun Ulumu Hatai Mukaddimes*, i, Istanbul, 1939.

the Faramin Great Mosque built by Busayin Khan in 1322. Dragons are also common around various banknotes. The exorcism gate of Baghdad was built before 1225, with a dragon embossed on the arch of the gate. Dragon decorations are also typical in various paintings.

The pattern of Iranian blankets also imitates the Chinese weaving of hunting and animal patterns and is good at using clouds to show the Chinese style. The clouds and thunder in Chinese paintings, the transformation of Tai Chi diagrams into curves, as well as golden plates, flying geese, and apples symbolizing peace, as well as peaches symbolizing longevity, have all become commonly used themes in Iranian painters' works.

During the Ilkhanate period, Chinese hats, thrones, and clothing were popular in Iran. In short, Chinese fashion has become the most popular and infectious new thing in Iranian society during the Mongol rule.

CHAPTER 4

# Foreign Trade in the Early Ming Dynasty and Zheng He's Voyages to the West

## 4.1 International Trade Across the Indian Ocean and Zheng He's Seven Voyages

In June 1402, the Emperor of Ming, Zhu Di, ascended the throne in Nanjing and decided to utilize the substantial royal treasury as a foundation for active state-sponsored trade. This approach aimed to establish connections with various foreign monarchs and secure dominance in overseas commerce. Consequently, starting in 1405, he issued orders for Zheng He and Wang Jinghong to lead an extensive fleet of treasure ships, initiating the grand endeavor of the Seven Voyages to the West.

In order to establish comprehensive government control over overseas trade and effectively curtail the operations of illicit trading groups, Zhu Di spared no expense. He utilized the fleets led by Zheng He to repeatedly venture to the West, employing the government's political authority and economic assets to assert dominance over the entire international trade network in the Indian Ocean. This era marked an unparalleled period in Sino-foreign relations and the history of Chinese maritime trade, propelling China's maritime endeavors to a pinnacle that astounded the world.

From December 1405, when Zheng He led his fleet on its first voyage, to July 1433, when the final return took place, there were a total of seven journeys to the West. The concept of the "Western Ocean" during the early Ming Dynasty had dual implications. On the one hand, it referred


F. Shen, *Cultural Communications Between China and The Outside World Throughout History*,
https://doi.org/10.1007/978-981-97-4696-5_4

to the sea west of the southern portion of the South China Sea (as defined in the "*South China Sea Journal*"), and on the other hand, it pertained to the Western Regions around Calicut. The term "seven voyages to the West" specifically denoted the expeditions to Calicut. Hence, Zheng He asserted that he completed seven voyages, discounting the 1424 voyage to the Old Port for sealing purposes. During this span, Chinese navigators, at the helm of their fleets, traversed vast sea expanses extending from the Ryukyu Islands, the Philippines, and the Maluku Sea in the east to the Mozambique Channel and the coast of South Africa in the west. These regular voyages fostered friendly diplomatic relationships with numerous countries and regions across Asia and Africa. They facilitated the development of mutual trade, the exchange of cultural and technological knowledge, the enhancement of maritime transportation, and the stimulation of socioeconomic advancement in various overseas locales. Within a concise 28-year timeframe, this era brought about a flourishing political landscape in the South China Sea and the Indian Ocean region.

Zheng He and his colleagues embarked on numerous diplomatic missions to various Asian and African countries, significantly elevating the international stature of the Ming Dynasty. Each of these journeys was backed by a robust national treasury, and through the exchange of gifts and trade, highly sought-after Chinese-colored coins, porcelain, valuable medicines, and bronze items were introduced to the international market, captivating ruling elites and merchants from different nations. The fleets that ventured on these expeditions also supported Chinese merchants residing overseas, who possessed significant fleets and maritime weaponry. This collaboration transformed their stance from passive resistance against foreign private traders to active support. This enabled the establishment of vital sea routes, including the Malacca Passage, circumnavigating the Indian Ocean via the Sunda Strait, and direct voyages to the Cape of Good Hope to explore the Atlantic Ocean—an unprecedented historical endeavor. These efforts led to substantial progress in enhancing diplomatic relations between the Ming Dynasty and various foreign countries. Consequently, many foreign emissaries flocked to China, resulting in a constant flow of diplomatic envoys.

While Zheng He embarked on his journeys to the West, the government concurrently promoted state-sanctioned trade. This initiative had dual effects: it expanded the scope of import and export transactions, bolstered the domestic market, and extended political support to overseas Chinese merchants in their imports of foreign goods. The government allowed leeway for intermediaries at coastal ports and among those actively engaged in overseas trade, ensuring continued profitability. This approach significantly augmented China's competitive position vis-à-vis Muslim merchants in the Indian Ocean region.

The cargoes transported by Chinese treasure ships to various countries were substantial in quantity and renowned for their distinctive qualities. China's unique products like brocade, sarong, damask, and silk, as well as the newly developed blue and white porcelain and glazed porcelain, stood out as special commodities on the global stage. A wide range of blue and white porcelain plates and bowls, glass beads, musk, rhubarb, cinnamon, iron tripods, iron mills, and copperware constituted major export goods. Consequently, a significant outflow of copper coins, gold, and silver became an unstoppable trend.

A diverse array of goods was transported from the Indian Ocean to China. According to records from sources like the *Statutes of the Ming Dynasty*, *History of Ming Dynasty*, *Ying Yai Sheng Lan (The Overall survey of the ocean shores)*, and *Hsing-ch'a Sheng-lan (Description of the starry raft)*, there were a total of 185 types of goods shipped to China from the Indian Ocean. Among them were 29 types of incense, 23 types of treasures, 22 types of medicines, 17 types of hardware, 51 types of cloth, eight types of products, and eight types of animals. Additionally, there were 21 types of goods, eight types of pigments, three types of food (salt, sugar cream, pepper), and three types of wood.

The bulk of imports included aromatic herbs, pepper, sorrel, treasures, and cotton products. Notably, there were goods such as frankincense, balsam, precious stones, coral, ivory, rhinoceros horn, renowned horses, and pigments from Africa and the Arabian Peninsula. Pearls and lapis lazuli were sourced from the Persian Gulf, while double-edged knives came from Syria. Sugar frosting was imported from Bangladesh, Russian gold lazuli ware from the Persian Gulf, pepper from Kozekhot, and a substantial amount of Indian cotton, including varieties like mahmal from Bengal, Sanah-baf, Kain ganeh, Percallen from Southern India, mousseline, five-colored cloth, Byrampaut, and other notable cotton fabrics. There were also items like Sakhlat and silk-cotton blends.

Thanks to Zheng He's voyages to the West, the Ming Dynasty government's practice of bestowing gifts through trade reached its anticipated pinnacle. This approach involved providing substantial economic support to neighboring nations, fostering a strong appeal that extended to countries adjacent to the Strait of Malacca and encompassed the entirety of the Indian Ocean region. These nations were keen on establishing direct trade relations with China through official means due to the highly advantageous nature of this engagement, which held significant benefits for the trading partners.

## 4.2 Zheng He's Voyages to the West and China's New Explorations in Science and Technology

### *Treasure Ship and Ship Design*

Zheng He went to the West, relying on a large and well-organized fleet of ships, often using hundreds or thousands of boats each time. Luo Maodeng's *The Adventures of the Eunuch Sanbao in the Western Seas: A Popular Fictional Account* (1597) divides the ships going to the West into five categories: treasure ships, horse ships, grain ships, seat ships, and warships according to their size and performance. The treasure ship is the leading trade ship for transporting treasure goods, also known as treasure ship, treasure boat, jewel ship, or dragon ship, with a vast establishment, complete equipment, good stability and seaworthiness, and able to sail long distances through storms. The 1st volume of Ma Huan's *Ying Yai Sheng Lan* described the treasure ship as "The biggest one is 444 feet long, 180 feet wide; the middle one is 370 feet long and 150 feet wide." Tan Qian's *Guoque* also said a treasure ship with "its biggest one is 440 feet long and 180 feet wide, the second largest is 370 feet long and 150 feet wide."[1] One foot in the Ming Dynasty was equivalent to 0.311 meters. It means the treasure ship's length reached 138.084 meters, 56 meters wide, and nine masts were erected on board. The smallest warship is also 180 feet long, 68 feet wide, and has five masts on board. The fleet consisted mainly of treasure ships. Therefore, it is called a treasure ship fleet.

The size of Zheng He's treasure ship, based on the shipbuilding tradition from the Song, Yuan, and Ming dynasties, was calculated using the concept of volume. In 1936, a monument with 148 character inscriptions was discovered at the Jinghai Temple in Nanjing, offering insight

[1] According to version research, the scale of the treasure ship was not found in the initial and final manuscripts of Ma Huan's book, but only appeared in the biographical notes of the final manuscripts after 1527. See Wan Ming's proofreading: *Annotations on the Ming Yingyai Shenglan*, Ocean Publishing House, 2005, "Research on the Origin of Ma Huan's *Yingya Shenglan*," pp. 26–28.

into historical ship dimensions. These inscriptions provide details about the ships of the third year of the Yongle reign (1405), mentioning "the general and official army riding 2000 volumes sea ships and eight sculling ships," as well as the seventh year of Yongle when they rode "1,500 volumes sea ships and eight sculling ships." Unfortunately, this monument was destroyed during the Japanese invasion, but its inscriptions were recorded in Zheng Hesheng's *Compilation of Zheng He's Legacy*. Researchers believe the 2000-volume ships represent some of the larger treasure ships in Zheng He's fleet.

During the 1980s, the length of the treasure ship's hull was once speculated to be comparable to modern 10,000-ton ships. The length-to-width ratio of the treasure ship was 2.46:1, similar to the enormous sea vessels constructed during the Song Dynasty. This type represented the tradition of crafting large maritime vessels along the southeastern coasts of Fujian and Zhejiang. The design of the treasure ship was suitable for anchoring in areas with rocky shores and deep waters, such as the East China Sea and the South China Sea. This design differed from the flat-bottomed maritime sand ships used for navigation along the sandy shores of the coast north of the Yangtze River's mouth. The construction of large treasure ships and most other sea vessels occurred at various shipyards, including the Longjiang shipyard in Nanjing, Liujiagang in Suzhou, and shipyards in Zhejiang, Fujian, and Huguang.

The earliest depiction of a Chinese ship in Western maps can be found on the Catalan Map of 1375, which portrays a Chinese sailing ship along the Malabar coast. The accompanying note states: "This ship is called Enchi. It has a keel length of 60 varas and a draft of 34 varas. It had at least four masts, sometimes up to ten, and sails made of bamboo and palm tree leaves."[2] This image represents a Chinese sailing ship of the Yuan Dynasty. The configuration of Zheng He's treasure ship during his time can be observed in the map compiled by Mao Yuanyi in the Ming Dynasty (commonly referred to as *Zheng He's Navigation Map*). These engravings depict the treasure ship as a three-masted sailing vessel. Furthermore, in the illustration at the beginning of the volume *The Sutra of the Heavenly Concubine* (included in *The Complete Collection of Chinese Art*, Painting and Printmaking Volume), which was engraved by Sheng Hui, a monk who accompanied Zheng He's voyage to the West in 1420, there are

[2] *Atlas de Santarem*, Amsterdam, 1849, reprinted in 1985, picture 35–36.

five rows of five sailing ships in a group. These sailing ships are of the Fujian type, characterized by their raised bow and stern. The sailing ship features six masts, with three masts in the front and middle sections and three auxiliary sails in the stern area, which can be lowered. The deck is divided into three levels, with the stern cabin being one level higher than the middle deck. The last auxiliary mast has a horizontal beam in the middle for hanging flags.

The shape and volume of the treasure ship are under continuous research. From the history of Chinese shipbuilding and world shipbuilding, it is estimated that the capacity of the treasure ship is between 800 and 1000 tons. Compared with the world shipbuilding industry, which was in the process of transformation at that time, Europe was still in the beginning. In Europe, the three-masted ship gradually replaced the two-masted boat around 1400, and a century later, the Basque shipbuilders of the Bay of Biscay revolutionized medieval sailing with the design of the full-rigged ship. The Atlantic type, with a bowsprit above the stern and a rudder on the side, was called a carrack. In the middle of the fifteenth century, the highest tonnage of Genoese galleys was only 1400 tons, and most ships were under 1000 tons. Until 1490, the capacity of most Venetian galleys was 1200 tons,[3] and the average tonnage of the Venetian merchant fleet in the fifteenth and sixteenth centuries was 600 tons. The English ship Gracedieu, built in 1418 in Southampton, was 60 meters long and over 1400 tons, but was not used. As for the sixteenth century, when the Portuguese fleet dominated the Indian Ocean, the cargo was lower, usually only 800 tons or less. 50-meter-long and 600-ton ships were rare. The galleon, called a building ship, only started to be built in Venice in the sixteenth century, initially around 100 tons, and only after half a century, it reached more than 1000 tons.

The highest number of treasure ships used in the seven voyages is 63, and the general number is more than 40. A large number of soldiers and crew members accompanied each expedition. In the first voyage, it is said that more than a thousand ships of all sizes were used (Luo Maodeng, *The Western History,* vol.15), and the *Hongzhi Taicang Zhouzhi (弘治太仓州志)* says that there were 208 ships in the fleet. This time, more than 27,870 people were employed in the mission (*Guoque*, Vol. 13), which became a model for all subsequent trips to the West. The third

[3] *Ships and seamanship in the Ancient World*, Princeton University, 1971, pp. 221.

and fourth times were more than 27,000 people, and the last time was 27,550 people, according to *Qianwenji*. The treasure ships could accommodate about 400–500 people each, and more importantly, they had sufficient compartments for transporting treasure and goods. Thanks to the improved design of the ships, the sea voyage was much safer and more comfortable than the Chinese sailing ships taken by Marco Polo and Ibn Battutah. The method of watertight compartments was of particular significance to the shipbuilding industry.

Zheng He's fleet sailed across various regions of the Indian Ocean, pioneering a new maritime route to the Mozambique Channel and the Cape of Good Hope through Java. This route predated European exploration by 300 years. This achievement not only garnered attention within the Arab maritime community but also captivated Venetian merchants navigating between Hormuz and Aden. Even European ship designers were enlightened. The caravel, which originated from Arab fishing boats in the thirteenth century, initially had a length-to-width ratio of 7:1. However, when repurposed for exploring new routes, the caravel's design was altered. For example, the "Leyden" caravel had its length-to-width ratio transformed to 3:1. By the sixteenth century, the Dutch further improved this design, creating a four-masted sailing ship with a front and rear structure. Moreover, the Dutch introduced curved side panels to reduce resistance, a concept possibly borrowed from China.[4]

### *Manufacture of the Blue and White Porcelain*

Blue and white porcelain is created by firing white porcelain with cobalt metal as the coloring agent. In 1957, fragments of blue and white bowls were unearthed at Jinsha Pagoda in Longquan, Zhejiang Province, revealing the existence of blue and white porcelain during the Northern Song Dynasty. In recent years, Yuan and Ming porcelain kilns have been discovered in Jiangshan, Zhejiang Province, and Yuxi, Yunnan Province, with blue and white porcelain production dating back to the Yuan Dynasty. Notably, Jingdezhen was a prominent representative of Yuan-era blue and white porcelain. During the Yuan Dynasty, imported cobalt

[4] Edward Bowness, Some Drawings of Old Ships, *The Mariner's Mirror*, 1956, XLII, 323–327. Jr. Edwin Do run The Origins of Leeboards, *The Mariner's Mirror*, 1967, L III, pp. 39–51.

materials were predominantly used for coloring blue and white porcelain, resulting in intricate patterns and sizable forms that were primarily exported abroad. As the Ming Dynasty dawned, blue and white porcelain reached its zenith, eclipsing Longquan's blue porcelain to become the most sought-after brand of export porcelain. During the early Ming Dynasty, the Yongle and Xuande periods produced exceptional blue and white porcelain with intricate patterns of intertwining branches, flowers, and fruits, often featuring motifs of clouds, dragons, and flying birds. The use of imported cobalt materials produced vibrant colors that penetrated the glaze and remained unfading, while the glaze surface often displayed characteristic brown speckles resembling orange peel. Throughout Zheng He's voyages to the West, porcelain played an essential role as an export commodity, and blue and white porcelain, in particular, held great significance.

The crafting and materials used in the firing of blue and white porcelain are intricately linked. The Yongle and Xuande periods are renowned for their use of imported cobalt materials, which became highly treasured. Wang Shimao's *Kui Tian Wai Cheng* written before 1589, noted: "The porcelain produced during the Yongle and Xuande periods within the imperial court has been highly valued. At that time, the colors commonly used were violet and sweet white. Su Ma Li Qing was often employed as decoration, and bright red was considered precious." The account of *Shi Wu Gan Zhu* (1591) conveyed similar information. Su Ma Li Qing is cobalt material produced in Somalia. They possess a higher iron content and a lower cobalt content. During the Zhengde reign, these materials were already referred to as "Huiqing." As stated in *Kui Tian Wai Cheng*, "Huiqing" was sourced from a foreign country. A eunuch of Yunnan acquired it during the Zhengde period and initially used it for refining stones, valuing it double the price of gold. It was known to be effective in kiln firing and produced good results. Huiqing, also known as "Huihuiqing," was imported from Yunnan during the same period. According to Yan Congjian's *Ming dynasty record (1574) of exploration and foreign relations*, tributes to the Ming Dynasty included Shiqing and Huihuiqing. However, after the Zhengde period, the term "Huiqing" began to be used more widely for the cobalt material used in blue and white porcelain production.

In the Yuan dynasty, Wang Dayuan observed various products along the coast of Benadir, among them being Qingmeng stone, which exhibited the color of the Bahri gemstone (sapphire). This cobalt material also served as a green pigment, commonly known as "Huihuiqing" or "Huiqing." The *Tian Gong Kai Wu* described it as: "Huiqing is a valuable green color from the Western Regions, and the exquisite type is sometimes referred to as 'Fotouqing.' When applied, it produces a fiery appearance." Sumatra could either be the source or a transit point for the cobalt material from the Middle East. However, the term "Fotouqing" likely referred to the cobalt material from Somalia. The best quality of this material was found in a region called Baduna in Idrisi's *Travelogue*, near Mogadishu in Benadir. The establishment of a direct route to Mogadishu and Brava during the early Ming Dynasty demonstrated the economic significance of this region, particularly concerning the materials used in blue and white porcelain. During the Zhengde and Jiajing periods, there was continued extensive use of imported Huiqing, which was employed in the official kilns in Jingdezhen. Jiajing's period saw a combination of imported and domestic cobalt materials, resulting in various shades of green. This practice continued until 1596 (Wanli's 24th year), when "Hui Green" was still being transported from the northwest to the mainland via Turpan.

During the era of Zheng He, blue and white porcelain was tailored for export and took into consideration the specific needs of various regions. Large quantities of these porcelains were sent to Southeast Asia, while many coarse bowls and plates were sold to East Africa. The affinity that Islamic countries had for blue and white porcelain played a pivotal role in opening up foreign markets. Notably, during the early Ming period, large plates and jars were produced with a Middle Eastern style specifically for export. Newly crafted items such as flat vases with two handles and squared vases with folded sides imitated the Iranian style from the thirteenth century. A unique piece created during the Yongle period featured a trumpet-shaped, hollow bronze seat with a slim waist at both ends, which was used to hold flower pots and jars. This design was rare in China and was influenced by a bronze seat style popular more than 3000 years ago in places like Palestine. Similar designs were later discovered through excavations in Syria. The Xuande period saw the production of Islamic-style cups, inspired by styles from the eighth-century Middle East. Additionally, items like pots, pencil cases, and goblets were crafted with Islamic influences. These rectangular pieces featured lids influenced by Chinese design, while their patterns and inscriptions were derived from Persia. Early Ming goblets were created in a style influenced by ancient Mediterranean or Near Eastern designs, differing from the Middle

Eastern style of the Yuan dynasty by featuring a more lavish mouth shape. In the Zhengde period, the "Hui ware" was designed with Arabic and Persian inscriptions. This collection included dishes, bowls, brush holders, deep-bellied jars, stoves, boxes, and candlesticks. Some of these pieces are now housed in the Palace Museum in Beijing, with a red-colored porcelain plate featuring Arabic inscriptions being a particularly rare treasure. The decoration of these exported blue and white porcelain items often featured Islamic binding patterns or Middle Eastern horizontal and straight overlapping decorative belt patterns. An example of the latter style is represented by blue and white jars unearthed from a Ming tomb in the Dongying Village in Beijing in 1965, showcasing a blend of Chinese and Iranian artistic styles.[5]

Blue and white porcelain has become a popular everyday utensil and a precious ritual vessel for all countries, regardless of rank. In the early fifteenth -century Persian codices preserved In Calcutta and Paris, many Chinese blue and white porcelain utensils and tableware appeared in the nobility banquet occasions in the illustrations.[6] Burma often buried Chinese porcelain jars in the ground or sent them to Buddhist temples for worship. The inhabitants of the Philippines used blue and white porcelain as an honored utensil and only took it out of their cellars during the holidays. Blue and white porcelain is also embedded indoors for viewing. The mosque excavated at Kilifi, north of Mombasa, Kenya, has a blue and white porcelain plate embedded above the gate. In Sarawak and along the Kenyan coast, distinctive pillar tombs can be seen, with quadrangular, hexagonal, or octagonal pillars standing five meters high in front of the tombs carved with floral motifs and inlaid with Chinese blue and white porcelain. Many exported blue and white porcelain wares were regarded as the most critical burial objects by Melanau People in Thailand, Calatagan in the Philippines, and Sarawak. In the South China Sea and around the Indian Ocean, the value of blue and white porcelain has long exceeded its economic meaning. Still, it has become a symbol of friendship and eternity with sacred spiritual connotations.

[5] *Antiquities*, 1979, No. 8, p. 75, Fig. 2.

[6] Basil Gray, Blue and White Vessels in Persian Miniatures of the 14th and 15th Century Reexamined, *JOCS*, 1948–1949, Issue 24.

The technique of manufacturing blue and white porcelain became a benchmark for porcelain artisans across different countries. In the fourteenth and fifteenth centuries, Egypt quickly adopted the newly emerged Chinese blue and white porcelain. The shapes and decorations were similar to those in China, with the exception that the porcelain body was of Egyptian origin. At the Faustat archaeological site, many pieces bearing the names of artisans in Arabic script were discovered. Before 1407, Vietnamese porcelain workers had already mastered the art of firing blue and white porcelain, exporting their products to Western Asia. Similarly, Bengal also produced blue and white porcelain, which was presented as a tribute to the Ming Dynasty. During the sixteenth century, the Persian king Abbasid attempted to establish a local porcelain industry by recruiting hundreds of Chinese porcelain artisans along with their families. These artisans moved to Iran and settled primarily in Isfahan. Some of these artisans might have also traveled to Turkey to assist local potters in their porcelain-making endeavors. This exchange of knowledge also benefited the porcelain industry in Syria.[7] The porcelain-making techniques reached Europe in 1470 when the Venetian alchemist Antonio learned from Arabs. Subsequently, Arab blue porcelain became a model for potters in Pisa, who began their porcelain production with blue and white pieces, imitating the Chinese style.

### *The Introduction of Cloisonné*

Cloisonné enamel is a type of Beijing specialty that gained renown during the Jingtai period of the Ming Dynasty. These delicate enamel wares bear markings such as "Jingtai Nianzhi" and "Da Ming Jingtai Nianzhi." In early Qing Dynasty literature, they were referred to as "Jingtai Imperial Workshop's enamel wares." The term "Jingtai Lan" first appeared in records and writings during the Guangxu period. It likely originated in places like Beijing and Guangzhou and spread from there. The name "Jingtai Lan" is thought to have been a long-standing term used among the people, where "Jingtai" signifies the time of production during the Jingtai years, and "Lan" represents "enamel."[8] The primary components

[7] Arthur Upham Pope, "*An Introduction to Seventh Century*", A. D., London, 1930, p. 94.

[8] Yang Boda: Cloisonné, *Cultural Relics*, 1981, Issue 1, Chapter 7.

of cloisonné enamel include quartz, porcelain clay, feldspar, borax, lead, saltpeter, and other silicate materials. During the casting process, fine copper wires are arranged to form a pattern following ink designs. These wires are then glued and welded onto the copper base, after which enamel is applied to fill the compartments. The piece is then fired, polished, and gilded. The resulting object is delicately crafted, but it may have visible joints due to the assembly process. Another technique involves chiseled enamel, where the pattern is drawn onto the copper base, and the areas outside the pattern are chiseled down. Enamel is then applied to the recessed areas, and the piece is baked, polished, and plated. While the lines in this style are rougher, there are no visible joints between sections.

The production process of cloisonné enamel is aligned with the "Daqi Kiln," which has some similarities with the "Folangqian," the latter being associated with the Arab countries. Arab countries were known for producing cloisonné enamel. After the Mongols' western expedition, both cloisonné enamel and chiseled enamel were introduced to China. The former was referred to as the "Dayi Kiln," while the latter was called "Folang Qian." The term "Daqi Kiln" was first mentioned in the Yuan Dynasty and was associated with products coming from various Arab countries. On the other hand, the term "Folang Qian" was likely introduced from Europe, possibly through the Levant and Asia Minor, and made its way to China. In summary, "Dayi Kiln" and "Folang Qian" represent two different techniques and styles of enamel work that were imported from different regions, with "Dayi Kiln" originating from Arab countries and "Folang Qian" potentially coming from Europe and the surrounding regions.

During the Ming Dynasty, cloisonné enamel exhibited a distinct Chinese style. Notable dated periods include Xuande (1426–1435), Jingtai (1450–1456), Jiajing (1522–1566), and Wanli (1573–1620). During the Xuande, Jingtai, and Jiajing periods, cloisonné enamel pieces typically featured a light blue base. Xuande pieces displayed alternating colors such as ochre, yellow, white, green, treasure blue, and yellow-green, while Jingtai pieces showcased ochre, yellow, white, blue, pink green, and yellow-green, demonstrating exquisite craftsmanship. This era also witnessed the height of development for blue and white porcelain and painted porcelain. In the Qing Dynasty, during the Kangxi and Qianlong reigns, cloisonné enamel experienced a period of prosperity. The Manufacturing Office was responsible for its production. Emperor Hongli (Qianlong) showed great admiration for Western-style cloisonné enamel designs. These designs were often imported from Guangzhou, and the Manufacturing Office also developed its own glaze colors. In total, there were 24 different colors used in cloisonné enamel, surpassing the variety seen during the Ming Dynasty. However, the glaze colors used in this period were not as vibrant as those seen during the Xuande and Jingtai periods, likely due to changes in the source of raw materials for the glaze formulas. During the Qianlong era, cloisonné enamel designs included imitations of ancient Chinese ceremonial vessel patterns and even European-inspired designs. The range of forms and types of objects expanded greatly, including everything from furniture and sacrificial vessels to Buddha statues and decorative items. Nevertheless, by this time, the production of cloisonné enamel was already in decline.

CHAPTER 5

# Western Transmission of the Four Great Chinese Inventions

## 5.1 The Paper and Paper-Making Techniques Transmitted from China to Europe

### *The World's First Plant Fiber Paper*

Paper is one of the greatest inventions in China, ranking alongside woodblock printing, the compass, and gunpowder as the Four Great Inventions of China. The invention and widespread application of paper have significantly advanced the development of world culture and produced an immeasurable and tremendous impact on human society. However, Europeans had long believed that ancient China only knew how to make paper from silk. Andrez of Balmades summarized the spread of papermaking techniques in 1782 as follows: "Ancient China made paper from silk, which was introduced to Persia around 652 AD and then to Mecca in 706 AD. The Arabs replaced silk with cotton and spread the technique of papermaking to Africa and Spain." In fact, before the advent of silk-based paper in ancient China, traditional paper was made from a thin film of plant fibers bonded together through physical and chemical processes.[1]

During the Western Han Dynasty, the standard writing materials were silk and bamboo slips. Silk was expensive, and bamboo slips were heavy,

[1] Pan Jixing, *History of Papermaking Technology in China*, Beijing, Cultural Relics Press, 1979, p. 3.



F. Shen, *Cultural Communications Between China and The Outside World Throughout History*,
https://doi.org/10.1007/978-981-97-4696-5_5

making them inconvenient for the transmission of text. Only paper made from plant fibers could be widely used as a writing material. The emergence of cotton paper marked the transition from silk to plant fiber paper.

In 105 AD, Cai Lun reformed the papermaking process, and before the mass production of finely crafted hemp paper, the true paper made from plant fibers had already appeared. Archaeological excavations have uncovered ancient paper artifacts from the Western Han Dynasty, such as the Lop Nur ancient paper, Baqiao paper, Juyan Jinguan paper, Fufeng Zhongyan paper, and Dunhuang Maquanwan paper.

The Lop Nur ancient paper was discovered by Huang Wenbi at the Western Han Dynasty beacon tower ruins in Lop Nur in 1933. It is a small piece of hemp paper measuring 4 cm by 10 cm, described as a square, thin slice with incomplete edges. The excavator described the hemp paper as "very coarse, uneven, and with hemp fibers still visible on the surface."[2] At the same time, wooden slips from the first year of the Huanglong era (49 BC) were also unearthed, so the hemp paper should belong to the Western Han Dynasty. However, the actual artifact was destroyed in the warfare of the 1930s.

In 1957, the Baqiao paper, which was discovered in a Western Han tomb in the eastern suburb of Xi'an at Baqiao, was once considered the earliest plant fiber paper in the world. The tomb dates back no later than the reign of Emperor Wu of Han (140 BC–87 BC), and the Baqiao paper was also considered to be fiber paper produced during the same period. Its raw materials consist of old hemp stalks and broken hemp fibers from cloth.[3] Baqiao paper is loose and rough, with a brownish-yellow appearance, and the direction of the fiber bundles can be seen when facing the light. Observed by solid microscope and scanning electron microscope, it has the characteristics of thick and coarse paper, loose fiber tissues, and unclear curtain lines.[4] In short, Baqiao paper is indeed an extremely primitive paper. However, a new discovery was made not long after. In 1986, after excavating the ancient tomb in Fangmatan, Tianshui, Gansu Province, a paper map with an inked surface was found on the chest of the

[2] Huang Wenbi, *Archaeological Records of Lop Nur*, 1948, p. 168.

[3] Pan Jixing, "The World's Earliest Plant Fibre Paper," *Cultural Relics*, 1964, p. 11.

[4] Wang Juhua, Li Yuhua: "*From the Analysis and Identification of Several Kinds of Han Paper Try to Discuss the Invention of Papermaking in China*", *Cultural Relics,* 1980, No. 1. See *Chinese Papermaking Technology History Manuscript*, p. 37.

tomb owner, and the artifact dates back to 141 BC. It appears that the earliest paper was used for military maps and the transmission of frontier reconnaissance information.

In addition to Baqiao paper, there are also Juyan Jinguan paper and Fufeng Zhongyan paper from the Western Han Dynasty made of plant fibers. In 1973, two pieces of paper were unearthed at the Juyan Jinguan site. The larger piece (21 cm × 19 cm) is white, with a fine and strong texture. Among the 20,000 bamboo slips unearthed at the same site, the latest one dates back to the second year of Ganlu (52 BC), so the paper should be from the same period. The other piece of paper (11.5 cm × 9 cm) is dark yellow, containing hemp fibers, thread ends, and broken hemp cloth pieces, with a loose texture. The unearthed stratum dates back to Emperor Ai's Jianping era (6 BC–3 BC). In 1978, the Fufeng Zhongyan paper was found inside a copper bubble attached to a lacquer ware storage in the Western Han Dynasty. It is milky yellow, made of hemp fibers, and belongs to the period of Emperor Xuan (73 BC–49 BC).

In 1979, eight pieces of Western Han Dynasty paper fragments were discovered at the Han Dynasty beacon tower ruins in Maquanwan, Dunhuang. The site is located 11 kilometers west of Xiaofangpan City. The unearthed Han paper is all hemp paper, with the largest piece measuring 32 cm in length and 20 cm in width. It dates back to the Ganlu era (53 BC–50 BC). The paper fibers are uneven, coarse, and yellow. Four other pieces have a finer and more even texture, dating back to the Ai and Ping eras (6 BC–AD 5). The remaining three pieces are white, with a fine and even texture, and are relics from the period of Wang Mang's rule.

The hemp fiber paper produced during the Western Han Dynasty was not suitable for writing and could only be used for wrapping, padding, and other miscellaneous purposes. From 1990 to 1992, the Gansu Institute of Cultural Relics and Archaeology excavated the Han post station site at Xuanquan, where they discovered a batch of hemp fiber paper from the reigns of Emperor Xuan and Emperor Yuan of the Western Han Dynasty. Four of these pieces have ink-written characters. This batch of well-defined hemp paper provides physical evidence of the papermaking process.[5] By around the first century BC, paper was already being used for

[5] *China Heritage News*, 5 January 1992.

writing in the imperial court. The use of paper for writing first appeared in ancient books during the Eastern Han Dynasty.

*Fengsutong* states that when Emperor Guangwu (AD 25–57) moved the capital to Luoyang, "a total of 2,000 carts were loaded with plain silk, bamboo slips, and paper scrolls." In addition to plain silk and bamboo slips, there were also ancient books written on paper, some of which might have been relics from previous dynasties. The "*Biography of Jia Kui*" in the "*Book of the Later Han*" records that in the first year of Jianchu (AD 79), Jia Kui taught twenty people to study the "*Zuo Zhuan*" using both bamboo slips and paper scrolls. These examples of plant fiber paper suitable for writing already existed before Cai Lun's invention of papermaking.

The *Biography of Cai Lun* in the *History of later Han dynasty* first recorded Cai Lun's invention: "Cai Lun, the eunuch, was in charge of making paper at the Imperial Workshop, which was called 'Cai Hou Paper'." By the fifth century, when Fan Ye wrote the *Book of the Later Han,* it had transformed into Cai Lun inventing plant fiber paper: "In the ninth year of Yongyuan (AD 97), he supervised the production of secret swords and various weapons, all of which were exquisitely crafted and became the standard for later generations. Ancient books and documents were mostly compiled on bamboo slips, and those using silk were called paper. Silk was expensive and bamboo slips were heavy, both inconvenient for people. Lun then creatively used tree bark, hemp fibers, worn-out cloth, and fishing nets to make paper, and presented it to the Emperor in the first year of Yuanxing (AD 105). The Emperor appreciated his ability, and from then on, everyone followed suit, so the world called it 'Cai Hou Paper'." (Volume 78) Fan Ye's definition of "Cai Hou Paper" is paper made from plant fibers such as tree bark and hemp cloth, which could replace silk and bamboo slips. Although the claim that only floss paper existed before the appearance of Cai Hou Paper has been proven wrong, it is a fact that in AD 105, Cai Lun established a new process for making plant fiber paper based on the papermaking techniques of previous generations, adding new raw materials such as tree bark and fishing nets, and creatively producing hemp fiber paper and bark paper suitable for writing.

In 1942, a paper bundle with writing was excavated at the southern Suku Tai beacon tower in Bayan Bato Mountain near the Ejin River in Inner Mongolia. The hemp paper was made between the third and fourth years of Yongchu (AD 109–110),[6] approximately the time of "Cai Hou Paper." In 1974, the Eastern Han paper with writing unearthed at the Hantan Slope in Wuwei is a relic from the second half of the second century and is the earliest existing paper with writing in China. Upon appraisal, the raw material was hemp fiber, with evenly interwoven fine fibers, few fiber bundles, and minimal holes. The fiber's degree of fibrillation was relatively high, and the paper was single-sided coated processing paper. The paste-like substance remaining in the coating was adhesive, and the paper's thickness was equivalent to 20–30 grams of thin paper. The structure and characteristics of the Hantan Slope paper indicate that at the time, in addition to precise and meticulous work in washing, soaking, boiling, and papermaking processes, techniques such as pressing and flat drying were also employed to achieve a smooth and compact paper surface.

After Cai Lun's reforms, not only large quantities of hemp paper could be produced, but also leather paper was invented. Dong Ba of the Three Kingdoms had already pointed out that paper made from wood bark was called Ceres paper, paper made from old hemp was called hemp paper, and paper made from fishing nets was called net paper. At the end of the third century, Lu Ji's "*Mao Poems on Grasses, Trees, Birds, Beasts, Insects, and Fish,*" said that Ceres was known as Ceresan or Broussonetia by the people of Youzhou; Ceresan by the people of Jing, Yang, Jiao, and Guang; Ceresan, and Broussonetia by the people of Zhongzhou. "Nowadays, people in Jiangnan make its bark for cloth, and pound it for paper, which is called Ceres bark paper." During the Wei, Jin, and Southern and Northern Dynasties, the use of tree bark paper increased dramatically, making it an affordable and readily available paper product.

The secret of making paper from broken hemp cloth during the Han and Jin Dynasties was not revealed to the world until 1910. In 1907, Aurel Stein excavated many Han wooden slips, paper documents, and nine Sogdian letters at the Great Wall beacon tower site near Dunhuang. After examination by Wilhelm Wessely in Vienna, it was found that the paper's raw material was ramie (Boehmeria nivea), made by pounding broken cloth into pulp and then producing paper suitable for writing. According to research by Henning, these letters were written by a Liangzhou merchant named Nanai-Vandak during the reign of Emperor

[6] Lao Gan, "*On the Academicians of Chinese Papermaking*", in *Collected Works of the Institute of Historical and Linguistic Studies*, No. 19, p. 496.

Huai of Jin (311–313) to his friends and family in Samarkand.[7] The identification of these paper documents has debunked the outdated claim that the Arabs were the first to invent paper made from broken cloth, and has unveiled the historical facts of Cai Lun's papermaking once again.

### *Ancient Paper Unearthed in Xinjiang*

Paper began to be mass-produced in Luoyang in the early second century, and was promoted nationwide, marking the beginning of the westward spread of paper. Due to military and foreign trade needs, the newly invented paper quickly spread to the Hexi Corridor and Xinjiang.

Ancient papers dating back earlier than the Han Emperor Xuan's time were unearthed at the Han Dynasty beacon tower site in Lop Nur in the early twentieth century, along with ancient papers with writing from the late Eastern Han period. In 1959, paper with vermilion powder was discovered in an Eastern Han tomb in Minfeng.

A large number of ancient papers from the Wei, Jin, Sixteen Kingdoms, Northern and Southern Dynasties to the Sui, Tang, and Five Dynasties have been unearthed in various places in Xinjiang, with official and private documents, contracts, and classic texts being the majority, and Buddhist scriptures being relatively few. The unearthed documents include those written in Chinese, as well as Old Uyghur, Turkic, Tibetan, Western Xia, Sogdian, Tocharian, Syriac, Greek, and Indian Sanskrit. In addition to manuscripts, there are also some printed religious classics.

Numerous Wei and Jin official documents, commercial letters, and Northern Dynasty paper documents have been unearthed at the Loulan site. In 1900, Sven Hedin discovered fragments of public and private documents here, with both fine and coarse paper quality, including gray mesh paper and pale-yellow letter paper with fine texture. Documents with dates range from the fourth year of Jiaping (252) to the fourth year of Yongjia (310). At the same time, a fragment of the *Strategies of the Warring States* was discovered, written in late Han clerical script, close to Wei and Jin regular script, and dated to the first half of the second-third

[7] W.B. Henning, *The Date of the Sogdian Ancient Letters*, *Bulletin of the School of Oriental and African Studies* (BSOAS), University of London, 1948, vol. XII, pp. 601–615.

centuries.[8] In 1914, Stein discovered Han Chinese paper-written fragments with the years Taishi 6 (270) and Yongjia 6 (312) at the northern end of Lop Nur. Along with the documents unearthed at the Loulan site, there were also wooden slips belonging to the Taishi period (265–274). Like the Han Dynasty barrier sites in western Gansu, Loulan used both paper and wooden slips during the Han and Jin periods. Wooden slips were phased out earlier in the inland areas than in the northwest border regions, with Loulan still using wooden slips as late as the mid-third century. Due to the low cost, easy portability, and ease of writing on paper, it was widely spread by domestic and foreign merchants, as well as border officials and soldiers. From the fourth century onward, paper completely replaced wooden slips, becoming the universally used writing material, and expanded from official documents and letters to the copying of classic texts.

The Gaochang ruins in Turpan are another site in Xinjiang where a large number of ancient papers have been unearthed and are also the outpost for the westward transmission of paper and papermaking techniques. A manuscript of the *Collection of Essential Teachings of All Buddhas* from the Western Jin period, written in the sixth year of Yuankang (296), was discovered in Tuyugou, making it the earliest existing Buddhist scripture.[9] Also found at the same site is the scroll 9 fragment of the Western Jin *Prajna Paramita Sutra* and other Buddhist scriptures from the Six Dynasties and Tang Dynasty. Numerous Six Dynasties and Tang Dynasty Buddhist scriptures have also been found in the ancient city of Harahwaz (50 km southeast of Turpan), including the fragment of *Ten Stages of Discourses* during the Northern Liang period. Tuyugou also contains many Han Chinese manuscript works, with the earliest being the annotated text of *Sun Tzu's Art of War- The Forms of Strategies and Attacks* from the Six Dynasties, and a significant number of Tang Dynasty annotated texts such as *Chu Xue Ji, Tang Yun, Zuo Shi Chuan, and Shennong Ben Cao Jing*. Gaochang also discovered paper versions of Manichaean scriptures translated into Sogdian, Middle Persian, and Uyghur languages.

[8] August Conrady, *Die Chi-nesischen hand-schriften und Son-stigen Kleinfun de Sven Hedins in Lou-Lan*, Stockholm, 1920, Tafel I; A Hermam, *Lou-Lan, China, Indien und Rome im Lichte der Aus-gratungen am Lob-nor*, Leipzig, 1931, pp. 109–110.

[9] Kagawa Akira, *Atlas of Western Regions Archaeology*, Volume 1.

Ancient paper fragments written in Kharosthi, Kuchean, and Old Hotanese have been unearthed in Yanqi, Kucha, Bachu, and Hetian. Yanqi and Hetian also discovered fragments of Sanskrit manuscripts and Kharoshti before the seventh century.

While transmitting and using paper, Xinjiang also began to manufacture paper. From the second half of the sixth century, there was bast fiber paper made from raw fibers, mainly mulberry bark paper. In recent years, many studies have been conducted on the ancient papers unearthed in Xinjiang, from raw materials and shapes to papermaking techniques, and the paper made in Xinjiang has its own characteristics. The raw materials are mostly broken hemp cloth, similar to those in the Central Plains. However, the paper curtains used for papermaking are not the bamboo curtains commonly used in the Central Plains, but grass curtains made from locally grown wild Juncus effusus stems. Therefore, the heavy papermaking method is often used to obtain thicker paper. This type of paper made in Xinjiang probably began in the early fifth century, and the documents from the eleventh year of Jianchu of Xiliang (415) and the twelfth year of Xuanshi of Beiliang (423) were both made in Gaochang.[10] By the seventh century, Gaochang indeed had papermakers known as "paper masters."

### *The Spread of Paper West of the Pamir Mountains*

Not long after the invention of paper, it crossed the Pamir Mountains and spread to the western regions of Asia. The Persians and Sogdians were the most widely traveled merchants in Central and West Asia, and the word "paper" appeared in their vocabulary quite early. In Persian, paper is called kāgad or kāgid, which can also be written as kāyad, kāyid, kāyaz, or kāgiz; in Sogdian, it is written as kāygdi. In the Uyghur language, it transformed into kegat or kagas. In the old Uyghur text fragments unearthed along the southern route of the Tianshan Mountains and in Central Asia, the word for paper was initially called Kägdä. Among these three major languages, the Uyghur language appeared later. The Uyghurs revered the Manichaean religion, which used the Sogdian and Middle Persian languages. It is possible that they borrowed the word directly from the Sogdian language. The pronunciation of the word for paper in

[10] Pan Jixing, *History of Papermaking Technology in China*, Beijing: Cultural Relics Press, 1979, p. 138.

Persian and Sogdian has similarities to that in Chinese. Friedrich Hirth believed that the Persian word for paper was derived from the Chinese term "穀纸" (cereal grain paper).[11] To be more precise, the word should have been translated from the Chinese term "穀楮纸" (Ceres kozogami). Jia Sixie's *Qi Min Yao Shu,* completed between 543–559 AD, discusses the planting of Chu in Chapter 48, saying, "*Shuo Wen* (a Chinese dictionary) states: Gu (穀) is Chu (楮). However, today people mistakenly call it "Jiao Chu (角楮)," which is incorrect. The confusion arises because the pronunciation of "Gu" and "Jiao" are similar. The bark of the plant can be used for making paper." It also mentions, "Being able to make paper has many benefits." This shows that the term "Gu Chu" has been used in the northern regions of China for a long time. Chu has been used for papermaking since the Eastern Han Dynasty, and since it is grown in various places, can be harvested every three years, and is relatively easy to cultivate, it has become one of the most widely used raw materials for papermaking.

The nine Sogdian documents written by Nanai-vandak to his friends in Samarkand between 311–313 AD are the works of Sogdian merchants who were among the first to enter the interior of China and use paper. Therefore, paper is correctly translated as "grain paper" in Sogdian text, and it was only later in Persian text that it was translated as "grain bark paper" or "grain bark." Persian documents unearthed in Xinjiang were dated to around 450 AD, later than Sogdian documents. However, the actual process of paper spreading to the West was very early. In the third century, when paper was being used in the Loulan area, China's paper had already spread westward from Persia and Vietnam to West Asia and the eastern Mediterranean, possibly due to the introduction by Persian or Sogdian merchants. The Qin envoys and merchants who came to China from the Red Sea to South China in the second and third centuries were the mediators of this long-distance transmission. According to Ji Han's "*Southern Plants and Trees Description*," the Qin people (Egyptian) who arrived in Jiaozhou even purchased honey-scented paper locally and offered it as tribute to the rulers of the Jin Dynasty. At that time, Egyptian papyrus and parchment were used along the Mediterranean coast. Persia and other Central Asian countries, such as Kangju, used parchment for official and commercial documents or for copying books. In

[11] Friedrich Hirth, *Chinesische Studien (Studies in Sinology)*, 1890, p. 269.

Syria, where Arabic was spoken, used bark to write books. Compared to Egyptian papyrus and Persian parchment, the plant fiber paper invented in China was cheaper, lighter, and softer, and it was finer, whiter, and more durable than papyrus. As a result, it drew the attention of Asian countries and became a new product distributed by Persian and Sogdian merchants with a wide footprint and diverse experiences. According to the records of Al Masudi, China's paper was already being used to write palace documents in Sassanid Persia, although the quantity was still small. However, it is a fact that China's paper was already being used for writing in Persia and Mesopotamia after the third century.

China's new invention of paper also spread to India at an early stage. Indian expatriate merchants living south of the Tarim Basin were probably the earliest mediators of this dissemination.

Which route was the earliest for the introduction of papermaking techniques to the Indian subcontinent? Kashmir was a center for the spread of this technology. Kashmir, located near Hetian and at the strategic point of the ancient Sino-Indian trade routes, established a paper mill in Nowshera at an early stage. Later, the technology spread to nearby Sialkot, which was the site of ancient Taxila. It is said that papermaking in various parts of India originated from Sialkot.[12] In 1931, a Sanskrit ancient paper was discovered in Gilgit, Kashmir, by the French Citroën Expedition. French Indologist Sylvain Lévi believed it to be from the sixth century.[13] This Sanskrit fragment is likely related to the Sanskrit fragments discovered by Aurel Stein in Hetian (no later than the seventh century) and the Kharosthi-script fragments found in Yarkand (sixth century). The paper used for these manuscripts should be produced in Xinjiang. The development of Kashmir's own papermaking industry was probably influenced by the Hetian paper.

Another factor that facilitated the spread of paper into the Ganges River Basin was the unimpeded Sino-Indian-Tibetan route from 639 to 670, through which paper was introduced to the southern foothills of the Himalayas via Tibet. In 641, Princess Wencheng married the Tibetan King Songtsen Gampo (Gyisong Ngonzan). In 649, Tibet requested from the Tang Dynasty "silkworm seeds, craftsmen for wine making, grinding

[12] Dard Hunter, *Papermaking by Hand in India*, New York: Pynson Printers, 1939, pp. 13–14.

[13] Andr Blum, *On the Origin of Paper*, 1934, p. 54.

saltpeter, and paper and ink making, which were granted." The paper produced in Tibet was made from Stellera chamaejasme L., a plant of the Daphne family. A Tibetan fragment of the "*Sutra of the Buddha's Teachings on Rice Straw*" made from this type of paper was found in Andeyue, Xinjiang, dating from the late eighth century. It is possible that this Tibetan papermaking technique was also introduced to Central India.

Not only was paper transmitted westward on land, but the maritime route has also been a means of paper distribution since ancient times. Egyptian and Syrian merchants may have become the main actors in transmitting this new product by sea as early as the third century. In the early Jin Dynasty, the southern coastal areas of China were not only capable of producing paper but also produced two unique types of paper: honey-scented paper (fragrant bark paper) and side-ridged paper (moss paper). In the Jin Dynasty's *Southern Plants Record,* a honey-scented paper was made from the bark and leaves of the honey-scented (agarwood) tree, with fish roe patterns in the paper and a strong texture. "In the fifth year of Taikang (284), Daqin (Rome, referring to Egypt at this time) presented 30,000 sheets of it." This type of paper continued to be produced in Guangdong during the Tang and Song dynasties, known as fragrant bark paper. In the third century, when Daqin (Egypt) merchants and envoys traveled to and from the Jin Dynasty, they took the maritime route. As Guangdong's honey-scented paper became a gift for Egyptian envoys traveling north to the Jin capital, it must have also spread to Egypt when the envoys returned to their country.

During the Tang Dynasty, ships regularly sailed the Persian Gulf routes, and Chinese paper was continuously exported from Guangzhou. In the mid-tenth century, al-Istakhri's *Book of Routes and Realms* noted that paper was a significant import for Siraf. The Tang Dynasty not only produced a wide variety of parchment papers in large quantities but also improved hemp paper and saw the flourishing of rattan paper production in the south. Moreover, Shaozhou in Guangdong even invented an unprecedented type of bamboo paper. These papers could all potentially be exported. Al-Istakhri's account reveals at least one fact: while the Arab world had already been widely producing plant fiber paper, they still imported a large amount of paper from China. This was because the locally made paper could not meet the rapidly increasing demand, and Chinese-made paper was of higher quality, with a rich variety, diverse colors, and larger sizes, all of which were unmatched by Arab paper.

Papermaking technology also spread to Japan and Java via sea routes. Paper was introduced to Japan from the Korean Peninsula. During the Three Kingdoms period on the Korean Peninsula, paper was made from hemp and paper mulberry, and locally made paper could be found in the relics of the National Scholar Peak at Daechong Mountain in Pyongyang, dating back to the sixth century at the latest. It then spread to Japan. According to the "*Nihon Shoki*," in the 18th year of Empress Suiko (610), "In the spring of March, the king of Goguryeo presented monk Damjing and Dharma regulations. Damjing was knowledgeable in the Five Classics and skilled in making colored paper and ink; he also introduced the production of grinding stones at that time." Damjing (579–631) was the first person to introduce papermaking and ink production to Japan.

Japanese scholars have identified remnants of household registers from the Dabao era (701–703) in the collections of Horyuji and Todaiji temples, as well as ancient papers from around the Tenpyo era (729–748). The ancient papers were mostly made from torn hemp cloth and paper mulberry bark, with a pulp-making process that used plant ash lye for cooking and added starch adhesive, similar to that in China.

Java began papermaking around the tenth century, and *Fuxuan Yelu* from the Song Dynasty already recorded Java's paper production. Bengal learned to make paper in the fourteenth and fifteenth centuries. Ma Huan, who accompanied Zheng He on his voyages to the Western Ocean, highly praised a type of white paper produced in Bengal in his book *Yingyai Shenglan*: "This white paper, also made from tree bark, is smooth and delicate, just like deerskin."

### *Papermaking in Central Asia, West Asia, and North Africa*

In the early stages of the paper's introduction to Central Asia, West Asia, and South Asia, these regions were still using tree bark, animal skins, palm leaves, and Egyptian papyrus as writing and recording materials. To meet the growing demand and surpass these traditional writing materials, it was necessary for these regions to produce paper themselves. In July 751 (the 10th year of Tianbao), a battle between the Tang Dynasty and the Abbasid Caliphate took place at Talas, which first introduced the novel papermaking technique to Central Asia, China's close neighbor. Talas is now located near the Talas River in Kyrgyzstan. During the Battle of Talas, Gao Xianzhi, the Tang's Governor of the Western Frontier, led a coalition of Tang troops and Central Asian nations, such as the Türgesh, against the Abbasid forces commanded by Ziyad ibn Salih. The Tang forces suffered a defeat, and 20,000 Tang soldiers were captured. Arab historical records specifically mention the capture of papermaking artisans among the prisoners. These artisans introduced papermaking techniques to Samarkand, laying the foundation for the paper industry in the Arab Empire.

Samarkand was the most important paper supplier in the Arab world for several centuries after the establishment of papermaking in 751 AD. Ibn Haukal mentioned the origin of Samarkand's paper industry in 976 AD, stating that paper was brought from China to Samarkand by prisoners captured during the Battle of Talas. These prisoners were owned by Salih's son, Ziyad, some of whom were papermakers. After the establishment of the paper industry, "the manufactured paper not only met the local needs but also became a major trade commodity for the people of Samarkand, thus meeting the needs of various countries and benefiting humanity."[14] Even earlier, in the first half of the ninth century, Tamīm b. Bahr wrote in his "*Turkic Chronicles*" that Abu al-Fadl al-Vāsjirdī, an Arab writer, stated that Muslims had recently won a victory and captured many people, some of whom were children who were now making high-quality paper, weapons, and tools in Samarkand. In 1030, the Arab scholar Al-Biruni also made a similar statement in his *Chronicles of India*, claiming that "the Chinese were the first to make paper," and that "the Chinese prisoners of war introduced paper-making techniques into Samarkand. Since then, many places have started making paper to meet their practical needs at the time."[15]

Once established, the paper industry in Samarkand produced high-quality paper due to the skilled Chinese craftsmen. Its sales quickly expanded, attracting the attention of al-Fadl ibn Yabyā, who was the governor of Khorasan. With his patronage, a new paper mill was established in 794 in Baghdad, the capital of the caliphate. After the paper mill was put into operation, the vizier Ja'far, brother of the Khorasan governor Harun al-Rashid (786–809), ordered the government to officially use paper for official documents instead of the extremely expensive parchment, which could be easily altered.

[14] Carter, *The Invention of Printing in China and its Transmission to the West*, 1955, p. 134.

[15] Saka ed., *Al-Bīruni's "Chronicles of India,"* (Ta'rikh al-Hind), 1914, London, p. 171.

The demand for paper sharply increased in the Abbasid caliphate, leading to a paper-making craze in the ninth century, which resulted in the establishment of two new paper mills in West Asia. First, a paper mill was established in Tihāmah, in southern Arabia, using plant fibers as raw materials. Soon after, a large-scale paper mill was established in Damascus. The paper continued to be exported to various parts of the Mediterranean for several hundred years and was even called "Charta Damascena" by Europeans. However, due to continuous technological improvements, Samarkand's paper continued to hold a high position in the Arab world until the early eleventh century. Paper was still considered a specialty of China and Samarkand by Arabs. Samarkand paper enjoyed the highest reputation in Central Asia and West Asia.

These paper mills established in Arab territory played a decisive role in promoting the use of paper. The early Arab paper-making process was entirely based on the Chinese method, from raw material selection to the manufacturing process. The raw materials used in later Arab paper-making were slightly different from those used in China. Firstly, although Arabs also used hemp cloth to make paper, they mostly used Linum usitatissimum due to its abundant production in Central and West Asia whereas China used hemp and ramie as raw materials. Secondly, in addition to using recycled hemp cloth, Arabs also directly used raw hemp fibers to produce paper, which was more expensive and consumed more raw materials.

Egypt began to have its own paper mills at the end of the Tulunid period (868–905) and the widespread use of paper began during the Fatimid dynasty (909–1171). Through Egypt, papermaking spread to Sicily and then continued westward along the Mediterranean, with Fez, the capital of Morocco, beginning papermaking around the second half of the eleventh century or in 1100, followed by its transmission to Spain.

The application of paper propelled the further development and prosperity of Arab science and culture. During the reign of Mamun (813–833), Baghdad established the "House of Wisdom" in 830, and the translation work actively initiated by the Caliph Harun al-Rashid of the Abbasid dynasty was carried out on a large and systematic scale. The "House of Wisdom" was composed of an academy of sciences, a library, and a translation institute. The lightweight paper produced by the paper mills in Samarkand and Baghdad provided the most convenient conditions for the development of the translation industry. The translation institute gathered scholars and translators of various faiths, including

Islam, Nestorian Christianity, Judaism, and Zoroastrianism, who specialized in translating various specialized works written in Greek, Syriac, Persian, and Sanskrit, widely absorbing scientific heritage from various countries around the world. Among these translators, the three most famous were Hunain and his son Isaac and nephew Hobaish. Isaac, known in Europe as Joannitius, was himself a physician who was proficient in Greek and had translated the works of Aristotle, Plato, Hippocrates, and Dioscorides into Syriac, which were then translated into Arabic by his students. The *Anatomy* of Galen, the famous Greek physician, and many other works were translated directly into Arabic by him. Many scientific works from ancient Greece were preserved almost entirely by their Arabic translations. Chinese scholars living in Baghdad also benefited from these translations. The Baghdad stationery dealer Abu'l-Faraj ibn Abu Ya'qub al-Nadim, in his *Catalogue of the Sciences (Kitāb al-Fihrist al'ulūm)* compiled in 988, mentioned that the physician and alchemist Razi (850–925) had taught a Chinese scholar, who used shorthand to transcribe a sixteen-volume Arabic translation of Galen's monumental work in a month through oral teachings.[16] By the tenth century, paper had become so widespread in the Persian Gulf and the Tigris-Euphrates region that multiple volumes of scientific works could be copied in a short period of time. The use of paper made it possible for the translation industry of the eighth and ninth centuries to develop, and paved the way for the creative development of Arabic science from the tenth century onwards. Caliph al-Hakim of the Fatimid dynasty (996–1021) established the "House of Science" in Cairo, which promoted teaching, research, and the development of translation. Arabic scientific works became the basis for various Latin translations from the eleventh century onwards. In the Middle East, Arabic paper ultimately replaced traditional papyrus and parchment, becoming the most convenient writing tool and packaging material.

### *The Rise of the European Paper Industry*

The introduction of Arab paper into Europe began around the ninth century. According to traditional accounts, paper was introduced into Europe through the spread of the Fatimids from North Africa, passing

[16] Ferrand, Gabriel, *Relations de voyages et textes géographiques arabes, persans et turks relatifs à l'Extrême*, Paris: E. Leroux, 1913, vol. 1, p. 135.

through Morocco and arriving in Al-Andalus under the rule of the Umayyad Caliphate of Cordoba in the mid-tenth century, and later widely adopted in France. However, there is evidence that Abdul Rahman of the Cordoba Caliphate (756–788) had already used stone printing (tab‘) to reproduce official documents and distribute them to various departments. Printing could not be achieved without paper. Abdul Rahman escaped the Abbasid family's search in Syria, fled to Maghreb, and rebuilt a new dynasty in Spain. He was likely already aware of paper production in Samarkand during his time in Syria. During his reign, he managed to obtain paper from West Asia, thus introducing paper into Europe.

After the establishment of the Damascus paper mills, Damascus paper became the main commodity exported to Europe for centuries. The Europeans used to call the paper "Charta Damascena" (Damascus paper).

Arabia became the nearest source of paper imported into Europe. Damascus, in particular, became a center for the export of paper to Europe.

In 1109, paper was brought to Sicily from North Africa. King Roger II of Sicily issued a decree written on paper in both Arabic and Latin, which became the oldest known paper document in Europe. Paper was then brought from Sicily to Genoa and it took more than 40 years of effort. The archive of Genoa still preserves a manuscript written on paper, some parts of which date back to 1154.

Although paper continued to flow into Europe from Damascus, Cairo, and Fez, the secret of papermaking remained in the hands of the Arabs. In 1224, Yakut mentioned in his *Geographical Dictionary* that Xātiva (Shātibah or Ja'tiva) was the papermaking center of Spain.[17] The papermaking industry in Spain was also one of the earliest in Europe. Xātiva in southeastern Iberia may have begun making paper as early as the 1220s. The St. Giles' monastery codex from 1129 contains both parchment and a new type of paper, which may have been locally produced in Spain rather than imported from North Africa. The paper mills in Spain were built by Moroccan Moravid people during a period of instability in the Islamic regime of Cordoba in North Africa. From 1086 to 1121, Christian forces from Castile and Leon invaded Spain, resulting in over 30 years of war.

The spread of paper in Europe facilitated the cultural prosperity of Cordoba under the rule of Abd al-Rahman III (912–961). His successor,

[17] Yaqut, *Mjan al-buldon*, edited by Husdenfeldt, Leipzig, 1866–1873, reprinted 1925, vol. 3, p. 235.

Al-Hakam II, placed great emphasis on the acquisition of books and sent people to bookstores in Alexandria, Damascus, and Baghdad to collect various books and manuscripts, totaling over 400,000 volumes, including some paper copies. Cordoba had 70 libraries, of which the library in Toledo was the largest and most renowned in Western Europe, with an extensive collection. After the Christian conquest of Toledo in 1085, the city became a place where Europeans went to study and absorb advanced Arab cultural and scientific knowledge. By the tenth century, Cordoba had become a cultural center that could rival Baghdad and Constantinople.

The widespread production of paper also led to a translation boom of classical works in Spain, which was only matched by the translation activities in Baghdad during the ninth century. Cordoba became a bridge between Greek classical culture and modern European science through the translation of works from Arabic to Latin. The Ibn Tibbon family of Jewish scholars in Provence played a significant role in translating important Jewish works into Arabic and then into Latin, including Judah ben Saul (1120–1190) and his son Samuel (1150–1230), who was born in Spain and moved to Lunel. Scholars from Italy, Germany, England, and France also contributed to this translation effort. Notable translators included Gerard of Cremona from Italy, Hermann from Germany, and Adelard of Bath from England. Some scholars directly translated Greek into Latin, such as James of Venice in the twelfth century and William Moerbeke of the Flemish people in the thirteenth century. Many important Arabic scientific works were translated into Latin and other European languages during the eleventh and twelfth centuries, allowing European scholars to access the ancient Greek intellectual heritage through Arabic translations.

The paper-making technology spread to Christian countries in Europe and surpassed Islamic countries, marking the beginning of a new craft in the Christian world in 1189. The first paper mill in the Christian world was built in the city of Huelgas, Catalonia, at the foot of the Pyrenees that year. However, the paper-making technique was still operated and taught by Muslim craftsmen. The Kingdom of Sicily had already been using Arab paper, but due to political conflicts and the intervention of the Pope, this advanced invention from Islamic countries could not be fully developed in Italy for a long time. In 1221, Frederick II (1215–1250), the ruler of Italy, ordered the prohibition of using paper to write official documents under the Pope's decree in order to resist the import of paper from Islamic countries. However, the actual need for paper did not decrease. After the Crusades, the paper mills in the eastern Mediterranean continued to export to Europe and this trend continued throughout the thirteenth century. The early paper used in Europe was made from hemp fibers, which were part of the same system as Arab paper and the even older Chinese paper, as confirmed by scientific analysis. In 1276, Montefano founded the first paper mill in Italy, and it was also around this time that the first records of Christians in Spain beginning to make paper appeared. From then on, the Italian paper industry flourished and by the fourteenth century had become the main source of paper supply in Europe, surpassing Spain and Syria.

Although paper had been used in Europe since the late eighth century, due to the backwardness of scientific culture and the small number of literate people, the use of paper was limited for a long time. Paper did not surpass parchment until the fifteenth century and achieved overwhelming advantages. Although parchment was expensive and consumed a large number of lambs, under the rule of Christian obscurantism, advanced scientific knowledge could not be disseminated, and paper could not be widely adopted. Parchment still had considerable adaptability to the traditional European writing methods using goose feather pens and ink solutions. Therefore, paper that could adapt to European writing habits had to be relatively thick, smooth, and non-absorbent. Italian paper was quickly promoted in the Western European intellectual community because it had these advantages.

In the fourteenth century, the rapid growth of paper usage in Germany promoted the development of the Italian paper industry, as Germany relied almost entirely on Italy for its paper needs. Especially after the introduction of woodcut printing into Europe in the late fourteenth century, the Italian paper used for printing playing cards and color images became abundant and widely used. It is said that paper mills had already been established in Cologne and Mainz, in southern Germany, by 1320. Soon after, one of the early printing centers in Europe, Nuremberg, established a paper mill in 1391, and woodcut printing was also born in Nuremberg at that time. The fourteenth century was a significant century for paper production in Europe. At the beginning of the century, paper was still relatively rare in Europe, with only two or three paper mills producing paper in Italy, except for Spain. By the end of the fourteenth century, Italy, France, southern Germany, and Spain were all producing paper. Paper had largely replaced parchment as the prevalent writing material, except for the wealthy. Of course, most of the surviving hand-written and printed books from that time were produced for the wealthy, and a large portion of them were printed on parchment. For example, Johannes Gutenberg, the inventor of European movable type printing, needed 300 parchment sheets to print a *Bible*.[18] From the fifteenth century, the papermaking technique spread from Germany to both the east and the west. Krakow, Poland established a paper mill in 1491, while Vienna, Austria did not start papermaking until 1498. In England, books were first printed, and it was not until 1511 that the first paper mill was built. Russia had a paper mill in Moscow around 1576, Denmark did not have one until 1635, and the first Norwegian paper mill was built in 1690. By the seventeenth century, most European countries had adopted the Chinese-style manual production methods and technological equipment for papermaking, but during the fifteenth and sixteenth centuries, parchment was still commonly used for writing.

The American continent initially relied on imported paper from Europe, in addition to parchment and bark as writing materials. Mexico, through the spread of Spain, established the earliest paper mill in the Americas in 1575. Then, in 1690, papermaking began in Philadelphia, USA. Benjamin Franklin, an American politician and scientist (1706–1790) who had worked as a printer, wrote an essay entitled *Description of Making Paper in the Chinese Manner*. He proposed an improvement to the European method of gluing small sheets of paper together and polishing them with agate or flint stones to make large sheets and introduced the Chinese method of using two papermakers to produce a large sheet of paper 20 feet long and 6 feet wide, which could be dried on a heated wall to obtain a smooth, large sheet of paper. Oceania did not have a paper industry until 1868 when printing and writing paper began to be produced near Melbourne, Australia. The long journey of paper, which started from China, first spread to the West in the first century BC and traveled around the world for two thousand years before reaching all five continents.

[18] Theodore Low de Vinne, *The Invention of Printing*, New York: F. Hart & Co., 1876, p. 41.

## 5.2 The Spread of Woodblock Printing

### *The Invention of Woodblock Printing*

Woodblock printing originated in China and is universally recognized as such. The invention of printing greatly expanded the role of paper and became a tremendous driving force for cultural dissemination and social development.

There are various claims regarding the origin of woodblock printing, but it is generally believed to have started during the Tang Dynasty in China. In ancient times, inscriptions were carved on materials such as tortoise shells, animal bones, stones, gold, wood, and bamboo as tools for communication and preservation. Despite the use of silk and paper for writing, they still had limitations in terms of preservation and dissemination, which led to the invention of woodblock printing. There were three factors that directly led to the invention of woodblock printing: first, rubbing and reverse writing on steles. In 498, during the Northern Wei Dynasty, the inscriptions on the stone carvings of the Laojun Cave in Luoyang had raised characters, and since the Jin Dynasty, there had been raised reverse writing on bricks and tiles. In addition, the stele of the Divine Way at Xiaojing (died in 523) Pass in Danyang during the Liang Dynasty had reverse writing inscribed on it, which directly inspired woodblock printing. Second, woodcut symbols and seals. In ancient times, seals were written in reverse script. During the Han Dynasty, large peach wood seals were already in use, and alchemists in the mountains carried the "Exceeding Seal of the Yellow God" as a talisman to ward off evil spirits. In the Jin Dynasty, Daoist sorcerers used jujube wood to carve symbol seals, and during the Northern Wei Dynasty (532), vermilion seals were used to stamp paper books or stitch-bound documents.[19] Thirdly, writing and drawing on wooden boards was also a precursor to woodblock printing. Since the time of Emperor Huan of Han, records of ink inscriptions on door panels have been found, and simple books with written funerary objects have been found in tombs. During the Sui Dynasty, Daoists engraved the stars, the moon, and the sun on wooden seals. The woodcut reverse script was then used to create woodblock printing using the rubbing method (using a brush and ink) that had been used since the Six Dynasties period.

[19] *Biography of Lu Tong in the Book of Wei*, Volume 76.

The invention of woodblock printing was influenced by Taoism and Buddhism. According to *Fayuan Zhulin*, volume 39, *The Record of the Western Regions* recorded that during his third visit to India in 660, Wang Xuance received various gifts from the abbot of the Mahabodhi Temple, including "four Buddhist seals". In India, Buddhist seals could be printed on clay or silk. After his return to China in 645, Xuanzang used Buddha statues to print on paper, and Feng Zhi's *Yunxiansanlu* quoted *Sengyuanyilu* as saying, "Xuanzang used the regurgitated paper to print the image of Samantabhadra, spreading it around the world." The Buddhist seals in India were made of clay, but in China, woodblock printing was used, and the seals were printed on paper. By the late seventh century, India also learned to print clay Buddhist seals on silk and paper. Some woodblock prints of these Buddhist seals date back as early as the twelfth year of the Tianbao era (753).[20] Tang Dynasty Buddha's seal with a thousand bodies has been found at Dunhuang.[21]

In the literature, the most reliable account of the origin of woodblock printing is found in the *Hongjianlu* (*Records of the Grand Scribe*) by Shao Jingbang (1491–1565) of the Ming Dynasty. Volume 46 of the book mentions that when Li Shimin's empress died at the age of 36, she left behind ten volumes of "*Nyuze*" (Women's Admonitions) containing stories of exemplary conduct by ancient women. The emperor was impressed upon reading the books and ordered that they be copied and disseminated to future generations. Although the source of Shao Jingbang's information is unclear, his account supplements the historical facts of the early stages of printing. *Nyuze* marked the beginning of rulers using their political power to print books.

The once considered to be the earliest surviving woodblock-printed book in the world is the 704–751 AD printed *The Great Dharani Sutra of Unsullied Pure Light* scroll unearthed from the stone pagoda of the Buddhist temple in Gyeongju, South Korea, and Houki edition of *The Dharani Sutra* (770) in Japan. The Korean printed book used the four character types of Wu Zetian, and Chinese scholars believe that it was printed in Luoyang in 702 AD. 770 AD is the year of the death of Shōtoku Tennō of Japan.

[20] Zhang Xiumin, *The Invention of Printing in China and Its Impact*, 1958, p. 62.

[21] Luo Zhenyu, "Secret Records of the Stone Chamber of the Mogao Caves," *Oriental Journal*, vol. 6, 1909.

According to the *Continuation of Japanese Chronicles* compiled in 797 AD, Shōtoku Tennō had ordered the printing of one million sheets of sacred text, which were placed in ten temples including Horyuji, Todaiji, Yakushiji, Daianji, Genkoin, Fukushoin, Saidaiji, Kofukuji, Chofukuji, and Shitennoji in the Kyoto area. The printing of one million stupa and one million texts in each temple was completed in the fourth month of the year 770 and the emperor passed away. In October of that year, Emperor Kōnin Tennō ascended the throne and changed the reign title to "Houki," hence the name "Houki edition" of *The Dharani Sutra*.The printing technology of Japan was naturally introduced from China, and it is possible that the Japanese envoy to Tang requested woodblock craftsmen to teach them the technique. The beginning of Japanese printing technology was in 754 AD when Jianzhen brought the technology to Japan.[22] In 1974, a paper single-page print of the *Dharani Jing* in Sanskrit was unearthed from a Tang Dynasty tomb at the Xi'an Diesel Engine Factory. The thirteen-line text surrounding the page has been identified as a print from the early seventh century and is the oldest surviving single-page print.

The printing of Buddhist images and books began in the Tang Dynasty during the reign of Emperor Taizong (627–649). From 783 to 785, there were "printed papers" used as certificates of trade and taxation by merchants. During the Changqing reign (821–824) of Emperor Muzong, Bai Juyi's and Yuan Zhen's poetry collections were printed in Yangzhou and Yuezhou. The printing industry had already spread to Luoyang, Jiangxi, Jiangdong, and other places outside Sichuan and Huainan, with Chengdu as the most important printing center. By the time of Emperor Yizong's Xiantong reign (860–873), book printing had become popular and the printing technology had become sophisticated. The surviving earliest printed book in China is the *Vajradhara sutra*, carved by Wang Jie in the ninth year of the Xiantong reign (868), which was stolen and transported to London by Stein in 1907. The Buddha images on the title page, with vivid expressions and exquisite carvings, are among the finest works of early carving and printing.

In the early tenth century, Feng Dao supported the carving of the Nine Classics in Sichuan, which was completed in 953. Soon after, the

[22] 禿氏祐祥:《東洋印刷史序説》, (京都)平楽寺書店, 1951.

newly established Song Dynasty reprinted the *Pali canon*. These large-scale printing projects opened the way for the government to use its power to print Confucian and Buddhist classics. Since then, printing books has become a trend, and woodblock printing is no longer just a tool for spreading folk beliefs such as yin-yang, dream interpretation, the Nine Palaces and Five Wefts (Daoism), and books for literacy education.

### *The Spread of Printing Technology to the West*

After the invention of printing, it first spread to Japan, Korea, and Vietnam, neighboring countries with close relations to China, until the rise of the Mongols. However, in the regions of Central Asia and the Arab world, where paper was rapidly being used, printing made little progress due to discrimination against it by Islam. In 1880, about 100,000 sheets of papyrus, parchment, and plant fiber paper were excavated near the Arsinoë region of Egypt, and about 50 printed items were identified from these discoveries, now stored in Vienna, Heidelberg University, Berlin, and Cairo. Of the six typographic prints in Heidelberg, one was printed on parchment, and the rest were printed on paper. Based on the various Arabic typefaces found on the prints, experts estimated that they were produced between roughly 900 and 1350, mostly during the latter part of this period. The earliest print was a *Quran* manuscript, no. 946 in the Reina collection, which according to Gruendler's study, appears to date back to the eighth century based on the font, but the actual printing date could not have been earlier than the tenth century.

From a printing perspective, the Egyptian typographic prints varied in quality, representing printing technologies from different times and places. Despite belonging to different eras, these prints shared three common characteristics: (1) their content is either Islamic prayers or *Quran* verses, with some containing exorcism spells; (2) the printed text is all in Arabic, except for one prayer that has Coptic phonetic letters around the Arabic text; and (3) most importantly, they bear a striking resemblance in appearance to the printed pieces unearthed from mainland China and Turpan since the twelfth century until the Mongol period, with identical printing methods that involve lightly brushing the ink on the paper with a brush.

The connection between China and Egypt in the field of woodblock printing was clearly established through different ways and media in different periods. In the early eighth to tenth centuries, woodblock printing was introduced to Mesopotamia and Egypt through both land and sea routes. The method of tab, which was used by the Abbasid Caliphate in Cordoba, is the earliest record of printing in the Arab world and was inspired by Chinese printing techniques. It is possible that this dissemination was initiated by the Chinese craftsmen who were forced to relocate to Kufa after the Battle of Talas in 751. In addition, a large number of Arabs residing in Yangzhou were likely to be the early promoters of the westward spread of printing. Yangzhou was one of the earliest places in China to engage in printing. In the 1950s, during archaeological excavations in Fayyum, 30 wooden blocks carved with Arabic script were discovered. These objects date back to the tenth century and from their carving methods and frame styles, they appear to be no different from Chinese woodblock printing. Later, before the rise of the Mongols in the eleventh to thirteenth century, woodblock printing spread through the mediation of the Uyghurs to Central Asia and West Asia, across the Pamir Plateau and entered Cairo. During this period, the Uighur civilization was thriving, and Uyghur generals were the actual rulers of Baghdad and Cairo. From 1172 onwards, the Uyghurs established the Ayyubid dynasty in Cairo, establishing Kurdish rule. Large numbers of Uyghurs moved from Central Asia to Mesopotamia and Egypt, which also had a certain influence on the westward transmission of printing technology.

In the thirteenth century, due to the expansion of Mongol power, woodblock printing once again formed a wave of westward transmission. This time, Persia played an unprecedented role and pushed printing technology into Europe. Famous historians of the Ilkhanate such as Rashid al-Din's *Compendium of Chronicles* (1310) carefully described Chinese printing technology.

Despite their knowledge of Chinese woodblock printing, the Persians did not use it for printing books. Instead, they popularized printed playing cards, which quickly spread to Europe in the fourteenth century and led to the establishment of handmade card-making workshops there. Playing cards were invented in China and were initially called "golden leaf grids." By at least the time of the Southern Tang Dynasty, there were shops in Hangzhou specializing in the sale of playing cards, which were known as "fan cards." By the end of the fourteenth century, card games were popular throughout Europe. Germany and Spain banned playing cards in 1377. Luxembourg had cards in the account books of the Grand Duke of Brabant in 1379. In 1392, playing cards were first seen in the court of Paris, but they were paintings, not prints. Soon after, in 1397, the mayor of Paris issued an order prohibiting workers from playing cards during working hours. The popularity of playing cards, of course, was due to printing.

The popularity of playing cards in Italy promoted the papermaking and printing industry. In 1423, Saint Bernardino of Siena had to deliver a speech against playing cards in front of the Church of St. Peter in Rome, causing the audience to gather all the cards on the square and burn them. In 1441, the Municipal Council of Venice ordered a ban on the import of playing cards and printed images to prevent imported products from threatening the livelihoods of local printing and painting craftsmen.

In Europe, playing cards and religious images were probably printed by the same group of craftsmen, starting from the second half of the fourteenth century. The earliest surviving religious print with a date is the Saint Christopher image of 1423 in Nuremberg, which was the same year that Rome banned playing cards. From a printing technology perspective, image printing began even earlier. Records from the city councils of Augsburg and Nuremberg between 1418 and 1438 mentioned playing card manufacturers five times. The records from the city of Ulm showed that playing cards were loaded into barrels and transported by ship to Sicily and Italy. Venice had become the center of European printing in the mid-fifteenth century, with nearly a hundred new printing houses established between 1481 and 1500, and the quality of published books was excellent. According to the *Kairouan Introduction* published in Bavaria in 1539, Father Pagnini (Alessandro de Paginini), who worked as a printer in Venice from 1485 to 1499, had published an Arabic version of the *Quran*, which is the earliest Arabic print in Europe. The printing process in Venice was completely the same as in China, with ink made from soot and glue dissolved in water to form a brownish-yellow color, and only one side of the paper was printed. Italy, Germany, and the Netherlands became the earliest bases of woodblock printing in Europe, and the influence of woodblock printing's transmission from China to Europe was apparent, with Persia under Mongol rule and Egypt under Turkic rule playing an essential role in this process of dissemination.

As for movable type printing, it is also an invention of China. Bi Sheng, a commoner, had already used mud to carve characters and made movable type for printing books during the Qingli period (1041–1048). In the thirteenth century, tin movable type was also used, but it was not widely adopted. In 1298, Wang Zhen officially used wooden movable type, which had already appeared among the people, to print 100 copies of the *Dade Jingde County Chronicle*, totaling more than 60,000 characters, which were all printed within a month. During the Yuan Dynasty, wooden movable type printing was also used in Anhui and Zhejiang. Soon, this wooden movable type was also used by the Uyghur people in the northwest. Aurel Stein found hundreds of ancient Uyghur woodcut movable single characters in the Mogao Caves in Dunhuang, which were sawn from hard wood and varied in size. According to research, they were relics from around 1300.

The woodcut single characters of the ancient Uyghur language had an inspirational effect on the origin of movable type printing in Europe. Missionaries, astrologers, and travelers who traveled between the Yuan Dynasty and Europe were the ones who spread this wooden movable type, bringing it to Germany, the Netherlands, and France, thereby promoting the experiments of movable type printing in these countries. The Augustinian monk Mendoza, in his book *The History of the Great Chinese Empire*, published in Rome in 1585, not only mentioned that the printing technique was transmitted from Rus and the Grand Duchy of Moscow to Germany but also believed that "some merchants from Arabia Felix (referring to Yemen) may have brought some Chinese books, so Johannes Gutenberg, who is known as the inventor, based his invention on these books."

During the time when Johannes Gutenberg (1397–1468) was engaged in typography, a Dutchman named Laurens Coster von Haarlem (1370–1443) had already printed a Christian pamphlet using moveable type in 1430, though the typeface was not very clear. Gutenberg had previously worked as a typographer and in 1450, he made a set of moveable types using an alloy of lead, tin, and antimony. He then invented a printing press and pawned it to a goldsmith. In 1453, Gutenberg printed the first Latin Bible using moveable type. This first Bible printed with moveable type, known as the *Gutenberg Bible*, had 42 lines per page and was printed on large-format paper or calfskin, with over 640 pages and 1200 sides, bound in two volumes. There are surviving copies of the Gutenberg Bible with the dates "1453" and "24 August 1456" inscribed on them. At the time, 180 paper copies and 30 calfskin copies were printed, for a total of 210 copies. Today, 32 complete copies of the *Gutenberg Bible* with 42 lines are known to exist, along with fragments. In addition, there are approximately 8 surviving copies of the later 32-line Gutenberg Bible, which was printed to imitate the style of hand-painted manuscripts of the time. Subsequently, typography workshops were established in Italy, France, the Netherlands, Hungary, Spain, England, Denmark, and Sweden by German typographers to print religious books such as *the Bible*. These fifteenth-century printed books, collectively called incunabula, meaning "swaddling clothes," marked the beginning of the era of printed books. Over a period of 50 years, approximately 30,000 different titles were printed in Europe, with about 200 copies of each title printed. These printed books had a significant impact on the cultural landscape of Europe.

The invention of printing revolutionized the cultural landscape of Europe, which had been severely underdeveloped. The printing press gradually changed the situation where only monks could read, write, and receive higher education, and put culture into the hands of the common people, which in turn promoted the religious reform launched by Martin Luther (1483–1546) and pushed the new cultural movement called the Renaissance to its peak in Western Europe.

## 5.3 The Use of the Navigational Compass

In ancient China, magnets and their polarity were already discovered. After the Tang Dynasty, for the needs of navigation, artificial magnetic compasses were created. This compass was made by magnetizing sewing needles, manually transmitting magnetism, and sharpening the needle's edge with a magnet. Shen Kuo (1032–1096), a great scientist in the Northern Song Dynasty, knew that the magnetic needle always deviated slightly to the east, and he recognized the deviation of the geomagnetic field. In his book *Dream Pool Essays*, Shen Kuo summarized the four methods of using the magnetic needle at that time: (1) the water-floating method, where the compass is horizontally placed on a wick and floats on the water surface, rotating freely and indicating the direction; (2) the fingernail rotation method, where the compass is placed on a finger nail; (3) the bowl lip rotation method; (4) the thread suspension method, where a thin thread is attached to the middle of the compass with wax and hung in a windless place. The last method is the most accurate, and the basic structure of modern magnetic variometers and magnetometers is based on this method.

In order to have a fixed device for the compass needle, so as to orientate during navigation, the compass, originally called the "di luo" (地螺), appeared. The *Yinhuahulu* (因话录) written by Zeng Sanyi in the

Chunxi period (1174–1189) mentions the "di luo," which was a transitional name from the traditional rectangular disk and circular luo jing pan (罗经盘) used by ancient feng shui masters. The luo jing pan is a circular disk made up of the traditional twenty-four geographic directions, composed of eight stems (甲乙丙丁庚辛壬癸), twelve branches (子丑寅卯辰巳午未申酉戌亥), and four gua (乾坤巽艮), with the compass needle indicating the direction. There are also two needle marks on either side of the needle, making a total of forty-eight directions.

The earliest recorded use of the compass in navigation was in the Xuanhe period (1119–1125) in Zhu Yu's book *Pingzhou Table Talks*. He wrote, "The boatmen know the geography. At night they observe the stars, and in the day they observe the sun, and in cloudy weather, they observe the compass." The book describes the experiences of his father Zhu Fu, who served as an official in Guangzhou between the Yuanfu and Chongning periods (1098–1102). By the end of the eleventh century, the compass was already being used in navigation. Twenty years later, in 1123 (Xuanhe 5), Xu Jing described the use of a compass in navigation in his work *Xuanhe Embassy To Koryŏ* referring to it as a floating compass needle: "After passing Penglai Mountain, the water is deep and blue, and the waves grow stronger. In the middle of the night, you can't stop, you can only look at the stars to move forward, or if it's cloudy, use a compass floating needle to tell north from south." In the first half of the twelfth century, this floating needle was likely the prototype of the water compass, and at that time, navigation relied primarily on astronomical navigation, with the compass being used only in cloudy weather.

In the thirteenth century, compass navigation had become the sole basis for maritime navigation. During the Xianchun period (1265–1274), Wu Zimu officially referred to the maritime compass as the "needle disk" in the *Mengliangl u*. The use of this water compass has a long history, and Joseph Needham even speculates that it may have been used for navigation as early as the ninth or tenth century.[23]

The use of the compass greatly improved the accuracy of navigation and allowed ships to safely navigate along fixed routes, providing assurance for regular voyages between ports of departure and destination. Once the compass was used, it was adopted by Arab navigators, and Persia and Arabia used the forty-eight-point method based on the Chinese

[23] Joseph Needham, *Science and Civilisation in China*, Cambridge, 1971, vol. 4, p. 563.

compass. The Arabic word for compass direction, Khann, is derived from the Min Nan Chinese word for compass, "Zhen." According to Léonard's research, the compass is mentioned in Arabic literature in Abu al-Fida's *"Geography"* in the early thirteenth century.[24] According to legend, the compass had already been called the "sailor's friend" in 1180.[25] As for Europe, Latin literature has already seen the use of the compass by the end of the twelfth century. It is a known fact that the compass had already been introduced to the Mediterranean in the twelfth century and was first adopted by Italian merchant ships.

The compass had enormous economic value and provided favorable conditions for the further development of maritime trade, so it was soon adopted by the Indian Ocean shipping industry. Europeans, who initially used the compass, were no match for Chinese navigators in terms of navigation technology. Even as late as the end of the fourteenth century, the compass used in Europe was not as accurate as that used in China. In the year 1391, the English poet Chaucer recorded the use of a thirty-two-point compass, which was different from the Chinese compass.

The use of the nautical compass led to the advent of the needle route and the nautical map, which further perfected maritime navigation. It can be seen that the compass helped calculate sailing schedules. Zheng He's voyages to the West relied entirely on routes for navigation, enabling them to sail freely and deeply into the Indian Ocean.

The use of navigational compass greatly changed the maritime industry, especially with the emergence of practical navigation maps. These maps were used to ensure safe navigation along a designated course, calculate the sailing schedule and determine a ship's location. In the latter half of the thirteenth century, there were many Han Chinese and Mongols who lived in India and had contact with Christian navigators from Venice, Genoa, and other Italian city-states in places such as Quilon and Kochi in southern India and Aden in Yemen. European navigators were able to interact directly with Chinese navigators and obtain practical sea maps of the Indian Ocean drawn by the latter. As a result, practical navigation maps emerged in Europe around 1300, breaking away from the traditional T-O world map system. The earliest surviving navigation maps

[24] Lerner, *Geographie d'Abu Fida*, Introduction, Paris, 1848, vol. 1, pp. CC III–CC IV.

[25] *Lslamic Qulture*, 1942, vol. XV, p. 404.

appeared around 1280, and the oldest extant map with a confirmed date is the 1311 map drawn by Pietro Vesconte, which features intersecting compass lines, rectangular grids, and calibrated edges. Rather than being influenced by Arab navigators, the European map-making tradition was likely influenced by navigators from farther east, which is more in line with practical reality. The appearance of grid-style navigation maps in Europe was not a simple imitation of Chinese navigation maps that had been using grids and compass directions. Italian and Spanish navigators not only recognized the economic benefits of compass navigation but also reintroduced quantitative cartography, which had been dormant for more than a thousand years among the Greeks, into the making of navigation maps. In other words, it was the practical use of the compass that prompted European navigators to take notice of Ptolemy's coordinate system again. The use of compass direction lines or rhumb lines on navigation maps allowed navigators to explore the coastline accurately and made the maps more precise. By the fifteenth century, European navigation maps had re-established Ptolemy's cartographic principles. When Gerardus Mercator (1512–1594) used cylindrical projection to create a world map in 1538, he laid the foundation for modern cartography. All of this was brought about by Chinese navigators who had been using the compass since the eleventh century and later relied on rhumb lines for navigation.

After the adoption of the Chinese compass in Europe, it was improved by using a pivot at the center of the magnet, allowing the magnet to rotate horizontally. It is referred to as a "dry compass." By the twelfth century, this type of dry compass was already in use in European navigation. The pivot-mounted magnetic compass appeared in the Southern Song Dynasty in the thirteenth century but was mainly used for divination rather than navigation. In 1745, the British physicist Gowin Knight developed the steel needle compass, which became widely used in the maritime industry. In practice, the dry compass was still not as accurate and practical as the liquid compass with the 48-point system at that time. However, with the development of history, the top-mounted magnetic needle has finally become the basic device of modern aviation and marine compasses.

## 5.4 The Invention of Gunpowder and Firearms and Their Impact

### *The Invention of Gunpowder*

The invention of gunpowder is closely related to alchemy and traditional Chinese medicine. Black powder is a powder mixture composed of saltpeter, sulfur, and charcoal in proportion, also known as medicine. The invention of gunpowder was caused by alchemists controlling the temperature when refining medicine. The eighth century Taoist alchemy book *Zhenyuan Miaodao Yaolue* mentioned the Redox reactions of saltpeter. A mixture of sulfur, saltpeter, and carbon—with honey acting as the source of carbon may cause an accidental fire, which drew the attention of alchemists to the explosive combustion of gunpowder.

Once gunpowder was invented, it was first used as a tool for traditional fire attacks. *The Biography of Li Xilie* in *New Tang Book* recorded that during the reign of Emperor Dezong (780–804), Li Xilie burnt down Liu Qia's barracks and defensive structures in the city. It implies that Li Xilie had used the newly invented gunpowder fire attack method by alchemists. *Jiu Guo Zhi* by Lu Zhen says that during the beginning of the Tianyou period (904) of Emperor Aidi of Tang, Zheng Fan attacked Yuzhang and ordered his subordinates to "launch machine-fired flying fire" to burn down Longshamen. "Flying fire" could be said to be the earliest name for gunpowder, and also the earliest general term for firearms, specifically referring to artillery launched by catapults or rockets launched by crossbows.

The spread of "Greek fire," a type of incendiary weapon, in West Asia and Europe, stimulated the use of gunpowder in military applications after it entered the coasts of Jiangsu and Zhejiang. "Greek fire" was invented by the Byzantines in the seventh century, using lightweight petroleum (naphtha) from the Middle East mixed with sulfur, asphalt, resin, charcoal, and lime, which was then stored in tanks or hand-held flamethrowers and used to attack enemy troops and ships.[26] In volume three of *The*

[26] W.Y. Car man, *A History of Firearms from Earliest London*, 1955, pp. 4–6; Pan Jixing, *A History of Chinese Rocket Technology*, Beijing, Science Press, 1987, pp. 29–31. A tenth-century silk painting from Dunhuang (now in the Musée Guimet, Paris) depicting a demon holding a long pole with a cylinder spouting fire at the top was considered by Joseph Needham to be the earliest tubular firearm (*Science and Civilisation in China*, Cambridge, 1986, Vol. 1, p. 4, p. 6). Similarly, the bottle held by the god of the Song

*Chronicles of Wu and Yue*, it is recorded that in 919 AD, the Wu and Yue navy had already loaded this incendiary oil, which was imported from Arabia, into iron barrels and attacked the army of Southern Tang at the mouth of the Yangtze River.

During the Northern Song Dynasty, the use of firearms was highly emphasized. In 970, Feng Jisheng and others developed rocket technology, and in 1000, Tang Fu invented rockets, fireballs, and spiked rollers. In 1002, Shi Pu produced fireballs and rockets, and Emperor Zhenzong witnessed the experiment. The same year, Liu Yongxi, a military commander, created hand cannons, and the court ordered their production for border defense. These hand cannons were an improved version of gunpowder bombs. Later, the catapult used to launch gunpowder incendiary projectiles was called "fire cannon,"[27] but the gunpowder incendiary projectiles launched by the catapult were also called "fire cannon," and the two became inseparable.

Rockets and cannons were common firearms of the Song army. Song Minqiu's *Dongjing Records* records that the capital of the Northern Song Dynasty, Bianliang (Kaifeng), had a large-scale arsenal dedicated to the production of gunpowder. They produced poisonous smoke balls, thorn fireballs, and cannons, which were all incendiary firearms and the most widely used firearms in the Song Dynasty.

The improvement in the manufacture of firearms involved packaging gunpowder in a system, igniting it, and causing an explosion, creating explosive firearms. In the eleventh century, the "thunderbolt fireball" appeared in the *Wujing Zongyao*, which can be considered as the precursor of explosive firearms, later modified into more powerful thunderbolt cannons. In 1125 (the first year of the Jingkang period), when the Jurchen attacked Kaifeng, Li Gang, a Song general, ordered the use of thunderbolt cannons and repulsed the Jurchen. The power of the cannon is closely related to the shell, with paper shells being inferior to leather shells, and leather shells being inferior to iron shells. In 1221, when the

dynasty in Cave 149 at Fowan, Beishan, Dazu Rock Carvings, Sichuan, was once identified as an ancient artillery piece (Joseph Needham, op. cit., pp. 580–581). The former has been identified by Chinese scholars as a Greek fire-breather (see Zhong Shaoyi, "Study of Early Tubular Firearms," in *Studies on the History of Gunpowder and Firearms in Ancient China*, Beijing, China Social Science Press, 1995, pp. 101–105), and the latter as a wind-bag used by the king of the wind goddesses (op. cit., pp. 136–138).

[27] *Wujing Zongyao*.

Jurchen attacked Qizhou (Qichun, Hubei), the defending Song army had a large number of leather cannons, while the Jurchen used more powerful iron cannons to attack the city. In 1274 and 1281, the Yuan army twice attacked Japan and used iron cannons.

### *The Earliest Tube-Shaped Firearms*

The use of firearms had to go through a necessary step from medieval projectile weapons to modern guns, which is to keep the cannon loaded with gunpowder on one's own position while using the flames or the spray of debris generated by the combustion of gunpowder to strike the enemy. In 1132, Chen Gui invented more than 20 long bamboo pole guns while defending De'an (Anlu, Hubei). Two people used one bamboo gun, and gunpowder was loaded in the bamboo tube, which was ignited on the spot. This was the world's first batch of tube-shaped firearms. In 1232–1233, the Jin army used "flying fire guns" to resist the siege of the Mongol army in Kaifeng and Guide. Later, the invention of the fire tube followed. In 1259, a sudden-fire gun was created in Shouchun (Shouxian, Anhui), using a huge bamboo tube with a debris chamber inside. It made a loud sound like a cannon and could be heard from a distance of more than 150 steps. This was the world's earliest tube-shaped firearm using a debris chamber, which paved the way for the future use of bullets in shooting tube-shaped firearms.

The fire tube (or "fire rocket") is a tube-shaped firearm that is wide at the top and narrow at the bottom and capable of firing flames. In the early Yuan Dynasty, the tube was made of metal, and the "fire gun" appeared. There were two types: a hand-held gun called "shou chong," and a bowl-mouthed gun placed on a wooden frame, known as the cannon. In 1970, a copper barrel measuring 34 centimeters in length and weighing 3.55 kilograms was unearthed in Banlacengzi, Acheng County, Heilongjiang. According to research, this was a firearm used to suppress the rebellion of Naiyan in 1287.[28] It is the earliest surviving tube-shaped firearm in the world. A bronze matte unearthed in 1974 in Jinglongchi Lane, Dongguan, Xi'an[29]; A bronze cannon excavated in Tong County, Beijing, and

[28] Wei Guozhong, "*Bronze Singijin Unearthed in Half-La Chengzi, A County, Heilongjiang Province,*" *Cultural Relics*, 1973, p. 11.

[29] *Archaeology and Cultural Relics*, 1981, No. 3. Cheng Dong and Zhong Shaoyi: *Atlas of Ancient Chinese Weapons*, PLA Publishing House, 1990, pp. 223–224.

a bronze cannon excavated in Zhangjiakou in 1961 also date from the late thirteenth and early fourteenth centuries.[30] These artifacts demonstrate that the Yuan Dynasty had begun manufacturing the world's first batch of metal-cast tube firearms. Most of the cannons produced from the thirteenth to the fourteenth century were bowl-mouthed firearms, also known as "zhan kou chong." In 1287, Dalimashili used a fire gun to deal with Bulutemür's army in Yanjing, which is the earliest record of the term "chong" referring to a firearm in literature.[31] After this event, copper guns were made in increasing numbers. The earliest known copper gun with an inscription was cast in the second year of the Dade era (1298). It was discovered in 1989 in Shiliin gol aimag of Inner Mongolia.

During the Ming dynasty, the technology for manufacturing bronze cannons had become more advanced. Surviving artifacts include a bronze gun from Hongwu year 5 (1372) in the collection of the Chinese Military Museum, two bronze cannons with inscriptions of Laizhouwei in Hongwu year 8 (1375) weighing 120 jin (73.5 kg) each and measuring 63 cm in length and 23 cm in caliber (Unearthed in Yingzili Village, Penglai, Shandong, 1988), and 25 bronze hand cannons from Hongwu years 10–12 (1377–1379) ranging in length from 32–44 cm. These include the 1378 bronze bowl-shaped cannon unearthed in Guanxian, Shandong in 1964 and in Hezhang, Guizhou in 1977. In 1958, Lulong in Hebei Province unearthed a Ming Dynasty "Yan" copper bowl-mouth cannon, as well as the "golden spear" modeled after the Ah Cheng bronze gun unearthed in Inner Mongolia. Three iron guns from the early Ming period were unearthed in Gaoyao, with a caliber of about 2 cm and a length of 26.5 cm, known as the "invincible hand cannon." The casting of iron guns lagged behind that of bronze guns due to the greater technical skill required to melt and cast iron. The earliest surviving iron cannon is the three-door iron cannon from 1377 (now in the Shanxi Provincial Museum), with a length of 100 cm and a caliber of 21 cm. Another iron cannon was made in November 1426 (Xuande year 1).

[30] Hebei Provincial Museum: *Selected Unearthed Relics of Hebei*, Cultural Relics Press, 1980, p. 232.

[31] *History of Yuan—Biography of Dalimashili*, Volume 145.

### *The Mediating Role of the Arabs*

Upon its invention, gunpowder and firearms quickly spread to Islamic countries. These countries originally produced sulfur but did not know how to use saltpeter, which was a knowledge passed down from China. It wasn't until the eighth century, when alchemy entered the Islamic cultural realm, that the Arabs discovered saltpeter. To Islamic countries, saltpeter was a Chinese specialty, so when it first entered West Asia, it was called "Chinese salt" by Persians and "Chinese snow" (talj al-sīnī) by Arabs, as saltpeter is white like snow and tastes salty like salt. Initially, saltpeter was used as a medicine to treat epilepsy and as a laxative. The Persian word šora originally referred to saltpeter or niter and was derived from the Sanskrit word sorāka, which broadly referred to niter, nitre, and potassium nitrate. The term "Chinese salt" in Persian was probably a reference to a type of sodium nitrate found in Zezhou, Shanxi province. Natron (nitrum), found in Greek and Arabic literature, was often mistaken for saltpeter. For example, in 1926, translator Goddard incorrectly translated natron from Strabo's book as "saltpeter," when in fact, natron is a type of salt compound, most commonly naturally occurring sodium carbonate (Na2CO3·10H2O). In the sixteenth century, Europeans still confused saltpeter (potassium nitrate), sodium carbonate, and sodium chloride, and the names were often mixed up. Thus, both Islamic and European knowledge of saltpeter was introduced from China. Saltpeter was initially used for medicine and alchemy in Islamic countries, but with the development of saltpeter mining, it was also introduced into the glass industry.

From the early thirteenth century, saltpeter began to be used to make gunpowder. This can be seen from the changes in its name. Ibn al-Baytār (1197–1248), a doctor who had traveled to North Africa and the Tigris-Euphrates river valley, had a note under the "bārūd" entry in his *Compendium of Medicine*: "This is what the old Egyptian doctors called Chinese snow. The common people and doctors in the West (North Africa and Spain) call it 'bārūd' which is 'flame saltpeter.'"[32] "Bārūd" is the Arabic word for gunpowder nowadays, but in medieval times, it referred to saltpeter.

Saltpeter, transformed from Chinese salt and Chinese snow to bārūd, not only makes it clear how saltpeter spread from China to Persia and Egypt, but also reveals the clear process of how it changed from a

[32] Leclerc, *Traite des simples*, 1883, p. 71.

medicinal and chemical agent to a material for preparing gunpowder and making firearms. Among various methods of arson and igniters, no saltpeter can be found in a book discussing the method of fire attack written in 1225, which is stored in Leiden University. It can be inferred that saltpeter had not been used to make gunpowder before that year. The appearance of the name "bārūd" in the thirteenth century is the best evidence that saltpeter had been used for combustion, just as during the Five Dynasties and Northern Song periods, saltpeter was used in gunpowder, and the new term "flame saltpeter" emerged. However, initially, this saltpeter that caused combustion was only used for fireworks, firecrackers, and fire performances. It is likely that it was directly introduced to Egypt through the sea route from China's southeast coast. At that time, Chinese sailing ships often sailed to Aden, equipped with firearms, shuttling between the Arabian coast and Quanzhou. Egyptian immigrants also lived in various parts of Hangzhou and Quanzhou, and they were the most reliable conveyors of this new invention. According to an Arabic manuscript from 1249, Oumlay, the Minister of State of the Ayyubid dynasty in Egypt, presided over the first production of gunpowder in Islamic countries. Under the rule of Sultan Ayyub (1172–1249), Egypt completed the initial experiments of using saltpeter to prepare gunpowder and make firearms. Gunpowder was indeed introduced to Islamic countries between 1225 and 1248. This was the first stage of the introduction of gunpowder to Islamic countries. In this stage, the manufacturing methods of fireworks and gunpowder were first introduced to Egypt through the sea route from the Southern Song dynasty.

The second stage of the introduction of gunpowder to Islamic countries occurred after the fall of Baghdad in 1258 when various firearms were introduced into Arab countries from the Yuan Empire. According to Al-Rammah's *The Book of Military Horsemanship and Ingenious War Devices (Kitab al-Furusiya wa'l-Munasab al-Harbiya),* we can learn that not only did gunpowder originate from China, but fireworks and firearms were also transmitted from China.

Chinese rockets and firearms also became the earliest firearms in Islamic countries. In "*Kitāb al-furūsīya wa'l munāsib al-harbīya*," there is a Khitan fire wheel, and the gun head is called "Sahm xatai," which uses the Jin dynasty's "flying fire" with rockets as the combustion body. An anonymous Arabic military book from the early fourteenth century, "Kitāb al-firussié birasm al jihad fī sabil allah," also mentioned firearms used in land battles and rockets used in naval battles, both called Khitan rockets. These two types of Khitan rockets, the former being firearms used in land battles, and the latter being rockets fired from tubular firearms in naval battles, are similar to the sub-chamber of the matchlock firearm because they are fired from a tubular firearm.

From the use of Khitan rockets in combat in Arab countries, two insights can be gained. First, the Arabs did not use bows and crossbows as tools for launching rockets from the beginning because they had strong petroleum machines and directly borrowed firearms from China, avoiding the detours that were necessary in the process of firearm invention. Second, around half a century after the introduction of gunpowder and fireworks to Islamic countries, i.e., the late thirteenth century, Arabs had already begun using small tubular firearms. Specifically, between 1267 and 1274, when the Mongol army besieged Xiangyang, two Muslim artillerymen were invited from Iran. These Muslims then transmitted the Khitan firearms and rockets used by the Mongol army to Islamic countries. In the thirteenth and fourteenth centuries, the West referred to China as "Khitan," so the firearms brought into Islamic countries also had Khitan names.

After the introduction of tubular firearms to Islamic countries, these new weapons were immediately put to heavy use due to their ease of use and powerful impact. In the late thirteenth and early fourteenth centuries, Islamic countries improved the firearms and matchlock firearms passed down by the Mongols and developed two types of "midfa'"(It means "firearms" in current Arabic), both of which were documented in the military books written by Shams al-Din Muhammad at the beginning of the fourteenth century. In the fourteenth century, the Mamluk army in Egypt was equipped with a batch of "midfa," and large copper cannons had already appeared in military use. Gunpowder and firearms were directly transmitted from China to Iran and India, so in modern Persian, Hindi, and Turkish, cannons are called "top."[33] The Arab countries with advanced science and technology in turn became the medium for the introduction of gunpowder and firearms into Europe.

[33] S. Mahdihassan, *Cultural Words of Chinese Origin, Bharatiya Vidya*, 1950, vol. 11, pp. 31 ff.

### *The World Significance of Gunpowder*

Europe's initial knowledge of gunpowder came from the Arabs, and the firearms they used were also introduced from the Arab world.

In the second half of the thirteenth century, Europe began to translate a book on fire warfare from Islamic countries called "*Liber ignium ad Comburendoshostes*" (The Book of the Burning Fire Attack) into Latin. This book was long attributed to the Greek author Marcus Graecus, but after meticulous research since the nineteenth century, it was discovered that the original author was someone from the Islamic world in the mid-thirteenth century. This is the earliest book on fire warfare in Europe.

Another Arabic manuscript called "*Eighty-eight Natural Experiments*" was also translated into Latin by Europeans, and many recipes are the same as those in "*Liber ignium ad Comburendos hostes.*" This shows that the earliest knowledge of gunpowder in European literature all came from Islamic countries.

The famous European scholars and alchemists Albertus Magnus (1193–1280) and Roger Bacon (1214–1292) both wrote about gunpowder, and their writings were based on *Liber ignium ad Comburendos hostes*. They played a role in spreading gunpowder knowledge in Europe, but Europeans learned to use gunpowder and make firearms during military conflicts with Islamic countries.

There are records that France, England, Germany, and Italy had used gunpowder, iron fire pots, and even had red copper guns and iron cannons before 1348, with France in 1338, Flanders in 1339, England in 1340, and Germany in 1346. In the 1420s–1440s, firearms spread rapidly throughout Europe because various Western European countries had learned sufficient lessons in fighting against Muslim armies. These lessons were most painful during three major battles. The first was the Battle of Akka in 1290, when the Mamluks used 92 catapults to continuously attack the city, throwing not only giant rocks but also firing fireballs, fire bottles, and fire pots, which eventually forced the Franks to withdraw from the last fortress on the Asian continent, marking the complete failure of the Crusades. The second was in 1325, during the Castilian revolt against Arab rule, when Arabs used catapults to fire fireballs and attack the city of Baza, causing casualties to both humans and animals and burning houses. The third was in 1342, when the Moroccans defended Algeciras with cannons against the invasion of Portuguese King Alfonso XI, causing significant casualties to the Portuguese. Like printing and papermaking, Spain and Italy were also the first places in which advanced firearms were introduced from Islamic countries. Just one year after the Arabs attacked Baza in 1326, the secret of firearms was mastered by Italians, and Florence ordered the manufacture of iron guns and iron bullets, making Europe's first batch of metal tube-shaped firearms, thus spreading firearms throughout Western Europe from Italy.

The earliest surviving European illustration of a firearm is a drawing of a bottle-shaped cannon with an arrow inserted in the mouth and a knight pointing a fuse, found on a wall in a chapel in Oxford in 1326. Another bottle-shaped cannon from the same year is depicted in the Holkham Bible. In a medieval Italian church, there are frescoes from 1345 and 1364 showing soldiers using hand guns in battle, and a bamboo-shaped fire gun outside a fortress with no bamboo node at the end and a fire hole at the back, with a stone ball inserted at the muzzle, being ignited by a person with a lit fuse. Its shape is similar to the English "telar," and is a good example of the "midfa'" firearm, which is derived from the same origin as the hand-gun.

England was one of the earliest European countries to use and manufacture firearms, as Derby Earl and Salisbury Earl of England learned how to use cannons in the 1342 battle of Algeciras and brought the knowledge back to England. In the 1345 Battle of Crecy between England and France, England used 24 iron cannons, 60 pounds of gunpowder, and the gunner Raymond Larchier received two iron cannons, 8 pounds of gunpowder, and 200 lead bullets from the King of Toulouse. In 1345, England also manufactured 100 ribaldos firearms, which were early versions of three or four-barrel guns. In 1357, England also replicated the "midfa" and produced a type of cannon known as the "telar."

Germany was the third European country to begin manufacturing and using firearms, after Italy and England. In 1348, a long red copper gun that could shoot clusters of arrows was recorded in Frankfurt's financial records. Russia also used explosive firearms in the fourteenth century. In 1382, during the attack of Токтамиш, Khan of the Golden Horde, on Moscow, the Russian defenders used a type of firearm called "tiufak" to repel the attackers. These firearms were found on the island of Khortize and consisted of iron cans filled with iron shavings, gravel, and gunpowder, wrapped with iron hoops.

Initially, most of the European tube-shaped firearms were made of cast iron, which was different from China's use of copper casting in the thirteenth and fourteenth centuries. Iron casting was difficult and the tubes were heavy and prone to cracking. By comparing the existing metal firearms, it can be seen that early Chinese firearms were superior in terms of technology and use. There are only three surviving fourteenth-century metal guns in Europe: one in the Bonn Museum, made of iron and weighing 4.25 kilograms with a length of 18.5 centimeters; one in the Völkerkundemuseum Herrnhut, made of iron and weighing 3.75 kilograms with a length of 16.7 centimeters; and one in the Germanisches Nationalmuseum Nuremberg, made of red copper and weighing 1.24 kilograms with a length of 33 centimeters and a diameter of 17 millimeters. The Tannenberg Castle, where the Tannenberg Handgonne was found in 1849, was destroyed in the 1399 war. The term "Handbüchsen" appeared in Nuremberg literature in early 1388, so the manufacturing period of the Tannenberg Handgonne was around 1390. The manufacturing of Western bronze guns is similar in structure to that of Chinese bronze guns, indicating a direct imitation of Chinese firearms.

In Europe, "firearms were first taken by the cities and the emerging urban-dependent monarchies as weapons against the feudal aristocracy."[34] Therefore, firearms played a decisive role in destroying the feudal knight class. In the process of primitive accumulation of capital, Western European colonial powers launched colonial aggression against coastal areas of Africa, Asia, and the Americas, and firearms were also introduced to these regions. Starting from the fifteenth century, Portugal's firearm technology developed rapidly, and the gun barrels were all made of copper, with great power. Therefore, firearms were not only a powerful catalyst in the process of overthrowing the feudal system in Europe, but also a powerful tool in the process of capitalist colonial expansion.

In Europe, "gunpowder, the compass, and printing - these are the three great inventions that herald the advent of the bourgeoisie. Gunpowder shattered the knight class, the compass opened up the world market and established colonies, and printing became a tool of the Reformation and, in general, the most powerful driving force for creating the necessary prerequisites for spiritual development."[35] These three great inventions played an unprecedentedly huge role in the development of history.

[34] *The Complete Works of Marx and Engels*, vol. 3, p. 207.

[35] K. Marx, *Machines: Application of Natural Forces and Science*.

CHAPTER 6

# The Early Contact Between China and the Americas and Western Missionaries in China

## 6.1 Colonial Expansion of European Powers and Sino-European Relations in the Late Ming and Early Qing Dynasties

### *Portuguese Invasion of Macau*

In 1453, the Ottoman Turks conquered the Byzantine Empire, and the Ottoman Empire occupied Western Asia, annexed Egypt and North Africa, and controlled the routes from the Red Sea, Persian Gulf, and Black Sea to the Mediterranean. They imposed significant taxes on merchants from various countries passing these routes. Meanwhile, on land, trade between European and Asian countries was heavily influenced by the Egyptian merchants and Arab camel caravans. This led to slow land transportation, high freight costs, and a lack of security guarantees, making it increasingly inadequate for the demands of the European market. Western European countries, particularly Spain, Portugal, France, and England, along with other emerging nations, had long sought to break away from the monopoly of the Italian city-states on East–West trade and to seek direct routes connecting Europe to India and China. Portugal and Spain were the first to succeed in discovering new ways. In 1498, a fleet of four ships led by Portugal's Vasco da Gama sailed around the Cape of Good Hope in Africa, arriving in the Indian city of Calicut. Subsequently, Spain's Columbus pioneered the North American route,



F. Shen, *Cultural Communications Between China and The Outside World Throughout History*,
https://doi.org/10.1007/978-981-97-4696-5_6

and Magellan and his companions completed a circumnavigation of the globe in 1522.

Portuguese colonial expansion in the East began with the opening of the New Route. Portugal occupied Malacca in 1511. On August 15, 1517, Fernão Peres de Andrade led a fleet and ran into Guangzhou with his envoy Thomé Pires. Since the Arabs called the Portuguese "al-Frandj," meaning Europeans, the Chinese called Portugal after "al-Frandj." This was an armed demonstration by the Portuguese to open the door of China with gunboats, and "the cannon sounded like thunder, in the name of paying tribute." After being rejected by the Ming Dynasty, they "retreated to the south end of Dongguan, built houses and fences, and relied on the cannon to secure themselves."[1] In 1554, in the name of trade, Portugal entered Guangdong's Lampacau by fraudulent means, and in 1557, it took advantage of the situation and forcibly occupied Haojingao (Macau). The Portuguese bribed the sea legate Wang Bo to gain a foothold in Macau, which became the base for Portuguese colonial trade in China and the Far East. Between 1557 and 1640, the Portuguese trafficked Chinese silk goods to Nagasaki, Japan, and Goa, India, where they were resold at a high profit. During the same period, Portugal also took over the triangular trade between Macau, Manila, and Nagasaki through Manila. Jesuit schools were also established in Macau, and missionaries followed the gunboats and began to knock on the doors of China.

### *Spain and the Netherlands Forcibly Occupied Taiwan*

Spain conquered the Philippines in the mid-sixteenth century, and Spanish Dominican priests entered Taiwan as missionaries from 1619 onward, opening the way for colonial expansion. Since the formal war between the Dutch and the Spanish in the Philippines in 1600, the contest in the Far East heated up.

The Dutch colonists invaded Asia after Portugal and the West, and in 1601, the Dutch asked Guangdong to open up to trade but were refused. In 1603, the Dutch attacked Macau but failed, and in 1607, they turned to Fujian, but the Ming government still refused to allow them to trade in the city. In 1622, the Dutchman Kornlis Rayerszon led 15 warships

[1] *Guangdong General Records*, 5.

to attack Macau again and was defeated. Dutch general Kobenloet led his remaining troops to Penghu. Then, the Dutch captured a branch of Taiwan's southern coast (now Anping), harassing the coastal areas of Zhangzhou and Quanzhou, attempting to use Macau as an example to occupy the land for trade. In 1624, the Ming Dynasty recaptured Penghu and captured 12 principal generals, including Kobenloet. The remnants fled to Tainan and have stayed there for 38 years.

Two years after the Dutch occupation of Taiwan, Spain sent troops to land in Keelung in 1626. In 1629, it occupied Tamsui and occupied northern Taiwan. In 1642, the Dutch occupied Tamsui and Keelung, driving away the Spanish.

In 1659, Koxinga decided to recover Taiwan and sent Huang Ting, the commander-in-chief, and Zheng Tai, the household official, to lead an army of 20,000 people. On February 1, 1662, the Dutch governor Coyett had to sign fifteen treaties and withdraw from Taiwan.

### *The Opening of Overseas Trade in the Early Qing Dynasty*

Following Portugal, Spain, and the Netherlands, England established the East India Company in 1600, dedicated to opening an Eastern Sea Route through the Cape of Good Hope. In 1620, the English ship Unicorn was rescued by the Chinese after a storm near Macau. With the permission of the Portuguese governor of Goa, England sent a boat to Macau on July 23, 1635. On August 12, 1637, two English ships bombarded Humen and forced their way into Guangzhou, following the example of Portugal and threatening Chinese sovereignty with the force of gunboats. However, the Ming government kept England from trading in Guangzhou.

In the early Qing Dynasty, overseas trade was still strictly prohibited in Guangdong. In 1646, Spanish emissaries arrived in Beijing, and in 1647, the Southern Ming regime banned Westerners from entering Guangzhou and only allowed foreign merchants to trade in Macau. In 1656, two Dutch envoys Peter de Goyer and Jacob dè Kegzer arrived in Beijing and were allowed to pay tribute once every eight years. Portugal sent envoys twice to Beijing in 1673 and 1676 to discuss trade, but the Qing government only temporarily allowed all countries to trade in Macau. In 1676, England set up a trading house in Xiamen, but it was closed after five years. In 1683, the Qing dynasty ended the ban on trade. In 1685, it officially announced the opening of trade, setting up customs offices in

Guangdong, Fujian, Zhejiang, and Jiangnan provinces to manage foreign trade. This allowed European countries to trade regularly with the coastal areas and to conduct extensive economic exchanges.

## 6.2 European Reports on China in the Sixteenth Century

In the mid-sixteenth century, two significant works focusing on China were published in Europe: *Treatise on Things Chinese* by Gaspar da Cruz and *The History of the Great and Mighty Kingdom of China and the Situation Thereof* by Mendoza.

Gaspar da Cruz, a Portuguese Dominican friar, along with ten other Dominican friars, went to Goa in 1548 and later traveled to Malacca. In the winter of 1556, he visited Guangzhou for several weeks and headed to Hormuz in 1560. He returned to Portugal in 1569 and died of the plague in 1570 shortly after. The same year, his book *Tractado em que se cõtam muito por estẽso as cousas da China* (Treatise on things Chinese) was published in Evora, Portugal.

Gaspar da Cruz had a clear intention of writing a dedicated book about China, which resulted in *Tractado em que se contam muito por estenso as cosas da China,* which became the first book on China published in Europe. In the preface, Gaspar da Cruz explicitly states that he wrote this book with the specific purpose of studying China's matters, covering both its geography and its people. He highly valued *Enformacao de China,* written by Galeote Pereira and other Portuguese individuals. With the assistance of Chinese sources, he gathered translations of official Chinese documents and private letters, thereby enhancing the book's value. Gaspar da Cruz's personal experiences in Guangzhou enabled him to make numerous observations about Chinese life and customs. These observations subsequently proved influential for the Jesuit missionaries, who are often recognized for introducing the realities of China to Europe. European academia unanimously considers Cruz's work an invaluable historical document coming after Marco Polo.

Based on Pereira's account, Gaspar da Cruz provides a detailed description of the 13 provinces of China, each having its own provincial capital. Drawing from his observations, he offers an intricate depiction of Guangzhou as a grand city. Gaspar da Cruz also informs his readers that China boasts numerous more significant and magnificent cities than

Guangzhou, characterized by robust city walls, expansive streets, and splendid architecture.

In Gaspar da Cruz's perspective, the Chinese people are characterized by diligence, with "everyone working to earn a living." He notes that in this country, laziness is despised, and those who are lazy are disliked, emphasizing the principle that one must work to eat. Consequently, the land is fruitful, and people have abundant clothing and food. Gaspar da Cruz regards the Chinese as being skilled and resourceful in various trades due to their inherent intelligence and capability. He also provides an account of the Great Wall of China and highlights how China excels in both farming and navigation compared to Europe.

Gaspar da Cruz provided a detailed account of various aspects of China's society. He discussed the administrative management system, palace life, supervision and appointment of officials, and the judicial system. Additionally, he covered topics such as Chinese cuisine, clothing, festivals, funerals, entertainment, and etiquette, ultimately concluding that "Chinese people are very polite."

Cruz also addressed the situation of overseas Chinese under the maritime trade ban. He noted that some individuals who went overseas for business chose to stay there. Some settled in Malacca, others in Siam (Thailand), and some found homes in Pattani. These Chinese communities were present throughout Southeast Asia. They often returned to China using their own boats under the protection of the Portuguese and would ask the Portuguese to help them pay their taxes.

Juan Gonzales de Mendoza, an Augustinian friar, was appointed as part of an embassy to China in June 1580. He carried a gift from Philip II to the Ming Emperor Wanli and embarked on his journey from San Luca to Mexico in February 1581. However, due to the obstruction of the Viceroy of New Spain, King Philip II eventually abandoned the plan to send the embassy to China. Mendoza returned to Spain in 1582. While in Mexico, Mendoza studied Gaspar da Cruz's *Tractado em que se contam muito por extensor as cosas da China* and Martin de Rada's account of China. He also discussed the state of China with a fellow monk who had previously been to Fujian with Rada. Mendoza also read the *Verdadera Relacion* written by Miguel de Luarca, who had been part of Gaspar da Cruz's mission. Mendoza held a special admiration for Gaspar da Cruz's book because Cruz had personally been to Guangzhou and meticulously documented many Chinese events in his account. Mendoza valued Cruz's writings so much that he frequently referenced them in his work.

Juan Gonzales de Mendoza's *The History of the Great and Mighty Kingdom of China and the Situation Thereof* is a substantial work that is divided into two volumes. The first volume consists of three parts, discussing China's geography, religion, politics, cultural customs, and social life. The second volume is divided into two parts, documenting the relationship between China and the Philippines. It contains extensive information about the social lives of overseas Chinese and includes firsthand accounts of the achievements of individuals like Limahong. Mendoza's descriptions of the Chinese empire are often exaggerated, portraying a kingdom that appeared grand and comfortable in that era. Through his imaginative commentary, Mendoza also offers insights into how Europeans might engage with this powerful nation across the Pacific in the future. Published in 1585 in Rome, this 440-page book was originally written in Spanish and was swiftly translated into various European languages.

In his work, Juan Gonzales de Mendoza compared the political and socioeconomic systems of the Spanish and Chinese empires. During the reigns of Charles V (1519–1556) and Philip II (1556–1598), Spain became a dominant force in Western Europe and established control over the Philippine Islands. By the time Mendoza published his work, Spain's influence was pervasive worldwide, and it had not yet faced significant setbacks. However, conflicts were ongoing in the Atlantic and Mediterranean regions, leading to frequent changes in the political landscape of Europe. In contrast, the Ming Empire, comparable in size to Europe, enjoyed political stability and effective governance, resulting in a state of national security. Mendoza delved into various aspects of the Chinese political system, including the imperial examination system that determined individuals' career paths, the hierarchical structure of officials, and the procedures for appointing and dismissing officials. He highlighted that the Chinese government deliberately avoided appointing local individuals as governors, regional administrators, or advisers (counselors), which helped prevent many problems related to unjust administration. This approach also facilitated the implementation of meritocracy. Mendoza's evaluation of China's judicial process emphasized that officials received sufficient salaries, and severe penalties were imposed on anyone involved in bribery or accepting gifts from litigants. He noted that judges were required to follow the emperor's orders, fast before court proceedings, and were forbidden from rendering judgments under the influence of alcohol. Violators of these practices faced strict punishments.

Additionally, Mendoza highlighted that judges couldn't solely rely on written documents or witness testimonies. Instead, they were expected to record witness statements and scrutinize the evidence personally. Through his comparative analysis, Mendoza sought to provide insights into Spain and China's governance and legal systems, offering readers a better understanding of their respective political and societal structures.

Mendoza believed that the Chinese were a talented nation, and as a result, their architecture was magnificent, and their artistic achievements were of high caliber. Each province had grand mansions resembling estates, with gardens, ponds, and wooded areas. Mendoza used various customs to illustrate that the Chinese were very polite and fond of feasting. The gentry and the wealthy frequently visited restaurants and entertainment venues, even more so than people from other cultures. They constructed houses and arranged living arrangements for the poor and disabled, eliminating the need for begging.

Regarding the Chinese character, Mendoza concurs with the anonymous author that the Chinese were deficient in bravery and martial spirit. He states, "If they could match our European nations in courage and bravery, they would have been capable of conquering the entire world. Despite their population and resourcefulness, they fall far short of us regarding courage and ambition." In his book, Mendoza expresses his disagreement with the idea of making efforts to conquer China through the grace of God, as it does not apply to the region. He believes seeking peace and adopting a missionary approach is a more prudent course of action.

## 6.3 Relations Between China and the Americas in the Age of Exploration

### *Zhangzhou and Manila Galleons in Pacific Trade Route*

In 1494, under the supervision of the Roman Catholic Church, Spain and Portugal ratified the Treaty of Tordesillas, which delineated their respective spheres of influence on a global scale. Portugal obtained control over the eastern hemisphere, while Spain took possession of the western hemisphere. In 1529, both countries signed the Treaty of Zaragoza, which led to a new adjustment of the prime meridian line. This expansion brought their influences to converge along the southeastern coast of China. Starting from 1565, Spain gradually occupied the Philippine Islands, thus opening up a maritime route encircling the Pacific Ocean.

In 1571, the Spanish quickly discovered that the Philippines had neither gold nor valuable spices like cloves, except for cinnamon. In the same year, Spain rescued a Chinese shipwreck off the coast of Luzon. When these sailors returned to Zhangzhou in Fujian, they spread the news about Spain's plans to build Manila City and actively sought trade. This led to Chinese merchants bringing substantial quantities of valuable commodities, such as silk, porcelain, benzoin, musk, meat, fruit, metalwork, and various goods, to Manila in 1572. The prices of these goods were surprisingly low. Using two galleons they constructed locally in Manila, the Spanish loaded them with Chinese goods and sailed to Acapulco, Mexico, in 1573. These ships carried 712 pieces of Chinese silk and 22,300 pieces of high-quality gilded porcelain and porcelain of various colors. The trade between Zhangzhou and Manila became firmly established by 1576, with annual sailings from Krabi, Manila Bay, to Acapulco. By the early seventeenth century, the number of sailing ships arriving from the Chinese coast in Manila increased significantly. In 1631, there were as many as 50 ships.[2]

Since the Manila sailboat trade route was established, a continuous stream of Chinese goods has entered the American market through the Pacific Ocean. Chinese silk and embroidered products, ranging from clothing and shawls to full-length stockings, quickly gained popularity in Mexico and the Americas. Some goods were transported from China's coastal regions to Manila and then resold in the Americas. Additionally, other products were manufactured on-site in Manila and the Philippines by the Minnan Chinese community. These skilled craftsmen engaged in various trades, including tailoring, weaving, dyeing, embroidery, leathercraft, furniture making, blacksmithing, silversmithing, locksmithing, coppersmithing, tinsmithing, harness-making, sewing, masonry, lime production, candle making, shoemaking, painting, felt production, winemaking, sugar refining, and oil extraction.[3] Although countless exotic goods were transported to Mexico, Chinese silk products consistently occupied a prominent position in the Manila sailboat trade. Chinese

[2] E. H. Blair and J. A. Robertson, *The Philippines Islands (1493–1898)*, Vol. 23, Cleveland, 1903–1909, p. 193.

[3] E. H. Blair and J. A. Robertson, *The Philippines Islands (1493–1898)*, Vol. 52, 1903–1909, Cleveland, pp. 212–238.

silk became the Spanish's most valuable merchandise to the American continent.

Before 1583, Chinese merchant ships typically brought goods valued at 200,000 pesos per year to Manila. However, after this time, the value of goods began to soar. Between 1587 and 1598, Chinese sea merchants brought goods worth 800,000 pesos per year, with some years exceeding one million pesos.[4] Most of these goods were redirected to Mexico, and Chinese products became the primary cargo source for Manila sailboats, while locally produced goods were scarce. Until the late eighteenth century, Chinese silk, porcelain, and other commodities made up 63% of the total value of Mexican imports.

The trade routes between Zhangzhou, Manila, and Acapulco brought economic prosperity to Latin America. Merchants from Mexico, Peru, and the Philippines engaged in galleon trade often enjoyed net profits ranging from 100 to 300%. In some cases, even as high as 600% to 800%.[5] Between 1571 and 1821, a total of 400 million pesos of silver coins were transported from Mexico to Manila, with at least one-quarter of this amount flowing into China.[6] The quantity of Mexican silver coins imported into China amounted to approximately 1/6 of China's original silver holdings, possibly even more. This development undoubtedly played a significant role in supplementing China's economy, especially since the country had progressively shifted toward silver circulation instead of relying heavily on banknotes since the mid-fifteenth century.

Amid internal and external challenges, the Manila galleon trade faced competition from the Royal Philippine Company, which annually transported 800 tons of Asian goods to New Spain. Ultimately, this trade was ordered to be closed on October 25, 1813. The Mexican War of Independence erupted in 1815, and the final Manila galleon closed its operations in Acapulco, marking its historical mission's culmination and signaling the Silk Road's conclusion on the Pacific Ocean.

[4] E. H. Blair and J. A. Robertson, *The Philippines Islands (1493–1898)*, Vol. 6, Cleveland, 1903–1909, pp. 269.

[5] William Lytle Schurz, *The Manila Galleon*, E. P. Dutton, 1959, p. 190.

[6] J. B. Eames, *The English in China*, London: Curzon Press, 1974, p. 63.

### *Introduction of American Crops*

With the establishment of formal connections between the Americas and the Old World, certain native American plants garnered attention and were introduced to Europe and the Philippines. From there, these plants entered various regions of China, profoundly influencing the country's economic activities and societal customs. Among the significant introductions from the Americas to China were vital food crops such as corn, sweet potatoes, and potatoes, essential cash crops like peanuts and tobacco, and notable fruits and vegetables including hot peppers and tomatoes.

Maize is a tropical American plant belonging to the grass family and an annual herb. Its history of cultivation in the Americas dates back over 5,000 years, as confirmed by Carbon-14 dating. Indigenous populations developed more than 200 varieties of maize, featuring kernels in colors such as red, white, yellow, blue, and purple. Columbus referred to this food as "sweet and delicious," known by the natives as "Mahiz." Subsequently, maize was introduced by Spain, France, and the United Kingdom, and it also made its way to Germany and Russia through Venice. In the sixteenth century, the cultivation of maize spread to Africa, Turkey, Iran, and Arabia.

The introduction of maize to China initially came from Western Asia. It is believed that maize was possibly cultivated in Yunnan as early as the fifteenth century, referred to as "Yumai," which is documented in the *Dali Fuzhi* (1563). By the late sixteenth century, the *Compendium of Materia Medica* mentioned that "corn species out of the West were rare in cultivation." *Xiangcheng County Zhi* (1551) in Henan and *Gongxianzhi* (1551) both documented the presence of Yumai. In the northern regions, it was known as "Fanmai" or "Xitianmai." The seeds were possibly brought to China by Muslims who had made the pilgrimage to Mecca. Due to its high yield, resistance to hunger, and ease of cultivation, corn was well-suited for hilly areas. By the early nineteenth century, corn had been planted in all mountainous fields in Sichuan, Shaanxi, Hubei, and Hunan, making it a staple for mountain farmers.

Sweet potatoes are also known as "Honshu," "Zhushu," "Shanyu," and "Digua." Gong Xianzeng, cited Su Yan's *Zhushu Shu,* argued that sweet potato seeds were brought back from the Philippines to Quanzhou for trial planting in the 12th to 13th years of the Wanli era (1584–1585), resulting in successful cultivation. Another record from the *Dongguan Fenggang Chen's Genealogy* states that Chen Yi, a resident of Dongguan, brought potato seeds from Vietnam to Dongguan for planting in 1582, two years before Quanzhou. Chen Zhenlong, a native of Changle, Fujian, learned to plant sweet potatoes in Luzon. He acquired cultivation techniques on-site and brought the seeds to Fuzhou's Nantai area in 1593 for trial growing behind his house. He concluded that sweet potatoes had "six benefits and eight advantages" and were no less effective than common grains. In 1594, during a severe drought, Jin Xuezeng promoted planting potato seeds all over Fujian to survive the drought, earning sweet potatoes the nickname "jin (golden) potato." New sweet potato species were later introduced from places like Brunei. Some of these sweet potatoes could weigh more than ten pounds. As a result, sweet potatoes gradually spread from Fujian and Zhejiang to areas north of the Yangtze River, including Shandong, Henan, and Hebei. In the North China Plain, sweet potatoes became commonly known as such to distinguish them from white and yellow potatoes that were available earlier.

Potatoes are native to South America, specifically Peru and Chile. The term "potato" originates in the native Peruvian term "papa," which was also borrowed into Spanish. Potatoes were introduced to Spain in 1525, where they were initially consumed in their raw tuber form. In 1565, the Spanish king presented potatoes to the Pope in Rome as a vegetable offering. During the reign of King Louis XVI, the agronomist Antoine-Augustin Parmentier introduced and popularized various delicious potato dishes, elevating the reputation of this crop. In the early seventeenth century, potatoes were cultivated in Taiwan. Upon encountering this crop in Taiwan, the Dutch called it "Dutch beans."

Peanuts are originally native to Brazil. In the 1950s, carbonized peanuts dating back 5,000 years were unearthed at archaeological sites in Zhejiang Province (Wuxing Qiansanyang site) and Jiangxi Province (Xiushui Shanbei site). This discovery led some experts to believe that China might also be one of the original regions where peanuts originated. During the Han Dynasty, peanuts were known as "Yu Ganzi," although their cultivation was not widespread then. The earliest records of peanuts in Chinese literature can be found in the works of that era, such as the *Changshu Xianzhi* from the Jiajing period and the *Jiading Xianzhi* from the Wanli period. In the Ming Dynasty, peanut planting was not widely planted. However, by the early Qing Dynasty, peanuts were already being produced in Huizhou, as mentioned by Ye Mengzhu in his *Yueshibian*.

Tobacco is native to the Americas. When Christopher Columbus arrived at a small island called Tobacco in Central America, he observed the indigenous people using pipes to ignite tobacco leaves and inhale the smoke. This led to the Spanish name "Tobacco." Initially, smoking was a social practice among Native Americans during gatherings and when welcoming guests. In 1560, Jean Nicot introduced tobacco to France, which was used as a topical pain-relieving treatment. Subsequently, tobacco cultivation spread to countries like England, Italy, Turkey, and Japan, and smoking became a widespread habit. Tobacco was also introduced to China, where it was translated as "淡巴菰" (dànbùguī) or "担不归" (dān bù guī). During the Wanli period (1573–1620), tobacco was brought from Luzon to Taiwan and then to Fujian and Guangzhou, gradually spreading throughout the country. In the northern regions of China, tobacco was introduced via Japan and Korea to the northeastern part before reaching the central areas. Tobacco use also had medicinal implications, as it was believed to help prevent miasma and dispel cold and dampness. This belief contributed to its rapid adoption from the upper classes to the general population. After a ban on tobacco was lifted in 1641, its cultivation and use became more widespread. By the middle of the nineteenth century, smoking was no longer limited to specific regions or social classes. Bao Shichen's work *Four Kinds of An Wu*, written during this time, highlighted the prevalence of tobacco use among people of all genders and ages.

Tomatoes are originally native to the northern valleys of the Andes Mountains in South America. They were introduced to China in the sixteenth century by Western missionaries. The earliest reference to tomatoes in Chinese records can be found in the *Yishi Xianzhi* of Shanxi in 1613, where they were referred to as "西番柿" (Xī fānshì) or "番柿" (fānshì). However, like the West, tomatoes were initially considered ornamental rather than a food crop in China until the nineteenth century. The tomato seeds found in a 2,000-year-old Han tomb on Mount Feng in Chengdu were cultivated and identified as tomato plants.

In the sixteenth century, peppers spread from the sea to the southeastern coast of China, known as chili peppers, spicy eggplants, sea peppers, and chili peppers. It was first recorded in the Ming Dynasty in the *Zunsheng Bajian* and the *Herb and Flower Book*. After drying, it can be used for years or processed into powder and sauce as an ideal condiment. Peppers are also commonly planted in inland mountainous areas and cold northern regions. Since then, the import volume of expensive spices such as pepper and cloves has dramatically decreased. Sichuan and

Hunan cuisine is also spicy, becoming a local characteristic among the eight major cuisines.

### *Sino-American Relations Before and After American Independence*

By the end of the sixteenth century, China had become a significant pivot point in global maritime trade. Heading westward, there was a route from Macao to Goa, India, which continued through East Africa, circumventing the Cape of Good Hope and crossing the Atlantic to reach Lisbon. In the eastern direction, ships departed from Zhangzhou Yuegang or Xiamen, heading toward Manila, and then traversed the Pacific Ocean to reach Acapulco, Mexico, forming the Pacific Silk Road.

In 1688, the Qing government lifted the maritime prohibition and opened ports like Guangzhou, Zhangzhou, and Ningbo as trading hubs for China and Western nations. Guangzhou reclaimed its status as a maritime gateway to the world, a role it had held for two millennia. By this time, the global landscape had been interconnected through trade routes spanning five continents. China was a prominent player in agriculture and manufacturing, and it occupied a strategic position at the confluence of the Pacific and Indian Ocean routes in the global maritime network. In 1757, the ports of Zhangzhou, Ningbo, and Yuntai Mountain were closed, leaving Guangzhou as the sole port for Western merchants to trade under the Canton System. As a result, Guangzhou became the exclusive nexus for Sino-foreign maritime trade routes.

Chinese silk, porcelain, cotton, and tea had already captivated the inhabitants of the thirteen North American colonies. Particularly significant was the tea trade, which kindled fervent sentiments for independence from Britain. On December 16, 1773, in the city of Boston, a pivotal locale among the thirteen North American colonies under British dominion, a gathering of eight thousand citizens vehemently protested the Tea Tax Act enacted by the British Parliament that year. They adamantly prevented the tea-laden ships of the British East India Company from docking and unloading. Starting with the Sugar Act in 1764 and furthered by the Quartering Act and the Stamp Act in 1765, followed by the Townshend Act in 1767 that imposed tariffs on essentials such as paint, glass, paper, and tea—items integral to the daily life of the colonists—these measures sparked the flames of anti-British sentiment across the colonies. In March 1770, violent clashes erupted between Boston citizens and British troops, forcing the repeal of the Townshend

Act. Yet, the costly tea tariffs persisted, making the tea tax a potent symbol of the British government's attempts to assert control over its North American colonies.

Dutch immigrants introduced the practice of tea drinking to the North American colonies. In 1640, Dutch settlers in New Amsterdam, a Dutch colony, were the first to adopt tea consumption. In 1664, the territory came under British control and was renamed New York. The local population quickly embraced the habit of drinking tea. By 1690, Boston had established the first tea dealership on the North American continent, three decades after direct tea imports from China to Britain had begun. Starting from the 1820s, tea was consistently shipped from Guangzhou to Britain and then transported to the North American continent. The tradition of consuming tea after lunch and in the evening also spread to many households in the thirteen North American colonies of the United Kingdom. By 1721, the four New England states (Massachusetts, Rhode Island, New Hampshire, and Connecticut)—boasting well-developed shipping industries among the thirteen colonies—had widely embraced tea consumption. By 1760, tea consumption in the thirteen colonies had reached 200,000 pounds. When the Townshend Act was introduced, most Americans were already accustomed to drinking tea at least twice daily. Before the Revolutionary War, Philadelphia, then known as North America's largest city with a population of 23,000, saw its residents develop a strong affinity for tea. Tea was shipped from the northernmost part of North America, New York, to the Appalachian Mountains in North Carolina, inhabited by Native Americans.

Records from British Customs between 1750 and 1774 indicate that Pennsylvania annually imported 40,000 pounds of tea from the United Kingdom, constituting one-fifth of the continent's average yearly tea imports from the UK. Even Boston, with a population of 22,000, witnessed the emergence of tea parties and groups dedicated to the transportation and sale of tea. However, tea consumption necessitated the payment of substantial commercial profits and taxes to Britain by the inhabitants of the thirteen colonies. Although approximately one-third of the ships flying the Union Jack flag were manufactured in the American colonies, British navigation laws precluded the direct involvement of the North American colonies in Asian trade. In 1773, the British Parliament passed the Tea Act, providing a convenient means of offloading the East India Company's surplus tea. This legislation transformed the import tax into a business tax equivalent to 1/80 of the tea's value. This imposition ignited the ire of Boston residents who, on the evening of December 16, dumped the tea from East India Company ships into the sea, causing a loss of tea valued at 18,000 pounds for Britain. To quell this anti-British movement, Britain responded by ordering the closure of the Port of Boston, revoking Massachusetts' autonomous charter, and deploying military forces in 1774. However, the anti-British sentiment continued unabated. On July 4, 1776, the Declaration of Independence was passed, officially marking the birth of the United States of America. After the Seven Years' War and with the assistance of European volunteer forces, the United States emerged victorious, compelling Britain to acknowledge its independence in 1783.

In 1784, shortly after its founding, the United States sent its first merchant ship, the Hartley, on a voyage to the East. The mission was to exchange a cargo of ginseng for Chinese tea valued at double the amount. After acquiring tea at the Cape of Good Hope, the Hartleys swiftly returned to Boston in July, marking the year when Americans first enjoyed tea transported by American ships. As the Hartley departed Boston and reached South Africa, Philadelphia businessman Robert Morris, with other investors, raised funds for an expedition to China. They built a 360-ton wooden sailing ship named the Empress of China. Led by 48-year-old Irishman John Green as captain and with a crew of 35 sailors, the ship embarked on its journey from New York on February 22, 1784. The Empress of China crossed the Atlantic, rounded the Cape of Good Hope, and navigated the Sunda Strait with guidance from a French ship. After 188 days of sailing, it entered the Pearl River on August 28, ultimately docking at Huangpu Port in the southern suburbs of Guangzhou. The ship was warmly welcomed by foreign traders from various nations. American merchants in Guangzhou sold their cargo of ginseng, feather yarn, fur, lead, and cotton and, in turn, purchased tea, silk, porcelain, Nanjing homespun fabric, and other desired items. On December 27, the Empress of China departed Guangzhou and retraced its route back to the United States. After stopping at the Cape of Good Hope, the ship arrived in New York on May 10, 1785, completing a journey of 13,700 miles that took one year and three months. The ship carried 4,498 units (each weighing 133.5 pounds) of Chinese goods from Guangzhou. Among them, 2,460 units were black tea, and 562 were green tea, accounting for two-thirds of the cargo. The remaining shipment included 24 pieces of cotton fabric, 962 units of porcelain, 490 pieces of silk, 21 units of cinnamon, black pepper, silverware, lacquerware, carved ivory, silk gloves, and clothing.

The total value was 71,767 Chinese taels of silver. Captain Green also acquired many Chinese goods, including black tea and lacquerware. In addition to popular items, he had more than 300 pairs of high-end satin pants tailored for American men and 600 pairs of women's long-sleeved gloves made by Guangzhou tailors. He also acquired craftsmanship such as ivory fans, dressing boxes, and snuff bottles from small workshops. Notably, this journey introduced the Shanghai rooster to the United States and became a popular breed. These Chinese goods exposed Americans to a new world and foreshadowed a later Chinese craze in American society, which occurred much later than in the European continent. The voyage was a success both in navigation and business. Investors reaped a quarter of the profits, equivalent to a total investment of $109,000, yielding a net profit of $30,727. New England merchants were elated by the success, prompting merchant ships from Salem, Boston, Providence, New York, and Philadelphia to embark on voyages to Canton, sparking an enthusiasm in Chinese trade conducted by American merchant ships.[7] Meanwhile, the tea tax collected by the government increased from 12.5% to 119% that year.

The commencement of navigation between Philadelphia and Canton marked a remarkable achievement in American public discourse, often paralleled with the third Pacific expedition conducted by British navigator James Cook from 1776 to 1781. It's worth mentioning that during Captain Cook's third voyage, there were already two native American sailors who accompanied the expedition and arrived in Canton. They were among the earliest Americans to set foot in China. One of these sailors, John Ledyard, documented his experiences in the *Journal of Captain Cook's Last Voyage to the Pacific* upon his return. This journal provided a firsthand account of China for Americans.

By 1803, the total trade value between the United States and China surpassed that of Britain and exceeded the total trade value between European countries and China. By then, the United States had become as important as Britain, emerging as a significant nation facilitating China's global connections through trade relations. This development all began with the arrival of the Empress of China in Guangzhou. Accompanying the Empress of China to Guangzhou was the influential business agent

[7] K. S. Latourette, *History of the Early Relations between the United States and China, 1784–1844*, New Haven: Yale University Press, 1917.

Samuel Shaw, whose role was pivotal in fostering U.S.–China trade relations. A mere ten days after the Empress of China's return to New York, Samuel Shaw submitted a comprehensive sailing report to Secretary of State John Jay, expressing a highly optimistic view regarding the bright prospects of trade with China. In 1786, Samuel Shaw embarked on another journey to China and was appointed the U.S. Consul in Guangzhou, dedicated exclusively to fostering business ties with China. The communication between the two nations flowed seamlessly during this period. At that juncture, American agriculture had already made strides by successfully introducing three high-yielding crops from China—rice, sorghum, and peas. The American Philosophical Society, headquartered in Philadelphia and led by figures such as Benjamin Franklin and Thomas Jefferson, was a reflection of the intellectual elite of the United States. As early as 1771, the society had expressed aspirations to import Chinese techniques and technologies, intending to bolster the production of cotton, raw silk, and tea and the advancement of porcelain manufacture within the United States. The society consistently gazed eastwards toward China, a nation renowned for its manufacturing prowess and technological progress. After the United States achieved independence, the American Philosophical Society committed itself to introducing Chinese expertise in sericulture, silk reeling techniques, and innovative livestock breeding methods. They were eager to learn from China's agricultural practices and cultivation technology, particularly interested in China's papermaking and printing technologies. This mutual curiosity paved the way for productive cultural exchanges between the two nations throughout the nineteenth century.

## 6.4 Jesuit Missionaries in China in the Ming Dynasty

### *Christianity Entered China Three Times*

The spread of Christianity in China is usually said to be Nestorianism in the Tang Dynasty for the first time and the second time in the Yuan Dynasty. In the middle of the Ming Dynasty, with the expansion of Portuguese colonial power, Catholicism regained the opportunity to preach in China. This was the third time that Christianity was introduced into China.

The Pope, utilizing Portuguese and Spanish naval power, extended religious influence to the East, leading to a completely different dynamic in the third propagation of Christianity in China since the sixteenth century. With the rise of the Reformation and Calvinist Protestantism, the Catholic Church, under papal control, sought to establish new religious dominions in areas reachable through new maritime routes. The Holy See's aspiration to develop China as a subordinate diocese, driven by political motives, rendered the Catholic missionary activities in China during the mid-Ming Dynasty more hazardous and with more conspicuous setbacks than at any prior time.

In the mid-sixteenth century, Saint Ignatius of Loyola from Spain organized a missionary group to accommodate the expanding influence of the Roman Church. On September 27, 1540, this group received approval from Pope Paul III and was named the Society of Jesus (Societas Jesu). On April 13, 1541, Loyola was appointed the first Superior General. He actively organized overseas missionary endeavors. King John III of Portugal promptly sought the Pope's authorization to send six missionaries to the East. One of the co-founders of the Jesuits, Saint Francis Xavier, was dispatched to India and Japan as the first step in his missionary journey. He was also the pioneer of Jesuit preaching in China. Departing from Lisbon on April 7, 1541, he reached Goa on May 6 of the following year and arrived in Kagoshima, Japan, on August 15, 1549. However, Japan had long adhered to Buddhism and Confucianism from China. In Japan, Xavier encountered many culturally advanced Chinese individuals, motivating him to commit to preaching in China. He believed that "once the Chinese are truly converted, the Japanese will naturally abandon the erroneous teachings derived from China."[8] Upon returning to Goa, he set out for China again on April 14, 1552. In August of that year, he stopped at Shangchuan Island in Guangdong Province. Due to the tightening of the Ming Dynasty's maritime restrictions, he was compelled to remain on Shangchuan Island, situated 30 nautical miles from Guangzhou. He passed away on December 20 in a thatched hut on the island.

Saint Francis Xavier can be attributed as the pioneer of the Jesuit mission to China. However, it was Matteo Ricci (1552–1610) who laid the foundation for the Jesuit mission in China. Before and after

[8] Cros, *Saint Francois de Xavier*, Vol. II, pp. 103–104, addressed to Simon Rodriquez on January 29, 1552.

his arrival, Macau remained the sole stronghold for Jesuit activities in China. In 1557, just five years after the Portuguese occupied Macau, Fathers Pérez and Texeira established residence there in 1562. Jesuit Carneiro was appointed as the first bishop of Macau. Through missionary work, they compelled the local Chinese to assimilate into Portuguese culture—adopting Portuguese names and living Portuguese lifestyles.

In order to conduct missionary activities in the Chinese mainland, the Jesuits chose to send four Italian Jesuits—Alexander Valignani, Michael Ruggieri, Franciscus Fassio, and Matteo Ricci—who were well-versed in the Chinese language and culture. All four adopted Chinese surnames. Valignani spent an extended period in Macau and passed away there in 1606. Fassio was occupied with missionary work in Japan and died shortly after returning to Macau in 1612. Only Michael Ruggieri and Matteo Ricci engaged in missionary activities among those sent to China. In 1582, Ruggieri gained permission from Chen Wenfeng of the Guangdong Provincial Council to reside in the Tianning Temple in Dongguan, Zhaoqing. He began formal preaching there, but this endeavor was brief. Ruggieri later ventured to Guangzhou, Shaoxing, and Guilin. After returning to Italy in 1588 and retiring to Sarano, he passed away in 1607. Ruggieri was the first Jesuit to enter the Chinese mainland in the late Ming Dynasty. He was also the first to compose the Catholic Catechism in Chinese. The book was subsequently revised by Matteo Ricci and printed in Zhaoqing in 1584. One thousand copies were distributed across various circles in society.

### *Matteo Ricci's Sermon*

Matteo Ricci's arrival in China opened up a new dimension of cultural exchange between China and the West in the late Middle Ages. He was the first Jesuit missionary to enter Beijing, the capital of the Ming Dynasty, and the most influential figure in Chinese society at the time, in terms of the upper echelons, cultural and academic circles, and later in Jesuit missionary work.

In 1582, Matteo Ricci arrived in Macau and studied Chinese there. With Ruggieri, he traveled to Guangdong's major cities, Canton and Zhaoqing. In 1583, Ricci and Ruggieri settled in Zhaoqing. Ricci began to preach in Zhaoqing, interacting with officials and literati at all levels, promoting European cultural relics and rules and regulations, introducing Western knowledge of astronomy, arithmetic, science, and chemistry, and displaying self-timing bells, maps, celestial instruments, and Venetian trigonometry. Ricci stayed in Zhaoqing from 1583 to 1589, when he was expelled by a new viceroy to Shaozhou. Ricci continued to befriend officials and the gentry, studied Confucianism, and preached Western science and technology to evangelize in Shaozhou. He befriended Qu Taisu of Suzhou, who took him as his teacher, and he listened to Qu's advice and changed from Western monk dress to Confucian clothing.

Matteo Ricci's ultimate goal was to reach Beijing and pay respects to the Wanli Emperor, aiming to facilitate the acceptance of Catholicism in China. In May 1595, he arrived in Nanjing. However, due to Nanjing's reluctance to host him, he returned to Nanchang. He befriended the King of Jian'an there, discussing Western friendship practices, which he compiled into his work *On Friendship*.

On May 18, 1600, Matteo Ricci set out northward, accompanied by the Spanish cleric Didacus de Pantoja. Upon their arrival in Beijing on January 24, 1601, Matteo Ricci offered gifts to the Emperor and submitted a memorial. The gifts included an image of God, two pictures of the Mother of God, a copy of the Book of God, a cross adorned with pearls, two self-ringing clocks, an atlas of the nations, and a Western zither. The Emperor accepted the various gifts, summoned Ricci for an audience, accorded him the treatment of an esteemed guest, and directed his courtiers to learn the Western zither from him. Through the efforts of Zhao Bangjing from the Ministry of Rites, Matteo Ricci was ultimately granted permission to select a residence at the eastern end of Xuanwu Gate (initially named Shuncheng Gate).

*The True Meaning of the Lord of Heaven* was composed by Matteo Ricci between 1593 and 1596, consisting of two volumes. Upon its completion, the book gained popularity and spread widely through copied editions. Xu Guangqi received a copy from Luo Ruwang in Nanjing, and after reading it throughout the night, he resolved to be baptized. Feng Yingjing, an imperial censor of Huguang who had been imprisoned for accusing the eunuch Chen Feng, also became a believer in Catholicism after obtaining a copy of this book. *The True Meaning of the Lord of Heaven significantly represents* Matteo Ricci's promotion of Catholic philosophical thought. It is a crucial work that harmonizes Confucianism and Catholicism, substantially influencing Ming Dynasty officials and intellectuals.

In Beijing, Matteo Ricci propagated Western scientific works to garner the trust and cooperation of officials who advocated practical learning and opposed abstract discussions about clarity of mind. He collaborated with Xu Guangqi to translate the first six volumes of Euclid's Elements, participated in the calendar revision, created translated and illustrated maps of the five continents, and acted as an intermediary for introducing Western disciplines like painting, music, architecture, and surveying. Over his lifetime, he authored and translated more than twenty works. In his later years, he dedicated considerable effort to his missionary work. In 1609, he established the Catholic Society, which outlined guidelines for members to gather monthly for charitable activities, focusing on aiding the impoverished. He also personally supervised the construction of the Beijing Cathedral. Matteo Ricci passed away on May 11, 1610, in Beijing. By the code of the Ming Dynasty, foreigners who died in China had to be buried in Macau. Diego de Pantoja made a special plea to the court, requesting a burial plot in Beijing in the light of Ricci's contributions to China. The Wanli Emperor granted this request and designated a Buddhist temple. In October 1610, Ricci's remains were transferred there.

During the late Ming Dynasty, the three key figures of Catholicism—Xu Guangqi, Li Zhizao, and Yang Tingyun—all directly embraced Matteo Ricci's missionary efforts and academic guidance, subsequently receiving baptism. Xu Guangqi (1562–1633), hailing from Shanghai, achieved the position of Minister of Rites in 1630. He was baptized by Jean de Rocha in 1603 in Nanjing, adopting the name Paul, and is recognized as a trailblazer of modern science. Originally from Hangzhou, Li Zhizao (1565–1630) learned about geography and almanacs from Matteo Ricci in Beijing. He received baptism from Ricci in February 1610 during severe illness. Yang Tingyun (1557–1627), another native of Hangzhou, was deeply influenced by Ricci's teachings and was baptized by Lazzaro Cattaneo in Hangzhou in May 1611. Moreover, there were ten other notable Catholic believers in the Ming Dynasty, including Qu Taisu, Feng Yingjing, Li Tianjin, Zhang Tao, Sun Yuanhua, Wang Zheng, Han Lin, Duan Gun, Jin Sheng, and Qu Shisi. All of them were renowned officials and literati.

At the time of Matteo Ricci's death, the Catholic Church had already been established in Zhaoqing, Shaozhou, Nanchang, Nanjing, Beijing, Shanghai, and Hangzhou. There were approximately 2,500 Catholic believers in China.[9]

[9] Pasquale M. D'Elia ed., *Storia dell'introduzione del cristianesimo in Cina*, p. 60.

Ricci played an enlightening role in Xu Guangqi's and Li Zhizao's study of European astronomy, almanac, geography, and surveying, making them pioneers of modern Chinese science. The exchange of ideas between the Chinese and Western academic achievements while compiling scientific works like *Euclid's Elements* and *Tongwen Zhisuan* was collaborative. These works were completed through the cooperation of Matteo Ricci with Xu Guangqi and Li Zhizao, enabling the introduction of Western science to China.

Matteo Ricci imparted Western astronomy, geography, and mathematics knowledge to Chinese scholars, with Xu Guangqi among the earliest and most accomplished recipients. The *Biography of Xu Guangqi* in the *History of the Ming Dynasty* notes that "Xu Guangqi learned astronomy, calendrical calculations, and firearms from Matteo Ricci and diligently advanced these subjects." Under his guidance, the *Chongzhen Calendar* was developed, and advancements were made in producing Western artillery and calendrical systems.

Matteo Ricci's efforts paved the way for Jesuits to study China and introduce Western science and technology. Figures like Diego de Pantoja, Sabbatino de Ursis, Joannes Terrenz, Jacques Rho, Julius Aleni, Johann Adam Schall von Bell, Ferdinand Verbiest, Martinus Martini, Joachim Bouvet, and Joannes-François Gerbillon inherited the tradition established by Matteo Ricci. They achieved significant milestones in promoting scientific endeavors and contributed to the cultural exchange between China and the West.

### *Catholicism in the Late Ming and Early Qing dynasties*

In the year before Matteo Ricci's passing, Nicolaus Longobardi, an Italian from Sicily, assumed the leadership of the Society of Jesus in 1609. However, he did not show respect for Chinese social traditions. Consequently, opposition to missionary activities also emerged. This was first witnessed in the Nanjing case of 1616, followed by the Astronomical Bureau(Qintianjian) event of 1659.

At the recommendation of Xu Guangqi, Western missionaries became involved in revising the Chinese calendar and producing artillery. Figures such as Joannes Terrenz, Julius Aleni, P.Francois Sambias, Johann Adam Schall von Bell, and Jacques Rho served in the Ming court, obtaining legal recognition and providing them with a basis for their missionary activities. During the early years of the Chongzhen period, Johann Adam

Schall von Bell, Nicolò Longobardo, and Jacques Rho were permitted to enter and leave the palace freely. Influenced by the persuasion of Xu Guangqi and Johann Adam Schall von Bell, Catholicism made significant inroads into the imperial court.

To expand their missionary efforts in China, the Society of Jesus sent Nicolas Trigault to Rome in 1613 to request more priests. At that time, there were fewer than 20 priests in China, only five residences, and less than 10,000 followers. As a result, China was assigned to a sub-province, equal to Japan. However, it wasn't until 1618 that China became an independent province within the Jesuit organization. During this period, the mission was divided into two major regions: North China and South China. Francisco Furtado served as the head of the North China mission, overseeing activities in Beijing, Shandong, Shaanxi, Shanxi, and Henan, with a total of five priests. Meanwhile, Julius Aleni served as the head of the South China mission, stationed in Fujian, and managing affairs in Nanjing, Jiangxi, Hubei, Sichuan, Zhejiang, and Fujian, with a total of thirteen priests. Starting from 1575, the Guangdong and Guangxi regions were overseen by the Macau diocese. Catholicism was introduced to Hainan Island in 1632, and in 1641, Francesco della Penna started missionary work in Sichuan. Consequently, by the end of the Ming Dynasty, missionaries and believers were present in thirteen out of the fifteen provinces, with Yunnan and Guizhou being the exceptions.

During over 50 years spanning the late Ming Dynasty and early Qing Dynasty, the Catholic Church experienced significant growth in the number of believers. The count of believers gradually increased from 2,500 in 1610 to 5,000 in 1615, 13,000 in 1617, 38,000 in 1636, and even 150,000 in 1650.

According to a report written by Father Galiani in prison in Guangzhou in 1667, there is a summary of the Jesuit missions in China from 1581 to 1664. This summary includes 246,000 baptisms over 83 years. During this period, 82 Jesuit missionaries were sent to the Chinese province and 12 to the Japanese province. Within the same timeframe, the Jesuits were involved in printing 131 Catholic books, 100 books on arithmetic, and 55 books on various scholarly subjects, ethics, and physics. The Catholic community also published 14 doctrinal books.[10] Notably, during this time, Li Zhizao's *Tianxue Chuhan*, a

[10] Augustin Colombel, *Histoire de la mission du Kiang-nan*, p. 122.

series consisting of 52 volumes, was published in 1629. This compilation included eight doctrinal works and 12 scientific and technical works. These translated works were published and reflected how early Jesuit missionaries utilized the dissemination of knowledge in natural sciences and technology to promote Catholicism. This dual focus resulted in a symbiotic relationship, which became a distinct feature of the missionary efforts during the late Ming and early Qing dynasties.

During this period, the Franciscan and Dominican orders also actively preaching and converting in various regions. They conducted missions in Fujian, Zhejiang, and Shandong. In 1590, the Pope dispatched four Franciscan friars to China, and in 1633, a group of ten clergypersons from the Philippines arrived and conducted preaching activities in Shandong. Between the years 1633 and 1660, a total of 3,500 individuals were baptized as a result of their efforts. The Dominican order also played a role in the missionary landscape. In 1631, ten Dominican priests arrived in Fujian to carry out their missionary work. Between the years 1650 and 1664, they baptized 3,400 people. At this time, the Dominicans had 30 priests involved in their mission activities.

## 6.5 Catholics in China Before the Middle of the Qing Dynasty

### *Johann Adam Schall von Bell and Missions*

In the mid-seventeenth century, the most renowned Jesuit missionary in China was Johann Adam Schall von Bell (1592–1666). He joined the Jesuit order in 1611. In 1622, he arrived in Macau and then proceeded to Beijing to further his studies in the Chinese language before being assigned to preach in Xi'an. In July 1629, under the leadership of Xu Guangqi, he was involved in the revision of the calendar. Shortly afterward, Li Zhizao and Xu Guangqi passed away, and the task of revising the calendar was continued under the guidance of Li Tianjing, with the participation of Johann Adam Schall von Bell, Nicolò Longobardo, and Giacomo Rho.

In May 1644, as Qing forces entered Beijing, the residence of Schall von Bell and Longobardo was within the imperial city and should have been vacated within three days. However, Schall von Bell, wishing to safeguard the calendar plates, religious texts, books, and astronomical instruments, submitted a petition and received special permission to keep them. Thus, Schall von Bell gained access to the newly installed Shunzhi Emperor and became one of his trusted counselors. In November 1644, he assumed control of the Astronomical Bureau's seal, integrating the calendar bureau into the Astronomical Bureau. By 1645, Schall von Bell had become the Qing Dynasty's first foreign head of the Astronomical Bureau.

Due to Johann Adam Schall von Bell's favor and position in Beijing, the Jesuits were able to prosper smoothly. In 1650, the Shunzhi Emperor granted Schall von Bell a piece of land adjacent to the Catholic Church in Xuanwumen to rebuild a new church. Empress Dowager Zhaosheng issued a grant of silver, and nobles and officials joined in donating to the project. The new church was completed in 1652, and it bore a plaque inscribed by Shunzhi himself, reading "Qin Chong Tian Dao." In 1654, Shunzhi Emperor also bestowed upon Schall von Bell the land adjacent to Matteo Ricci's tomb outside Fucheng Gate, which was used to construct a church dedicated to the Virgin Mary. In 1657, the Shunzhi Emperor personally composed the inscription "Tongweijiajing" for the Xuanwumen Catholic Church. He even provided an imperial tablet for the Catholic Church, showing his support for Catholicism.

The Shunzhi Emperor died in 1661 and Schall von Bell's position started to erode. In 1664 he was accused of planning a rebellion and of having wilfully miscalculated the time and place of a funeral and in that way contributed to the death of Consort Donggo. Schall von Bell and other Jesuits, Ferdinand Verbiest included, were imprisoned and condemned to death. Schall von Bell had a stroke in prison. However, after a five-day earthquake in Beijing, the Jesuits were pardoned and only five Chinese Christians working in the astronomical office were executed. The Jesuits were exiled to Macau. Schall von Bell died within one year of his release. He was buried in the Jesuits' Zhalan Cemetery in Beijing.

During the Qintianjian incident, various provincial governors also apprehended missionaries. In May 1665, these priests were brought to Beijing for trial. Among those detained were individuals like Nicolas Trigault from Shanxi, Jean-François Ronusi de Ferrariis from Shaanxi, and Jean Valat from Shandong. The priests from the provinces who were later released included 25 Jesuits, 4 Dominicans, and 2 Franciscans, totaling 31 individuals. Eventually, these priests were released from detention, but 25 of them were confined within the Jesuit Church in Guangzhou and were prohibited from engaging in missionary activities.

During the Kangxi era, the resurgence of Catholicism in China was greatly facilitated by the Belgian Jesuit priest Ferdinand Verbiest (1623–1688), who played a pivotal role while serving in the Astronomical Bureau. In June 1669, Verbiest assumed the deputy head of the Astronomical Bureau, and in 1673, he was promoted to the head. In 1674, he successfully crafted astronomical instruments and was subsequently promoted to the title of secretary of the Court of Imperial Sacrifices. In 1678, the Kangxi calendar project was successfully concluded, and he was granted the title of minister of the Office of Transmission. Due to his successful cannon casting efforts, he was promoted to the vice minister of the Ministry of Works in 1675. In 1676, Verbiest also took on the role of vice president of the Chinese diocese. His efforts were instrumental in urging the Jesuit order to send missionaries to China to expand their missionary activities further. Five Jesuit missionaries arrived in Beijing on February 8, 1688. All of these missionaries were well-versed in various fields, including astronomy, calendar-making, and geography. They also maintained connections with the newly established French Académie des Sciences. In March 1699, an additional ten French missionaries made their way to China, contributing significantly to both missionary work and scientific advancement. The growth of missionary activities during this period led to a significant expansion in less than 40 years, from 1664 to 1701. Over the span of 1581 to 1712, a total of 249 Jesuit missionaries arrived in China (with 127 deaths en route), along with 48 Dominicans, 56 Franciscans, 17 Augustinians, and 30 non-affiliated priests.[11]

### *The Prohibition of Catholicism After Yongzheng*

During the reigns of Yongzheng, Qianlong, Jiaqing, and Daoguang, the Qing dynasty issued a series of orders to restrict Catholic missionary activities in defiance of the Roman Catholic Church's ban.

In the early years of Yongzheng's reign, the Fujian Church Case occurred. The Dominicans in Fuan were accused of building a church. The governor of Fujian and Zhejiang suggested to Yongzheng that all Western missionaries, except those required for service in Beijing, should be sent to Macau. He proposed that the Catholic churches be converted into public facilities and that strict measures be taken against those who

[11] Augustin Colombel, *Histoire de la mission du Kiang-nan*, p. 465.

mistakenly entered the religion. On January 11, 1724 (the 11th day of the 12th month of the first year of Yongzheng's reign), Yongzheng ordered all missionaries from different provinces to leave the country within six months. Over 50 clergy members (including 37 Jesuits) were expelled, along with five bishops. In August 1732, the 35 clergy members who had been gathered in Guangzhou were also expelled to Macau by the governors of the two provinces. However, the 20 priests working at the Astronomical Bureau were allowed to remain but were prohibited from preaching.

For more than 200 years, the Roman Catholic Church continuously held the authority of missionary activities in the hands of Western missionaries. However, this approach contradicted Chinese cultural practices and the path of intellectual achievement, leading to a decline in the influence of the Church. During the Qianlong era, various provinces continued to reject Catholicism. There were notable incidents such as the execution of the Dominican Bishop Mgr. Sanz in Fuzhou in 1747, the execution of the Jesuits António-José Henriques and Tristan d'Attimis in Suzhou in 1748, and the imprisonment of priests in Nanjing, Hubei, and Sichuan. In 1785, four Franciscan priests were arrested in Huguang for attempting to infiltrate the country's interior. Subsequently, priests from various provinces, including Sichuan, Shanxi, Shaanxi, Shandong, Jiangxi, Guangxi, and Huguang, were either apprehended or voluntarily surrendered themselves to officials. They were then sent to Beijing. However, 18 Western missionaries were eventually released with leniency. In Beijing, many priests also served in the imperial court. Those who served in the Astronomical Bureau included individuals like Ignatius Kögler from Germany, André Pereyra from Portugal, August von Wallerstein from Austria, Anton Gogeisl from Germany, José de Espinha and José Bernardo de Almeida from Portugal, and Castiglione from Italy. Castiglione was known for his paintings at the court, while Franz Stadlin was a Swiss watchmaker, Giovanni Giuseppe da Costa was an Italian pharmacist, and Etienne Roseset was a French doctor.

The dissolution of the Jesuits had a significant impact on the state of Catholicism in China, leading to a period of decline. Driven by religious convictions, Portugal, Spain, and France separately suppressed the Jesuit order. On July 21, 1773, Pope Clement XIV officially issued an order to ban the Jesuits. This order reached China by 1775, and the Jesuit order was disbanded in the same year. On December 17, 1783, responding to the proposal of the King of France, the Roman Propaganda Department dispatched members of the Lazarists to Beijing to assume the missionary role that the Jesuits had held. In April 1785, three Lazarist missionaries arrived in Beijing and settled in the North Church. This marked the conclusion of the Jesuits' 200-year presence in China as they withdrew from the historical stage.

CHAPTER 7

# Chinese Culture and Western Europe in the Eighteenth Century

## 7.1 The Translation of Chinese Ancient Books and Western Studies on the Chinese Language

### *Western Translation of the Four Books and Five Classics*

Western missionaries needed to study the classical Chinese texts, known as the Four Books and Five Classics, to adapt to Chinese society and understand it. The earliest translation of the Four Books into Latin and sent back to Italy was done by Matteo Ricci in 1593. In 1626, a Belgian missionary named Nicolas Trigault translated the Five Classics into Latin, and they were printed in Hangzhou, making it the earliest published Western translation of Chinese classics.

The formal Latin translations of the Four Books were the collaborative work of Prosper Intorcetta (1625–1696), an Italian Jesuit and Ignatius da Costa (1599–1666), a Portuguese Jesuit. They translated *Da Xue* (Great Learning) and titled it *Sapientia Sinica,* published in 1662 in Jianchang. They also translated *Lun Yu* (Analects). Intorcetta further translated *Zhong Yong* (Doctrine of the Mean) as *Sinarum Scientia Politico-Moralis*, which was published in Guangzhou in 1667 (reprinted in Goa in 1669). The first European printing of Confucius' works occurred in 1687. That year, a Belgian priest, Philippus Couplet (1624–1692), published *Confucius Sinarum Philosophus* in Paris. The Chinese title of the book is *Xi Wen Si Shu Jie* (Explanations of the Four Books in Western Languages). It includes an introduction to Chinese classical



F. Shen, *Cultural Communications Between China and The Outside World Throughout History*,
https://doi.org/10.1007/978-981-97-4696-5_7

texts, a biography of Confucius, and Latin translations of *Da Xue*, *Zhong Yong*, and *Lun Yu*, using the translations by Intorcetta and da Costa, with annotations. Additionally, *Notizie Varie dell'imperio dell Cina*, an Italian work by Joannes Grueber, an Austrian priest who previously served the Roman Curia in exploring overland routes between Europe and Asia, was published seven years after his death. The book includes Italian versions of *Zhong Yong* and a biography of Confucius.

The complete translation of the Four Books was done by Franciscus Noël (1651–1729), a Belgian scholar, using a literal translation method in Latin. After arriving in China in 1687, Noël conducted missionary work in Huai'an, Wuhe, Shanghai, Jianchang, and Nanfeng. In 1702, he was sent back to Europe by the Church, and in 1711, he published his translation of the Four Books and *Philosophia Sinica* at the University of Prague. *Philosophia Sinica* provides a systematic introduction to China's Confucian classics and ancient philosophical thoughts.

The earliest attention to the Five Classics was focused on the *Book of Changes (I Ching)* and the *Book of Documents (Shu King)*. During the reign of the Kangxi Emperor, Joachim Bouvet, Claude de Visdelou, Joseph Henri Marie de Prémare, and Jean-Baptiste Régis were involved in translating and studying these texts. In 1710, Bouvet, who studied astronomy and calendrical calculations, focused on the *Book of Changes*. The Kangxi Emperor summoned a French Jesuit, Jean-François Foucquet, from Jiangxi to assist Bouvet in his research on the *Book of Changes*. This collaboration lasted for six years. Bouvet authored a Latin work titled *Idea Generalis Doctrinae Libri I-king*, which summarized the *Book of Changes*. Manuscript copies of this work are held in the National Library of Paris. Bouvet also had a manuscript on the study of the *Classic of Poetry (Shih-ching)*. Claude de Visdelou researched the *Book of Changes*, *Classic of Poetry*, and *Book of Rites*. There is also a Latin version of the *Book of Documents* in four volumes, six books, which are kept in the Vatican Library. The sections of the *Book of Rites* dealing with the rituals and sacrifices were also translated into Latin. Jean-Baptiste Régis also translated the *Book of Changes* and included research and annotations, many of which were contributed by Pierre Vincent de Tartre. However, it was not until 1834–1839 that these translations were published in two volumes, titled *I-King antiquissimus Sinarum Liber*. Joseph Marie de Premare (1666–1735) selected and translated the*Book of Documents* and eight odes of the *Classic of Poetry*. He was the first to translate the Chinese drama *The Orphan of Zhao* into a European language. His work was included by Du Halde in his *Description Géographique, Historique, Chronologique, Politique et Physique de l'Empire de la Chine*. There is also a manuscript titled *Essai d'introduction préliminaire à l'intelligence des King* held in the National Library of Paris. Additionally, Julianus-Placidus Hervieu (1671–1745) had a translation of selected works from the *Classic of Poetry*.

During the reign of the Qianlong Emperor, three French Jesuits researched ancient Chinese texts: Alexander de la Charme (1695–1767), Antoine Gaubil, and Joseph-Marie Amiot. Alexander de la Charme was well-versed in old classics and began his translation of the *Classic of Poetry* into Latin with detailed annotations in 1733. He also translated the *Book of Rites*. Antoine Gaubil's translation of the *Book of Documents* was published in Paris in 1770 and was known for its faithful and well-researched translation. Following Prosper Intorcetta's translation of the Analects, Joseph-Marie Amiot published the French work *Vie de K'ongtse* in Beijing in 1784. He also published *Abrégé de la vie des principaux disciples de K'ong-tse*, a brief account of the lives of Confucius's main disciples.

Indeed, Italian and French missionaries played a significant role in translating, researching, and introducing ancient Chinese classics and Confucian teachings to Europe. Through their efforts, these works were translated into Latin and French, allowing Chinese philosophical and cultural ideas to spread among Europe's intellectual and elite circles.

### *Compilation of Chinese Grammar and Dictionaries*

The European academic community began studying and compiling Chinese language and dictionaries from the Italian missionary Martinus Martini (1614–1661). During his travels in various European countries from 1653 to 1657, he wrote the *Grammatica Sinica* in Köln, Germany, which became an essential gateway for European scholars to delve into studying China. Later, Philippus Couplet made revisions to the book, and a manuscript of it is still preserved in the Hunterian Museum at the University of Glasgow. Under Couplet's guidance, the German physician Christian Mentzel began studying the Chinese language and wrote *Clavis Sinica*, although it was not published. In 1685, Mentzel compiled the *Sylloge minutiarum Lexici Latino-Sinico Characteristici*, a handbook of Latin-Chinese vocabulary, which was officially published. Around the same time, the Spanish missionary Francisco Varo authored *Faecilis et Perspicua Methodus ad Linguam Mandarinam Addiscendam*. Theophilus Siegfried Bayer's Latin work *Museum Sinicum* was published in 1730. The first volume discussed the progress of Chinese literature in Europe and Chinese grammar, while the second volume contained a Chinese dictionary and dialects. With these publications, the foundation of Chinese grammar and Latin-Chinese dictionaries had taken shape.

Subsequently, various Chinese-European language dictionaries were published. Joachim Bouvet authored the *Dictionarium Sinico-Gallicum* and *Méthode pour étudier la langue chinoise*. Most importantly, Joseph-Marie de Premare wrote *Notitia Linguae Sinicae* in Guangzhou in 1728, the first Western work to extensively study Chinese linguistics. It drew heavily from Bouvet's *Méthode pour étudier la langue chinoise* written in Latin and French and contained over 13,000 examples, showcasing a profound understanding of the elegance of the Chinese language. The book was reviewed by Étienne Fourmont (1683–1745), an advisor to Louis XIV of France. Fourmont translated Francisco Varo's *Faecilis et Perspicua Methodus ad Linguam Mandarinam Addiscendam* into Latin and incorporated research materials from Joseph-Marie de Premare to write *Grammatica Sinica*, which played a milestone role in the development of Sinology in France. Many of France's renowned Sinologists are Fourmont's students.

In 1732, Carolus Horatius a Castarano compiled a Latin-Italian-Chinese dictionary completed in Haidian, Beijing. In 1733, P. Bazilius a Glemona authored a Chinese-Latin dictionary. The following year, Dominique Parrenin translated Pierre Danet's Latin dictionary into Chinese, and the original manuscript is preserved in the Hunterian Museum at the University of Glasgow. Alexander de la Charme's *Han-Mongolian Comparative Dictionary* and *The Pentaglot Dictionary*, compiled by collaboration between Jean Joseph Marie Amiot and the Hanlin Academy, are monumental works in the same category.

## 7.2 Introductions About Chinese History, Geography, and Science

### *Writings on Chinese History and Geography*

After European missionaries began their activities in Asia, the history of China, a great Eastern country, attracted the attention of Western Europeans. In addition to the collection of 7000 Western books by Nicolas Trigault, it is worth noting that when Philippe Couplet arrived in Rome in 1682, he brought over 400 works by missionaries in China, later housed in the Vatican Library. In 1694, Joachim Bouvet returned to France and presented 300 volumes of Chinese books to King Louis XIV. Furthermore, half a century later, Joseph-Marie de Premare immersed himself in studying Chinese classics in Guangzhou and extensively collected Chinese works, which he sent to the Royal Library of Fourmont, the predecessor of the National Library of France. The thousands of volumes on China preserved in the Royal Library were initially collected by Joseph-Marie de Premare in China.

Due to this abundant collection of Chinese historical works, Europeans, who boasted of their emerging civilizations, began to understand the origins of an old Eastern country. One of these works is *Historia de las cosas más notables, ritos y costumbres del gran reyno de la China*, written in Spanish by Juan Gonzalez de Mendoza and published in Rome in 1585. Within five years, Italian, French, English, Latin, and German versions of the book were printed in various parts of Europe. This was one of the earliest systematic introductions to Chinese history and geography. In 1658, Martinus Martini published the first ancient Chinese history written in Latin, titled *Sinicae Historiae Decas Prima*, in Munich. However, he only completed the first ten volumes, covering the period from legendary times to the end of the Western Han Dynasty. The French translation of the book was published in Paris in 1667. Following that, the 13-volume *Histoire Générale de la Chine* by Joseph-Anne-Marie de Moyriac de Mailla was published between 1777 and 1785. This comprehensive history was primarily based on the *Tongjian Gangmu*. It provides a detailed account of the history of the Ming and Qing dynasties. As the author participated in mapping China during the Kangxi era, his discussion of Chinese geography is highly accurate. Of particular interest to Enlightenment thinkers was Athanasius Kircher's *China monumentis, qua sacris qua profanis, nec non variis naturae & artis spectaculis, aliarumque rerum memorabilium argumentis illustrata*, which is commonly known simply as *China Illustrata*. Originally published in Latin in Amsterdam in 1667, it was later translated into French and German, becoming the essential source for the European intellectual community to understand China.

The earliest systematic study of Chinese geography and the actual measurement of Chinese maps was carried out by Martinus Martini. In 1655, *Novus Atlas Sinensis* was published in Amsterdam. This work, a folio with 17 maps and 171 pages of text, was, in the words of the early twentieth-century German geographer Ferdinand von Richthofen, the most complete geographical description of China that we possess and through which Martini has become the father of geographical learning on China. Jean-Baptiste Bourguignon d'Anville, a prominent geographer at the French court who was considered an authority in European geography, produced various maps of China based on the *Huangyu Quanlan Tu*.

### *An Exploration of Chinese Astronomy, Biology, and Medicine*

In 1687, five French scientists arrived in China. They were Jesuit missionaries sent by King Louis XIV of France and funded by the French government to conduct a comprehensive survey of the country. This survey was part of a broader effort by European countries such as England, Italy, and Russia to expand their foreign markets. The French-led survey of China lasted for almost a century. During this period, the study and exploration of natural science and medicine in China were predominantly carried out by French Jesuits. One of the participants in this survey was Jean de Fontaney. In a letter dated November 8, 1687, addressed to the French Academy of Sciences, he provided an overview of the disciplines investigated and studied by their team in China. These included astronomy and geography, Chinese chronicles, sinology, natural science, medicine, and China's political, economic, and social conditions.

The research findings were compiled into three monumental works. The first is "*Description Géoographique, Historique, Chronologique, Politique et Physique de l'Empire de La Chine et de La Tartarie Chinoise*. This four-volume book was published in Paris in 1735 by Jesuit missionary J.B. du Halde. It was also published in The Hague the following year and quickly spread across Europe with an English translation by R. Brookes. This work, renowned as an encyclopedia of China, contains translations of various Chinese literary works, including *Guwen Guanzhi* and *The Orphan of Zhao*. It became essential for French Enlightenment scholars and those interested in China in German and English contexts. Meanwhile, between 1702 and 1776, a comprehensive collection of letters titled *Lettres édifiantes et curieuses écrites des missions étrangères par quelques Missionnaires de la Compagnie de Jésus* was published in 34 extensive volumes. Additionally, from 1772 to 1788, 15 volumes of *Mémoires Concernant l'Histoire, les Sciences, les Arts, les Moeurs, les Usages, etc., des Chinois par les Missionnaires de Pékin* were printed, with the 16th volume published in 1814.

The achievements of the French scientists who came to China in natural sciences primarily focused on astronomy, biology, and medicine. One notable accomplishment is the work of Antoine Gaubil. In 1732, under the editorship of E. Souciet, he published *Histoire Abrégée de l'Astronomie Chinoise* and *Traité de L'Astronomie Chinoise* in Paris.

French Jesuit missionaries paid great attention to Chinese plants and produced numerous works in this field. However, the most significant contributions came from Pierre Nicolas d'Incarville. In the book Voyage en Chine, he documented his activities in China from 1721 to 1722 and sent a collection of 4010 botanical and zoological illustrations to Adrien de Jussieu in Paris. The illustrations were once part of the French National Museum of Natural History collections but have been lost. The Adrien de Jussieu Scientific Library in Paris still preserves 72 botanical illustrations by Pierre Nicolas d'Incarville. The widely cultivated Reine-Marguerite in Western Europe was sent to the Royal Garden in Paris by d'Incarville from China in 1728. Initially, it only produced white flowers, but later, it bloomed in red and violet colors. By 1750, it had transformed into a multicolored variety. Pierre Nicolas d'Incarville was also one of the earliest Western researchers on Chinese silkworms and authored the book *Elevage des vers à soie* with 23 color illustrations. Pierre-Martial Cibot arrived in China in 1760 and studied Chinese geology, minerals, chemistry, animals, and plants. Volume 13 of the *Mémoires Concernantes L'Histoire* contains Cibot's observations on Chinese bees, swallows, and cicadas. His research on Chinese plants can be found in Volume 11 of the same book. The plant specimens he sent back to France in 1780 are still preserved in the French National Museum of Natural History, including 149 species collected inland and 144 species collected in Macau. Francois Bourgeois provided detailed information on the use of rubber in China. Jean-Paul Louis Collas (1735–1781) compiled a list of Chinese plants and trees that could be introduced to France. Joseph-François Charpentier de Cossigny introduced Chinese lychee to France.

Traditional Chinese medicine was a focal point of study for the Jesuit missionaries. The first European work on Chinese medicine was sent to Europe by an unknown Frenchman from Guangzhou. Published in Grenoble in 1671, it was a French version of the *Maijing* by the renowned physician Wang Shuhe of the Jin Dynasty. In 1658, Polish physician Michał Piotr Boym wrote *Clavis medica ad Chinarum doctrinàm de pulsibus*. The book was published in 1686 and translated into various European languages. In 1676, a pulse examination book in Italian was published in Milan, which was reprinted in Nuremberg ten years later and widely disseminated. An English translation was published in London in 1707. Chinese medicine thus became a new area of knowledge for Europeans.

The third volume of Jean-Baptiste Du Halde's *Description Géographique, Historique, Chronologique, Politique et Physique de l'Empire de La Chine et de La Tartarie Chinoise* is dedicated to traditional Chinese medicine. It includes translations of various classics such as *Maijing*, *Maijue*, *Compendium of Materia Medica*, *Shennong's Classic of Materia Medica*, *Mingyi Bielu*, and *Yiyao Huibian*. It provides numerous prescriptions and mentions medicinal substances like donkey-hide gelatin, gallnuts, and cypress bark. Joseph-Anne-Marie de Moyriac de Mailla's *Histoire Générale de la Chine* discusses traditional Chinese medicine and herbal remedies. Dominique Parrenin sent technical information and samples of donkey-hide gelatin to the French Academy of Sciences. Martialus Cibot discusses the pathogenesis of smallpox in China, while Dominique Parrenin translated the world's first-ever written book of forensic science, *Collected Cases of Injustice Rectified*. Joseph-Marie Amiot sent numerous Chinese medicinal herbs and medical instruments to Henri Bertin, who was fascinated by China. Two Chinese Christians, Louis Kao and Etienne Yang, also sent Chinese herbs upon Bertin's request. The fingerprint identification method used in China since the Tang Dynasty also made its way to Europe and became popular in the nineteenth century.

Chinese surgical techniques attracted the attention of the Royal Academy of Surgery in France, which appointed two members to gather information about them. Longevity techniques and qigong therapy in China also piqued Western interest. Amiot emphasized the respiratory and embryonic breath, referring to it as "cong-fou".

## 7.3 Confucianism and the Enlightenment Movement in Europe

The Enlightenment Movement that shook Europe in the eighteenth century advocated rationalism, attributing all phenomena to nature rather than miracles. It emphasized the legitimacy of human understanding of natural laws and rational principles, rejecting the notion of divine intervention and miracles. This social and intellectual trend originated in England, flourished in Germany, and peaked in France, becoming a precursor to the bourgeois social revolution. The advocates of the Enlightenment defended natural rights, promoted wisdom and education, and opposed the dominant Christian religious theology that characterized medieval European society. However, the tension between the abstract and the concrete, reason, and nature persisted, just as the uncompromising situation between the intellectual dominance of ancient Greek culture and the religious belief in divine purpose. The Enlightenment thinkers sought inspiration and knowledge from the non-Christian world of the East, drawing on ideas that aligned with rational principles.

China, a vast and glorious civilization in the distant East with a completely different temperament from Europe, became a source of spiritual inspiration for Enlightenment thinkers through the introduction of Jesuit missionaries. Ancient Chinese philosophy, represented by Confucius' Confucianism with "Heaven" as the natural law and Song Confucianism with "Dao" as the fundamental principle of reason, which they regarded as the "foundation of Heaven and Earth and the source of all things". Confucius' ethical and moral philosophy centered around "Ren" (benevolence), and his advocacy of education became a fundamental source for French philosopher René Descartes (1596–1650) in his promotion of rationalism. The traditional Confucian values of benevolent rulership and unified society, especially the stable and prosperous social conditions during the early Qing dynasty under Kangxi's reign, as reported by Jesuit missionaries, served as a model of enlightened absolutism that Enlightenment thinkers advocated against the expansionist aspirations of European monarchies. The wealthy Chinese culture during the era of the Enlightenment provided nourishment for Leibniz's classical speculative philosophy, Voltaire's natural theology, Quesnay and Turgot's physiocracy, and contributed to the birth of modern European civilization.

### *Song Confucianism and Leibniz's Philosophical Revolution*

Gottfried Wilhelm Leibniz (1646–1716) was a German polymath active as a mathematician, philosopher, scientist, and diplomat. He held great admiration for the natural theology of Chinese Confucian philosophy. In his letter "Discourse on the Natural Theology of the Chinese" addressed to Monsieur De Remond, Leibniz stated, "This philosophical doctrine or natural theology has been established for about three thousand years and has great authority, far predating the philosophy of the Greeks". He found a unity between the Chinese concept of the Mandate of Heaven and the Christian belief in God, viewing that the Chinese also unified God and matter. According to Leibniz, the belief of Chinese philosophers honored the highest rationality, which he believed was present everywhere. In his later years, Leibniz expressed his philosophical ideas of the "pre-established harmony system" through his concept of "monads," opposing Descartes' mechanistic materialism in explaining the world. He believed that the world consisted of a vast living organism composed of innumerable "monads," which are living organic entities, and he acknowledged that this philosophical thought was influenced by his appreciation of classical Chinese teachings. He stated that the Chinese concept of "li" is equivalent to the highest entity we worship as "God". Leibniz openly proclaimed that the Chinese concept of "li" and the European concept of "God" are in harmony.

Leibniz began studying China at the age of 21, and in 1687, he read the works and biography of Confucius published in Paris. After meeting Jesuit missionary Claudio Filippo Grimaldi in Rome, he published *Novissima Sinica* in 1697 based on the Jesuit writings and materials provided. In the same year, he corresponded with Joachim Bouvet, who was in Beijing, discussing binary arithmetic and the *I Ching* for six years. When the second edition of *Novissima Sinica* was published, it included his Latin translation of Bouvet's work *Histoire del Empereur de la Chine*. Over 25 years, Leibniz maintained correspondence with French Jesuit missionaries in China, including Claudio Filippo Grimaldi, Joachim Bouvet, Claude de Visdelou, Jean de Fontaney, Pierre Jartoux, and Italian Jesuit Giovanni Laureati, discussing Chinese Confucian natural theology. In 1990, Rita Widmaier published *Leibniz korrespondiert mit China* in Frankfurt, which included 71 letters written by Leibniz from July 1, 1689, to February 5, 1714. Most of these letters were in French, with a few in Latin, surpassing the previously published 12 letters by a significant number. In 1715, shortly before his death, Leibniz wrote a lengthy letter to French Regent Philippe d'Orléans' advisor, De Remond, titled "Discourse on the Natural Theology of the Chinese". In this letter, he argued that China's concept of destiny and the Mandate of Heaven represented the unchanging laws of nature, and to conform to rational laws, one must align with Heaven to achieve pre-established harmony. This expressed his continued pursuit of the natural laws revered in Chinese philosophy, which led him to a final rupture with the dogmas of revealed theology within the Church.

Leibniz was the first philosopher in the West to recognize the usefulness of Chinese culture in aiding the development of European culture. By comparing Chinese and European civilizations, Leibniz acknowledged that Europe excelled in logic, metaphysics, mathematics, and military technology and theory, while China excelled in ethics, practical philosophy, politics, social customs, and ceremonial systems. Leibniz firmly believed in the "complementary relationship" between Chinese and European civilizations. According to his understanding, China surpassed Europe in certain aspects because "their experience is much older than ours, their traditions are rarely interrupted, and they maintain them much better than we do". In the preface to his work *Novissima Sinica*, Leibniz used Chinese civilization to criticize European Christian culture mercilessly, stating, "We formerly believed that no nation could surpass us in ethics or progress in worldly affairs. But now, from the East, from China, we are awakened".[1] He warned European society, saying, "In my view, we are currently in a state of moral decline from which it is difficult to extricate ourselves. I even believe that we should invite representatives from China to come and teach us about the purpose and practice of natural theology, just as we send missionaries to China to teach them revealed theology". He made a fitting assessment comparing Chinese and Western cultures: "The strengths of European culture lie in mathematical and speculative sciences, including the military field, where China is not superior to Europe. However, in practical philosophy, Europeans are far inferior to the Chinese". Leibniz actively sought to implement the practical philosophy of China and advocated for establishing scientific academies in Berlin, Vienna, and St. Petersburg, with research on China included in their projects. His vision for the nation was to establish a

[1] Deutens, *Opera Omnia*, t IV, pt. I, 78–86.

unified country under the rule of an enlightened monarch characterized by benevolence, justice, perseverance, and extensive knowledge, and the Kangxi Emperor of China became a role model embodying these ideals.

Indeed, Leibniz made significant scientific achievements that aligned with certain Chinese philosophical concepts. In 1679, he formally presented his work *De Progressione Dyadica*, which is entirely consistent with the diagrams of the "Sequence of the Sixty-Four Hexagrams of Fuxi" and the "Orientation of the Sixty-Four Hexagrams of Fuxi". Furthermore, in 1714, he published his work *Monadologie*, which incorporated notions of the "Dao" from Laozi, Confucius, and Chinese Buddhism. These connections highlight Leibniz's interest in and engagement with Chinese philosophy and his incorporation of Chinese concepts into his philosophical framework.

The dual sources of Leibniz and his student Christian Wolff's ideas, which extend beyond Plato to include Chinese philosophy, were elucidated by the German philosopher Carl Gunther Ludovici in his work *Ausführlicher Entwurf einer vollständigen Historie der Leibnitzischen Philosophie*. The close relationship between Leibniz and Chinese Confucianism has reached a point where it is difficult to separate and untangle.

### *The Encyclopedists' View of Chinese Culture*

France's intellectual and political revolution began with the Encyclopedists, led by Denis Diderot. Among them were many individuals who enthusiastically studied Chinese thought and explored Chinese culture. Figures such as Montesquieu, M. de Voltaire, Jean-Baptiste le Rond d'Alembert, and François Quesnay conducted in-depth research on Chinese history, philosophy, criminal law, social customs, and political systems through the writings and reports of Jesuit missionaries who had been to China. In 1713, Montesquieu had an extensive conversation with a Christian with the family name Huang from Fujian, China, who had lived in Paris for ten years and worked at the Royal Library. He wrote detailed notes about their discussion.[2] Voltaire had multiple conversations with French Jesuit Jean-François Foucquet. The awakening effect of Chinese civilization was particularly championed by Voltaire, who

[2] *Oeuvres Completes de Montesquieu*, Paris, vol. 2, pp. 927–943.

advocated for natural theology and enlightened absolutism. As a follower of John Locke, Voltaire criticized Leibniz while also using the model of Chinese civilization to challenge and mock European society, which considered itself the true civilization of the world.

According to Voltaire, the "discovery" of ancient and esteemed Chinese civilization by Europe served as a powerful contrast to the arrogance of the Christian world. In his *Dictionnaire philosophique,* he praised China as the "most beautiful, oldest, largest, most populous, and best-governed country in the world". He used Chinese chronology to refute the biblical account of the creation of the world. The Old Testament recorded the creation date as 3761 BC, while the Enlightenment scholars of the Encyclopedists acknowledged that China had "populated and multiplied for more than fifty centuries". Voltaire mercilessly mocked those Christians who boasted European civilization was the most advanced in the world, pointing out that while China was vast and flourishing, with a well-developed and wise system of governance, "we were still just a small group of wild men wandering in the Ardennes forest!"[3] He believed that the history of human civilization, science, and technology originated in China and had long been far ahead. He praised the historical records of China that "hardly have any fabrication or fantastic tales, with no self-proclaimed representatives of God receiving revelations as the Egyptians and Greeks, and the Chinese history has been written rationally from the beginning".[4] He expressed astonishment at the greatness of Chinese civilization, noting that "among all the peoples of the world, only they have continuously recorded solar eclipses and planetary conjunctions. Our astronomers have verified their calculations and are amazed that almost all the records are true and reliable".[5]

As a proponent of natural theology, Voltaire also regarded Confucianism as a form of natural theology. It fundamentally differed from the prevailing superstitious revelation in Europe and served as an exemplary "rational religion" that embodied sublime reason, conformity to nature, and morality. This new "rational religion" characterized by elevated motivation and adherence to natural and moral principles was the faith that Voltaire sought. He believed that "Confucius enabled humanity to attain

[3] *Oeuvres Completes de Voltaire,* Paris, 1865, vol. 3, p. 76.

[4] *Oeuvres Completes de Voltaire,* Paris, 1865, vol. 3, p. 25.

[5] *Oeuvres Completes de Voltaire,* Paris, 1865, vol. 3, pp. 25–26.

the purest understanding of God" by emphasizing virtue as the means to educate people. According to Voltaire, individuals should cultivate themselves and govern their countries following the natural laws of reason, not relying on divine revelation.[6] A fundamental aspect of Voltaire's natural theology was its connection to innate moral principles. It adopted the ethical norm of "Do not do to others what you would not want them to do to you," attributed to Confucius. The unity of innate moral principles, reason, and the progress of civilization formed the basis of Voltaire's natural theology.

Voltaire and the Enlightenment scholars of the Encyclopédie expressed their opposition to the oppressive rule of monarchies under religious authority in Europe by praising Chinese thought and politics. They presented a political system that embodied sublime reason, morality, tolerance, and moderation as their ideal. Diderot praised Confucianism, stating that "governing the country and bringing peace to the world could be achieved through reason or truth alone". D'Holbach considered China as a model that combined politics and ethical morality, stating that "people feel that in this vast country, ethical morality is the only religion for all rational individuals". He remarked, "China is the only country in the world that combines politics and ethical morality. The long history of this empire has made all rulers understand that prosperity can only be achieved by relying on morality". He advocated governing through virtue and coined the term "ethocratie" in his work *Ethocratie, ou Le gouvernement fondé sur la morale* (*Ethocracy or Government Founded on Ethics*). He openly declared, "European governments must take China as a model".[7] The natural philosophy, moral philosophy, and political ideals of Chinese Confucianism became powerful weapons for the fervent atheists or natural theologians within the Encyclopédie movement.

### *Physiocratis' Adoration to China*

The Physiocrats, following Voltaire's enthusiastic praise of Chinese civilization, once again embraced Chinese culture as the beacon of European intellectual thought in the 1760s, particularly in politics and economics. The term "Physiocrats" derived from Greek and originally meant "the

[6] *Oeuvres Completes de Voltaire,* Paris, 1865, vol. 3, pp. 25–26.

[7] D' Holbach, *Système Social*, 1773,vol. 1, p. 86.

school of natural rulership," with its central premise being the substitution of natural laws and order for divine revelation. François Quesnay, the founder of the Physiocratic school, was hailed as the "Confucius of Europe" for his publication of *Le Despotisme de la Chine* in 1767. In the introduction, he presented China's political system as a legitimate form of despotism, with the Chinese emperor being a legitimate absolute monarch. He regarded China as a shining example of a state founded on the law of nature, stating that "the law of nature serves as the foundation for human legislation and the highest criterion for human behavior" and that "all nations have disregarded this, except China". He believed that natural order was the source of human reason and rationality was the basis of natural rights.[8] Quesnay advocated an agrarian-based economy, emphasizing that only agriculture could create wealth while downplaying the role of money and commercial capital. His ideas called for a change in the prevailing mindset that undervalued and hindered agriculture. These ideas are traced back to the philosophies of Socrates, Fuxi, Yao, Shun, and Confucius. Quesnay greatly admired China's emphasis on agrarianism and the historical policies of Chinese emperors regarding agriculture. As a royal physician, he even encouraged King Louis XV of France to imitate the Chinese emperors by participating in the "land registration" ceremony for spring plowing in 1756. After Quesnay's death, his students regarded his famous work *Tableau économique*, published in 1758, and his lifelong pursuit as the realization of Confucius' teachings and moral standards, guided by respecting heaven, loving humanity, and restraining desires.

Quesnay's advocacy for implementing a single land tax policy was also influenced by the ancient taxation system in China, particularly by the inspiration drawn from the equal-field system described in the *Rites of Zhou*. He believed that the fundamental principle behind China's ability to maintain social stability for thousands of years lay in collecting taxes. The Chinese people have always considered a tax on the income from labor and a head tax as "unreasonable and unfair". According to Quesnay, the only solution was to impose tariffs on the pure income derived from land ownership to provide the necessary revenue for the state. The ancient Chinese thoughts and political system have left an indelible imprint on the political ideals and economic theories of the Physiocrats.

[8] Auguste Oncken, *Oeuvres Economiques et Philosophiques de F. Quesnay*, Frankfurt, 1888.

Turgot (1727–1781), the reformer of the Physiocrats, served as the Minister of Finance from 1774 to 1776. His understanding of economic principles was based on the manifestation of individual psychology, thus paying more attention to the interrelationships of various aspects in the flow of wealth. During his tenure as the Intendant of Limoges, Turgot met two Chinese youths, Louis Kao and Etienne Yang, studying in France. He persuaded them to extend their stay in France for another year. Upon their return to China in 1764, Turgot wrote them instructions, entrusting them with collecting information on Chinese crafts and scientific techniques. He also specifically prepared the renowned work "Reflections on the Formation and Distribution of Wealth" for them to have a systematic understanding of the theories and doctrines of the Physiocrats. The instructions outlined a survey framework consisting of 52 questions, including 30 questions on agricultural economic issues related to land, capital, labor, rent, and taxation in China, 15 questions on craft issues such as papermaking, printing, and textile industry, and seven questions on natural geography, resources, and historical matters. The activities involving China were necessary for the reforms of the Physiocrats in France. After Gao and Yang returned to China, they regularly reported to Henri Bertin and assisted the Jesuits in compiling the "Beijing Missionary Reports". The French government funded the publication of this material.

In response to the Physiocrats' demand for research on Chinese agriculture and horticulture, the Jesuits sent the abundant materials they collected to France. The missionaries focused on cultivating rice, mulberry trees, and tea in China. They also examined Chinese farming tools and grain storage methods, as Henri Bertin requested. During the same period, Chinese plows were also introduced to Europe. Chinese floral and grafting techniques were also brought to Europe at this time, significantly transforming the appearance of European gardens in the eighteenth century.

## 7.4 The Rococo Style and the Chinese Cultural Boom in Western Europe

The art of seventeenth and eighteenth-century Europe underwent significant changes due to socioeconomic transformations, giving rise to two distinct styles: Baroque and Rococo. The Baroque style prevailed during the reign of Louis XIV, known for its grandeur and opulence. During the Enlightenment period, in harmony with philosophical thought, there was a call for personal liberation in art, aiming to break free from the constraints of artificiality and rigidity of tradition. This led to the emergence of Rococo, a new artistic style characterized by vividness, elegance, lightness, and naturalness. Rococo began to rise around 1690 and flourished after the death of Louis XIV in 1715, continuing until the late eighteenth century.

With its novelty, delicacy, softness, intricacy, and elegance, China's artistic style became a source of inspiration for the Rococo style, which sought beauty, variety, and novelty. In this artistic style, paintings emphasized light and subtle colors, while architecture favored rounded forms over sharp angles and corners.

Before the rise of the Rococo style, Leonardo da Vinci in Italy had already incorporated Chinese landscape painting techniques, evident in the subtle and profound mountainous background of his renowned artwork, the Mona Lisa. His Tuscan landscapes also showcased similar characteristics. Furthermore, The Temptation of Saint Anthony also exhibited the spirit of Chinese painting.

In the seventeenth and eighteenth centuries, there was a popular trend in France for the Chinese style and taste, which involved imitating Chinese objects and aesthetics. The term "Chinese style" originally carried the meaning of grotesque. The Rococo style prevailed in France and was also referred to as the "French Grotesque". France was the center of Europe's literature, fine arts, theater, etiquette, fashion, and interior decoration during this time. As a result, Chinese tea, silk, embroidery, porcelain, and lacquerware became popular in European society. The influence of the Rococo style extended beyond porcelain to lacquerware, silk textiles, and various other objects. The Rococo style and Chinese aesthetics were closely connected as if they shared a common origin.

### *Imitation of Chinese Arts and Crafts*

Since the seventeenth century, the French court and nobility have been competing to acquire Chinese porcelain, lacquerware, silk, and embroidered textiles. They took pride in adopting Chinese sedan chairs, mirrors, fans, and clothing styles.

Chinese porcelain was considered a scarce and valuable treasure in fifteenth-century Europe. The Landesmuseum in Kassel, Germany, houses a blue and white porcelain bowl, making it the earliest dated Ming dynasty porcelain in Europe that has survived. It has been a cherished heirloom of the Hessian family for centuries. The Warham Cup at Oxford University is another exquisite piece, a blue and white porcelain bowl with the Hongzhi reign mark, gifted by Bishop Warham between 1502 and 1532. There are also a pair of Xuande blue and white porcelain bowls, which were then presented to Sir John Trenchard. These four pieces of porcelain are all adorned with magnificent silver mounts and are highly treasured. At the time, the value of a piece of porcelain, as stated by Damião de Góis, was equivalent to that of seven enslaved people. After Portugal established new sea routes, porcelain became the most precious gift in European society. Both Vasco da Gama and Francisco de Almeida used porcelain to win the favor of King Manuel I of Portugal and obtain special privileges. Since the seventeenth century, Dutch and English merchant ships participated in trade with the Far East. In over 80 years, from 1602 to 1682, the Dutch East India Company alone imported around 16 million pieces of Chinese porcelain to various countries. In 1604, the Dutch looted the Portuguese carrack, Santa Catarina, carrying porcelain back to Europe. The Chinese porcelain from the shipwreck was named Kraaksporceleint and was auctioned in Amsterdam. King Henry IV of France and King James I of England competed to acquire these pieces, making Chinese porcelain highly sought-after throughout Europe.

During the Rococo era, Europeans' understanding of China was not based on texts but on light-colored porcelain and vibrant silk fabrics. When porcelain first arrived in France, it was referred to as "celadon" after the character Celadon in Honoré d'Urfé's famous novel *L'Astrée* in 1610, who was often depicted wearing a light green cloak. In seventeenth-century England, the term "Chinaware" was used to refer to porcelain from China. European monarchs during that time were particularly fond of Chinese porcelain. King Louis XIV of France commissioned his Prime Minister, Jules Mazarin, to establish the French East India Company to order porcelain from Guangdong with the French coat of arms. A special room was dedicated to displaying Chinese porcelain in the Palace of Versailles for the king's collection. Russian Tsar Peter the Great also commissioned porcelain from China, and the Kangxi Famille Verte porcelain with the double-headed eagle displayed in the Palace Museum in Beijing was one of them. Prussian King Frederick the Great considered the collection of porcelain from various countries his wealth. Under his patronage, exquisite Meissen porcelain was produced, inspired by Chinese porcelain. In times of financial difficulty, producing Meissen porcelain became a means for Frederick to repay his debts. Some of Germany's Rococo-style palaces, such as Charlottenburg, Schlossbrühl, and Oranienbaum, gained great fame for their porcelain collections. Queen Mary II of England also had a passion for Chinese porcelain and had numerous glass cabinets in her palace to display various types of porcelain.

Consequently, British society embraced the fashion of decorating and using Chinese porcelain, making it an essential part of living rooms and interiors. British Prime Minister Robert Walpole was even considered an expert in identifying Chinese porcelain. British newspapers frequently published advertisements for the sale of porcelain.

After the mid-seventeenth century, Chinese porcelain became a significant import in Europe. The Netherlands, France, and England all imported large quantities of porcelain. In 1722 alone, 400,000 pieces of porcelain were shipped to England, mostly customized tableware and teaware with family crests. King Louis XV of France even ordered the melting of all silverware in the country and replaced it with porcelain, making it a nationwide trend. By the eighteenth century, porcelain had become a common household item. With the introduction of tea, chocolate, and coffee to Europe, hot beverages became part of social life, increasing the demand for porcelain drinkware. The tea-drinking habit started in the Netherlands in 1610, spread to Paris in 1636, became a social custom in London by 1650, and reached Moscow in 1659. Tea drinking became particularly popular in England due to the large-scale importation of tea leaves. By the late seventeenth century, around 20,000 pounds of tea were imported into England annually, and by the early nineteenth century, this number skyrocketed to a staggering 20 million pounds per year. Tea had replaced chocolate as an essential beverage. Entire porcelain tea sets, like tableware, became common vessels in England and other European countries.

The increasing demand for Chinese porcelain stimulated and drove the establishment of Europe's porcelain industry.

The first successful imitation of Chinese porcelain in Europe was made by Venice, producing colored soft-paste porcelain. In 1580, Florence established a factory manufacturing blue and white soft-paste porcelain, which ceased operations in the early seventeenth century. In 1627, the ceramic workers of Pisa produced soft-paste blue and white porcelain bowls, initially in an Arabesque blue style and later imitating Chinese blue and white. The Netherlands learned the technique of making soft-paste porcelain in 1634 from Italy. Subsequently, France also attempted porcelain production, with yellow and transparent soft-paste porcelain produced in Lyon in 1673 and Saint-Cloud in 1695, using white porcelain from Fujian as a model. The Sèvres porcelain factory also had this type of soft-paste porcelain. It was not until 1709 that true hard-paste porcelain was officially manufactured in Europe. This invention was driven by Elector Augustus of Saxony in Germany, who discovered porcelain clay in the region. Johann Friedrich Böttger discovered the secret of the creation of hard-paste porcelain in 1708. In 1710, Böttger relocated the porcelain factory to Meissen, west of Dresden, and started producing polychrome porcelain. In 1713, the porcelain was publicly sold, gaining immense popularity. In 1717, blue and white porcelain was also made, leading the popularity of European porcelain to imitating Chinese colors and painting styles. In 1719, two employees of the Meissen porcelain factory, Samuel Stölzel and Christoph Conrad Hunger, established the Vienna porcelain factory in Vienna. In 1720, Stölzel returned to Meissen with painter Johann Gregorius Heroldt. Hunger went to Venice to spread the secrets of porcelain production and later traveled to Copenhagen and St. Petersburg to teach the craft. Viennese painter Johann Gregorius Heroldt devoted significant effort to producing polychrome porcelain, using Chinese polychrome, lacquerware, gold-layered vessels, and faience as his models. Meissen porcelain also adopted Chinese colors, depicting Chinese figures in gold and incorporating dragon motifs, although with a Westernized touch.

France also relied on Pere d'Entrecolles from Jingdezhen for the production of porcelain. In 1717, Pere d'Entrecolles sent samples of Gaolin clay from Jingdezhen to France. It wasn't until 1768, when a deposit of porcelain clay was discovered in Limoges, that the Sèvres porcelain factory successfully produced hard-paste porcelain.

The first porcelain factory in England was the Bow porcelain factory in Stratford. In 1750, it produced the first batch of soft-paste porcelain in England. Later, it incorporated bone ash and other elements to create bone china, a type of porcelain that fell between soft-paste and hard-paste porcelain. In 1768, the Plymouth porcelain factory successfully produced true hard-paste porcelain in England. These porcelain wares featured both Chinese-style colored patterns and distinct Rococo influences. Through the production of porcelain, the combination of the Rococo style and the ideal of transcendence and simplicity found in Chinese art was indeed realized.

In addition to porcelain, another aspect that embraced the "Chinese style" was using Chinese lacquerware in furniture and furnishings. During the reign of Louis XIV, Madame de Maintenon employed complete sets of Chinese lacquer furniture in both Versailles and Trianon palaces. In the Saint-Antoine region of France, lacquerware production imitating Chinese styles began in 1692, and it progressed even faster after 1730. The Martin family, particularly Robert Martin, excelled in the art of lacquerware reproduction, with a particular talent for depicting flowers and birds. The lacquer furniture in the residence of Louis XV's mistress, Madame de Pompadour, was created by him. French lacquerware primarily featured blue, red, green, and gold colors, and the interior cabinets were designed in the Chinese style, adorned with decorative patterns of peonies, birds, Chinese women, Chinese-style railings, and houses. Pierre Nicolas d'Incarville, a Jesuit missionary in China, also took an interest in Chinese lacquerware. In 1760, he published his work on "Chinese Lacquer" in Paris, which included exquisite illustrations.

In France, there was also a fashion for Chinese sedan chairs or sedan-style carriages. Sedan chairs were introduced to Europe in the early seventeenth century, and by 1644, they were already mentioned in French literature. During the reign of Louis XIV, nobles and officials would travel in sedan chairs. The materials and colors of the sedan chair canopies were strictly regulated according to the person's rank, and the body of the sedan chair was often decorated with lacquer paintings featuring popular Chinese flowers such as peonies and paeonies. Many women of high social status would travel in sedan chairs. Unlike in China, European sedan chairs were not carried on the shoulders but lifted by hand. In French, they were called "chaise à porter". These types of sedan chairs often appeared in the comedies of playwright Molière. For example, his play *Les Précieuses Ridicules*, published in 1659, featured such chaise à porter. The Elector of Saxony, Augustus II, also designed porcelain sedan chairs. This type of sedan chair remained popular in Germany and Austria until 1861.

The popularity of Chinese lacquer furniture in England began during the reign of William and Mary in the late seventeenth century. In the eighteenth century, it became fashionable for upper-class women in England to learn the art of lacquer painting, and lacquer painting became a subject in girls' schools. After 1680, British furniture makers began to imitate the patterns and colors of Chinese lacquerware, creating Chinese-style furniture. Publications such as *Designs of Chinese Buildings, Furniture, Dresses,*

*Machines, and Utensils* were released. Renowned furniture designers like Thomas Chippendale and George Hepplewhite designed and produced cabinets, tables, and chairs that imitated Chinese styles, using fine Fujian lacquer and carving dragons, pagodas, Buddha figures, and flowers. The furniture designed in eighteenth-century England, known as the Chippendale era, still bears the imprint of Chinese influence.

Wallpaper is an essential interior decoration that embraces the Chinese style. In ancient China, hand-painted wallpapers, known as painted panels(tie luo), were prevalent. In the sixteenth century, gold and silver stamped papers were introduced to Europe by Spanish and Dutch merchants from China. William and Mary of England were fond of Chinese wallpapers, and the trend also spread in France. The term "hang paper," which later evolved into "wallpaper," first appeared in 1700 and was similar to "tie luo". The designs of the wallpapers initially featured birds and flowers, but later, landscapes and figures were also depicted. Soon, Europe began producing wallpapers in various colors, resulting in a fusion of Chinese and English or Chinese and French styles. In 1638, Frankfurt imitated Chinese bird-and flower-patterns using gold and silver papers as substitutes for hanging parchment paintings. In 1688, France also successfully imitated Chinese wallpapers. The wallpaper industry in England began by ordering color-printed woodblocks from China and, upon arrival, pieced them together for sale. Despite the ability of Britain to produce its wallpapers, Chinese wallpapers continued to be imported to Britain for their exquisite craftsmanship.

Chinese silk fabrics also became a sought-after decoration among upper-class women in France and Germany. Madame de Pompadour, for instance, adorned herself with silk dresses featuring Chinese birds and flowers. By 1699, the English silk industry was already facing the risk of unemployment due to the importation of Chinese silk. In 1701, England even banned silk imports as a result. Lyon was able to produce silk fabrics from 1537, with the founders originating from Piedmont, Italy. Even in the eighteenth century, silk produced in Lyon still exhibited solid Chinese influences.

During the seventeenth and eighteenth centuries, Chinese folding fans became popular. In the French court, noblewomen would always carry a Chinese-style silk folding fan, regardless of the season, replacing the feather fans that were popular in the sixteenth century. Chinese influence could also be seen in architecture, with the introduction of "lattice fan" windows, "fan-shaped ventilators," and "fan-shaped domes".

During the Rococo era, Chinese fashion and aesthetics were highly sought-after and incorporated into various aspects of society. Chinese influence could be seen in advertisements, book illustrations, stage designs, and theatrical makeup. In the eighteenth century, Chinese shadow puppetry, known as "Ombres Chinoises," gained popularity, featuring plays with Chinese themes and characters. Chinese clothing became fashionable for tavern servers and masquerade balls featuring individuals dressed in Chinese attire became popular. Keeping goldfish and peacocks as pets also became a manifestation of Chinese fascination. Madame de Pompadour, for example, had kept goldfish imported from China.

### *The Promotion of Chinese Painting and Gardens*

The influence of Chinese landscape and figure painting on European art surpassed that of ceramics and lacquerware, becoming a subject of emulation for renowned painters. One notable artist is the French painter Jean-Antoine Watteau (1684–1721), whose works often employed light colors and depicted misty landscapes with serene mountains and flowing clouds. His painting "The Embarkation for Cythera," displayed at the Louvre Museum, exemplifies a Chinese aesthetic. Other painters with a Chinese sensibility include Berain, Gillot, and Pillement. The influence of Chinese painting also found its way into British landscape painting, particularly in the works of John Robert Cozens (?–1794). His watercolor landscapes resemble Chinese painting, often employing a brown-gray base with red and blue accents. During the early development of watercolor painting in Europe, many artists adopted Chinese ink techniques, reflecting the characteristics of Rococo-era art. The late works of the British landscape painter John Constable demonstrate a strong influence of Chinese painting, particularly in his masterpiece "The Hay Wain," which captures the idyllic beauty of the English countryside reminiscent of Jiangnan landscapes. In the nineteenth century, the French Impressionist painters were indirectly influenced by Chinese painting techniques and employed splashing ink methods in their artwork. Maurice Denis, in particular, excelled in imitating Chinese porcelain paintings.

Chinese garden art was highly esteemed for its harmonious integration with nature, starkly contrasting the symmetrical and ornate gardens of the Louis XIV era. British poets such as Joseph Addison and Alexander Pope created gardens inspired by Chinese landscapes, and this trend gained popularity throughout society. The Pillnitz Palace in Germany, located near the River Elbe, was the first to incorporate Chinese-style roof architecture. Subsequently, Potsdam in Germany, Het Loo Palace in the Netherlands, Chanteloup in France, and various places in Switzerland competed to construct Chinese-style clock towers, stone bridges, rockeries, and pavilions. The British architect William Chambers, who had visited China and later served as the King's architect, published *Designs of Chinese Building, Furniture, Dresses, Machines, and Utensils* in 1757, which became popular throughout Europe and served as a model for Chinese fashion. In his work *A Dissertation on Oriental Gardening*(1772), Chambers praised Chinese garden design: "The Chinese art of garden design is incomparable. Europeans cannot compare their artistic achievements with the splendor of the East". Chambers designed a Chinese-style garden called "The Chinese Garden" at Kew Gardens in 1757 for the Duke of Kent. The garden featured a lake with a pavilion and a 163-foot-tall ten-story quadrangular tower with dragons holding silver bells at the corners. There was also a Confucius Tower with depictions of Confucius' life. In 1763, Chambers published *Plans, Elevations, Sections, and Perspective Views of the Garden and Buildings at Kew in Surry*. W. Halfpenny also published *New Designs for Chinese Temples: Triumphal Arches, Garden Seats, Palings &c.* ... in 1750.

Chinese garden architecture became more refined in Britain and spread to France, where it became known as "Anglo-Chinese gardens". The Pillnitz Palace near Kassel in Germany, Chanteloup in France, and the Wooden Tower in Munich, Germany, all carried remnants of this style. Architects such as Le Camus, Belanger, and Jean Augustin were known for their designs in this genre. In 1773, the German landscape architect F. L. Sekell was sent to England to study Chinese gardens. The same year, Ludwig A. Unzer published *Über die Chinesischen Gärten*, expressing the need for Germany to catch up with this emerging architectural style in England. Following this, the Count of Kassel built the Mulang Village near Wilhelmshöhe, featuring a small stream named "Hu-Kiang," after a famous water town in southern Suzhou, China. The village was meticulously designed to resemble the gardens of Jiangnan in China, complete with black people dressed in traditional Chinese milking cows.

### *An East-Meets-West Fusion of Porcelain Styles*

In the eighteenth century, a large quantity of "Chinese export porcelain" was produced specifically for European markets, with the most conservative estimate being over 60 million pieces in 100 years.[9] These export porcelains, except for white porcelain and carved porcelains from Dehua kilns in Fujian, were predominantly produced in Jingdezhen, Jiangxi. Many of them were polychrome porcelains, including a significant number of Famille Rose pieces from the Yongzheng and Qianlong periods. Among the Qing dynasty's export porcelains, there were both general porcelain wares initially intended for domestic production and a considerable portion specifically tailored to the demands of foreign markets. Britain (1715), France (1728), the Netherlands (1729), Denmark (1731), and Sweden (1732) successively established trading agencies in Guangzhou. They placed orders with Jingdezhen for Chinese-style ceramics while providing samples (mostly made of wood, pottery, and paintings) for reference. Jingdezhen would then produce "blank wares" (unglazed porcelain) that were transported to special workshops on the southern bank of the Pearl River in Guangzhou. They were painted following Western painting techniques and designs, fired in kilns, and ultimately transformed into polychrome porcelain. These pieces were shipped to Europe and became known as "Henan polychrome" or "Guangzhou polychrome". There were tens of thousands of private kilns engaged in Guangzhou polychrome production on the outskirts of Guangzhou. Later, Jingdezhen also imitated Western porcelain and referred to it as "yangcai" or foreign-colored ware. In 1743, the superintendent Tang Ying established official kilns and compiled the book *Taoye Tushuo (Illustrated Treatise on Pottery Making)*, which included seven categories of imitated Western porcelain: replicas of Western sculpted vessels (such as plates, dishes, vases, and boxes) with painted renderings imitating Western brushwork; Western yellow-colored crafts; polychrome vessels imitating Western purple-colored ware; new imitations of Western enamel painting techniques featuring figures, landscapes, flowers, and feathers, all finely detailed and exquisite; Western red-colored vessels; Western green-colored vessels; and newly crafted Western black and gold colored vessels.

[9] Elinor Gorden, *Collecting Chinese Export Porcelain*, 1978, p. 22.

The porcelain produced explicitly for export to Europe was commonly referred to as "yangqi". Many traders were from the eastern part of Guangdong province and sold the porcelain to foreign merchants. These porcelain wares were known for their exquisite craftsmanship, and their styles varied annually. The production of "yangqi" was not limited to private kilns; even official kilns received significant orders from overseas, which were referred to as "Chinese export porcelain" in Europe. The painted decoration on the "yangqi" was called "yangcai" or foreign-colored ware. By the eighteenth century, most porcelain painting designs were in the style of "yangcai". Zhu Yan, in his book *Tao Shuo*, stated: "Regarding painted decoration on ceramics, it reached its peak during the Ming dynasty. Most of it was inspired by brocade patterns, with one-third or one-fourth imitating ancient styles. Nowadays, out of ten porcelain painting designs, four are 'yangcai,' three are sketchings, two are imitations of ancient styles, and one is based on brocade patterns". Yang Xiaogu, in his book "*Guyuexuan Cikao*," stated: "Western porcelain had already entered China during the Ming dynasty, and the Kangxi porcelain with painted enamel was an imitation of the Western porcelain, using the foreign colors technique. It was the first use of colored enamel on porcelain". The enamel-painted porcelain and enamel-painted bronze during the Kangxi period were used in the imperial court. The technique of enamel painting on porcelain arrived in China 300 years later than the arrival of enamelware in the early Ming dynasty. This type of enamel-painted porcelain was created by applying painted enamel on white porcelain and undergoing a second firing in the kiln. During the Kangxi period, the enamel pigments were imported, and it was not until the Yongzheng reign that the Chinese began producing them domestically.

The export of porcelain from Jingdezhen was influenced by European craftsmanship in terms of production techniques, types of wares, painting styles, and designs.

The types of export porcelain varied depending on the needs of the buyers, and they changed from year to year. The first batch of samples was shipped to the Netherlands in 1639, which included practical items in English styles. The painting techniques used for export porcelain also adopted Western methods. The depicted themes included armorial porcelain, figures from mythology, religion, and daily life, as well as ships and flowers. In addition to traditional Chinese landscape, sculpture, and bird-and-flower paintings, there were also a considerable number of

Western-style armorial designs and figure paintings. The book *Chinese Armorial Porcelain*, published in 1974 in the United Kingdom, collected around 2000 pieces of armorial porcelain used for Western military, nobility honors, and festive ceremonies.

British porcelain factories such as Worcester (established in 1752 and still in operation) and Lowestoft (1756–1802), as well as German Meissen and French Sèvres, have imitated Chinese porcelain forms and patterns. Both Germany and France were enthusiastic about following China's polychrome ceramics. The Sèvres factory, for example, replicated the coral-red glaze of the Qianlong period and created the "Pompadour Rose," a gem of Rococo art. In the eighteenth century, Chinese exported porcelain also produced products similar to Worcester and Lowestoft, to the extent that European and American ceramic historians have mistakenly identified Chinese products as Lowestoft porcelain.[10] Worcester porcelain factory was able to create dinnerware that closely resembled Chinese porcelain. In 1763, they boasted in the "Oxford Magazine": "Chinese porcelain and Worcester porcelain can be used together without any difference". The manufacturing and stylistic fusion of porcelain between East and West in the eighteenth century made it nearly indistinguishable.

## 7.5 Chinese Literature in Western Europe

### *Translations of The Orphan of Zhao and The Pleasing History*

Chinese literature caught the attention of Jesuit missionaries in the eighteenth century. In 1732, Joseph Henri Marie de Prémare translated the Yuan drama *The Orphan of Zhao*. The French translation was titled *Tchao-Chi-Cou-euih, ou L'Orphelin de la Maison de Tchao, tragedie chinoise*. In 1734, excerpts from the play were published in the Parisian magazine *Mercure de France*, and the complete space was included in the third volume of Du Halde's *Description géographique, historique, chronologique, politique et physique de l'empire de la Chine et de la Tartarie chinoise* when it was published the following year. As the first Chinese play translated into a foreign language, *The Orphan of Zhao* gained popularity in Europe with the English edition in 1736, the German edition in 1747,

[10] Elinor Gorden, *Chinese Export Porcelain: An Historical Survey*, Universe Publishing, 1977, pp. 49–50.

and the Russian edition in 1774. In 1741, a complete English translation was also published, including translations of the songs Joseph de Prémare had not previously translated. Voltaire adapted the play into a new script titled *L'Orphelin de la Chine* (The Orphan of China). It was first performed in Paris on August 20, 1755. Under the play's title, Voltaire added the text "adapted into a five-act play based on the teachings of Confucius". He hoped to use the power of morality to overcome the tyranny of monarchs. In this play, Voltaire depicted Genghis Khan, the conqueror, ultimately swayed by noble morality. In the preface, he declared, "This is a great demonstration that shows how reason and intelligence will inevitably triumph over ignorance and savagery,"[11] expressing his belief in the ideals of the Enlightenment and his continued struggle against authoritarian politics.

In 1761, the first English edition of a Chinese novel, *Hau Kiou Chuaan* (*The Fortunate Union* or *The Pleasing History*), was published in England. This edition was translated by a British merchant, James Wilkinson, a resident of Guangzhou (Canton). Wilkinson translated about 75% into English; another person translated the remainder into Portuguese. In 1761, Thomas Percy, the Bishop of Dromore, published this version anonymously; before Percy published it, the Portuguese portion was translated into English, and the entire version had been edited. This was the first-ever translation of a Chinese novel in a European language. In 1766, The Wilkinson/Percy English version was also translated into French by an anonymous translator identified only as "M." Christoph Gottlieb von Murr translated the Wilkinson/Percy first English version into German.

### *Goethe and Chinese Literature*

Johann Wolfgang von Goethe was initially exposed to Chinese literature like *The Orphan of Zhao, Gujin Qiguan,* and *The Classic of Poetry* through a German edition of the works of Du Halde. In August 1781, he started adapting a play titled *Elpenor* based on *The Orphan of Zhao* and *Gujin Qiguan,* but it remained unfinished. From Goethe and Johann Christoph Friedrich von Schiller's correspondence in January 1796, it is known that he has read the German edition of *Hau Kiou Chuaan*. In 1817, he read

[11] *Oeuvres Completes de Voltaire,* Paris, 1865, vol. 3, p. 680.

an English version of a Yuan drama titled *A Rich Merchant Distributed His Wealth to His Offspring (Laoshenger).* In 1827, he read The *Chinese Courtship in Verse*, translated by P. P. Thoms, and *The Two Fair Cousins: a Chinese novel* translated by Abel Rémusat.

Goethe, the great literary figure of the eighteenth century, had a unique and penetrating insight into Chinese literature. Johann Peter Eckermann once asked if *Hau Kiou Chuaan* was the best Chinese novel, to which Goethe promptly replied, "Certainly not. China has millions of novels like this. When they started writing, our ancestors still lived in the forests!" Regarding Chinese literature, Goethe made interesting comparisons with English and German literature. He said, "The characters, thoughts, actions, and emotions in their books are similar to ours. In no time, we feel like we are the same as them, except that in their world, everything appears more rational, pure, and moral than ours. Everything is reasonable, and there are no intense emotions or poetic sentiments, much like my work, *Hermann and Dorothea* and Samuel Richardson's novels. However, there is one difference: the external nature and the characters always coexist in their works".

Goethe's notable contribution to German literature is his efforts to transplant Chinese lyrical poetry into Germany from 1827 onward. He attempted to re-translate the five poems from the English version *of Chinese Courtship in Verse.* Goethe then incorporated these poems into his work titled *Chinesisch-deutsche Tages und Jahreszeiten*, which reflects his admiration for the philosophical teachings of Confucius and Laozi. These philosophical poems resonated with Chinese and Western literature in the early nineteenth century.

CHAPTER 8

# China and Its Delayed Industrialization

## 8.1 The Growth of Chinese Tea Export and the New Trends in Porcelain Export

### *The Export of Tea and European Maritime Trade*

Since the eighteenth century, the United Kingdom has maintained its position as the primary importer of Chinese tea in Europe, and later, with the addition of the United States, this continued for one to two hundred years. In the early eighteenth century, the British East India Company sold less than 50,000 pounds of tea annually. However, by the end of the century, the company's annual tea sales had surged to nearly 20 million pounds, marking a staggering 400-fold increase within that century. Throughout this period, the industrial cities and coastal regions of the United Kingdom, encompassing even the impoverished population, adopted tea consumption. The annual tea consumption for the general populace ranged from 5 to 6.5 pounds. In an average household, the consumption of Wuyi tea alone amounted to at least 15 pounds annually. For middle-class families, the annual consumption included 12 pounds of Kung Fu tea, new green tea, and 3 pounds of Gongxi tea. Wealthier households indulged in 24 pounds of Gongxi tea annually.

In 1699, the British East India Company obtained a charter for the tea trade, and in the early stages, the tea imported from China was predominantly green. In 1715, British ships entered Huangpu Port in Guangzhou, establishing direct trade. By 1721, British imports of



F. Shen, *Cultural Communications Between China and The Outside World Throughout History*,
https://doi.org/10.1007/978-981-97-4696-5_8

Chinese tea had exceeded one million pounds, prompting the British government to prohibit other European countries from importing tea. In 1725, the East India Company implemented a tea monopoly, resulting in a proliferation of adulterated tea in the London market. In 1757, the Qing government restricted foreign trade to Guangzhou alone, requiring British ships conducting business along the Zhejiang coast to relocate to Guangzhou. Starting from 1728, the Dutch began transporting tea from Batavia (modern-day Jakarta) to Guangzhou, and by 1795, they had withdrawn entirely from trade with China. France established a trading company in 1700 to conduct trade with China. In 1776, they found a consulate in Guangzhou to manage business affairs, with the volume of tea transported closely following Britain's. In 1732, Denmark dispatched ships to Guangzhou, and in 1733, Sweden sent ships for trade with China, which included transporting tea. The British East India Company maintained a leading role in the tea trade due to the British government's support. However, with the increase in the import of Chinese tea, tea smuggling also grew. Some tea imported by mainland European countries was smuggled back into the British market. It wasn't until 1784, when Britain passed the Commutation Act, reducing tea import taxes, that smuggling was effectively curbed. This ensured the East India Company's monopoly on tea sales. Subsequently, mainland European countries had to forego the benefits of smuggling trade.

Initially, the British East India Company primarily transported green tea. However, after 1760, due to rampant counterfeiting in the London market and the subsequent damage to its reputation, the company shifted its focus to importing black tea. This black tea primarily consisted of Wuyi tea, Kung Fu tea, small-leaf tea, and Gongxi tea. This change in preference among the British populace led to a strong demand for Fujian black tea, making it a best-selling product. By 1792, the British East India Company had shipped a significant quantity of black tea—approximately 156,000 piculs—worth 3,413,054 taels of silver to Britain.

In contrast, green tea only accounted for 1500 piculs, valued at 624,640 taels of silver. The proportions of black tea to green tea imports were vastly different. In 1813, the British Parliament abolished the East India Company's charter for trade with India, but trade with China continued until 1833. Between 1784 and 1833, the tea transported by the company constituted about 90% of the total value of imports. This proportion even reached as high as 94% between 1825 and 1833. The company's annual profits from the tea trade ranged from 800,000 to 1 million pounds. The British government also profited significantly from the tea trade by imposing the tea tax. This tax rate escalated rapidly, from 12.5% in 1793 to 100% in 1805, remaining unchanged for the following three decades until 1833. By 1833, the British government's revenue from the tea tax had grown to 3.3 million pounds, accounting for approximately one-tenth of Britain's national income.[1] Following the dissolution of the British East India Company in 1834, Jardine Matheson & Co. quickly assumed responsibility and continued the tea trade. After April 1834, British merchant ships began importing large quantities of Chinese tea from Singapore. The tea's quality was comparable to that from Guangzhou, but importing through Singapore saved time and bypassed the heavy taxes imposed in Guangzhou. This shift in import logistics contributed to a decrease in tea prices.

By the eve of the Opium War, tea import into Britain was still growing rapidly due to a thriving smuggling trade. From 1830 to 1834, an average of 32 million pounds of tea was transported yearly, and from 1835 to 1839, the annual average was 40 million pounds.[2] From 1838 to 1842, the average amount of tea imported into Britain was 42,355,000 pounds annually. Due to the war in 1843, this number fell to 17,727,754 pounds but rebounded to a high of 69,327,500 pounds in 1844. Consequently, the average annual amount from 1840 to 1844 was still as high as 36 million pounds. Guangzhou exported 350,000 piculs of tea annually, worth 94,450,000 silver dollars, accounting for 70% of China's exports.[3] The export of raw silk also increased similarly, reaching a peak of over 7900 piculs per year.[4]

Foreign traders in Guangzhou operated according to the regulations set by the Qing government, traditionally through a co-hong, an officially sanctioned guild of merchants, later known as the Thirteen Factories. Foreigners in Guangzhou were required to reside in the "foreign houses" of the Thirteen Factories. Goods had to be sold to these merchants, and the export of goods was also handled by them. In 1760, the Qing

[1] Michael Greenberg, *British Trade and the Opening of China 1800–1842*, Cambridge, 1951, p. 3.

[2] *Chinese Repository*, Vol. XIII, October 1843.

[3] Yao Xianhao, ed., *Materials on the History of Modern Chinese Foreign Trade*, vol. 1, Beijing, Zhonghua Book Company, 1962, p. 258.

[4] Yan Zhongping, ed. *Selected Materials on Modern Chinese Economic History*, China Social Sciences Press, p. 14.

court divided the co-hong into foreign, local, and Fujian-Chaozhou factories. The foreign factories became institutions solely for Western trade. The co-hong was temporarily abolished in 1771 but was reinstated in 1775, establishing a system of merchant protection. Foreign merchants were required to identify a guarantor among the Thirteen Factories to handle the payment of duties, trade in goods, and manage all activities of foreign sailors and other personnel. The foreign merchant ships arriving in Guangzhou numbered more than 30 yearly. They imported goods in late summer and early autumn, set sail for home in winter, staying for no more than four or five months. Resident foreigners continued the remaining trading affairs.[5] With the large-scale import of tea into Britain, beyond lead, tin, and cotton, the United Kingdom had no corresponding goods to export to China. In order to seek a balance in trade, the United Kingdom had to pay in silver. More than one million taels of silver flowed into China from the United Kingdom and India each year, with the highest annual amount (1820–1821) exceeding 5.56 million taels. It was not until after 1823 that payments in silver were no longer necessary.[6] To compensate for the deficit, the United Kingdom imported Indian cotton into China. From 1775 to 1833, the imports increased annually from 288,000 taels to more than 4 million taels, exceeding the value of woolen and metal products that Britain imported into China.[7] Another significant import was opium. It had been imported as early as the Ming Dynasty, known as "A Fu Rong" or "A Fei Yong," coming from Arabia and used as a medicinal ingredient, known as poppy or "foreign medicine." The method of consumption and production as "medicinal smoke" was introduced from abroad. Since the sixteenth century, it was brought in by the Portuguese from Turkey in tiny quantities. In the second half of the eighteenth century, the British promoted the cultivation of opium in Bengal. In 1773, Britain implemented a patent monopoly on opium, making the revenue from the opium trade a major source of financial income for India, which brought famine and disaster to the people of Bengal.

Starting from 1727 until 1767, the annual export of opium from India to China did not exceed 200 chests, each weighing approximately 133

[5] Xu Dishan, *Dazhongji*, The Commercial Press, 1931, p. 130.

[6] Earl Prichard, *The Crucial Years of Early Anglo-Chinese Relations, 1750–1800*, Washington, 1936, p. 397.

[7] Yan Zhongping, ed. *Selected Materials on Modern Chinese Economic History*, pp. 9–13.

pounds. At that time, the Portuguese mainly transported it from Turkey to China, with a customs duty of about 3 dollars per chest. After 1773, opium was mainly transported from India to Guangdong. By 1800, the opium imported into China had reached 2000 chests. Despite the Qing government's repeated orders to ban opium, smuggling was rampant, and the amount of opium being smuggled in kept increasing: between 1821 and 1828, an average of over 9700 chests of opium were smuggled in each year; between 1835 and 1839, an average of over 35,440 chests of opium were smuggled in each year.[8] Between the 13 years from 1814 to 1827, over 23.57 million taels were imported; in the nine years from 1830 to 1839, over 31.76 million taels were imported. Huang Juezi, the Chief Minister of the Court of State Ceremonial, lamented in his report: "The useful wealth of China is being used to fill the endless pit abroad, and this harmful substance is gradually becoming a worry for our nation."[9] In early 1839, Lin Zexu strictly enforced the ban on opium in Guangzhou, achieving noticeable success. This action infuriated the British government, which dispatched 16 warships, 20 transport vessels, four armed steamships, and 4000 soldiers, initiating the Opium War in June 1840. In March 1841, due to the war, 30 million pounds of tea couldn't be exported, resulting in a loss of 3 million pounds in tea tax revenue for the British government.

After the Opium War, with the opening of five treaty ports, the four major tea-producing areas in the Yangtze River Basin (Jiangsu, Zhejiang, Fujian, and Jiangxi) were able to transport tea without having to go through the inland waterways and over mountains to reach Guangzhou (where shipping fees were often one-third the value of the tea). Instead, tea could be exported through Xiamen, Shanghai, and Ningbo. In 1845, the British adopted the method of buying tea with foreign goods in Fuzhou, purchasing 72 million pounds of tea worth 15,825,954 yuan, which were shipped to the United Kingdom, 10 million pounds to Australia, and 3 million pounds to India. By 1851, tea exports had nearly doubled to almost 100 million pounds. However, until 1860, the tea imported by Britain still accounted for 90% of China's tea exports. After that, as various countries began directly purchasing tea from China,

[8] Yan Zhongping, ed. *Selected Materials on Modern Chinese Economic History*, pp. 22–23.

[9] Yan Zhongping, ed. *Selected Materials on Modern Chinese Economic History*, pp. 27–34.

Britain's share gradually declined. In 1865, tea exports exceeded 100 million pounds. According to the *Treaty of Yantai* between China and Britain, British merchants were allowed to purchase tea directly from places like Yichang and Wuhu in the Yangtze River area. Taiwan's tea also began to be exported to Britain in that year. Before 1893, Britain was consistently the largest market for Chinese tea, especially between 1880 and 1888, when it peaked. After 1888, Chinese tea exports began to decline year by year. While it initially remained at 1 million metric tons, it dropped to below 1 million in the 20 years following 1897. Since 1897, the export volume of tea, which was once China's leading export commodity, has declined year by year, falling from 17% to less than 13% and, in some years, even dropping below 4%.

From an external perspective, tea exports declined due to Japan's seizure of China's green tea market in the United States after 1886 and its entry into the market in northern China at low prices. On the other hand, India, Sri Lanka, and Dutch-controlled Sumatra quickly took over China's foreign market for black tea. With the support of the British, India imported black tea artisans and equipment from Fujian and Jiangxi in China.[10] Tea cultivation expanded in Assam and Bengal, commencing exports in 1857 with just 200 g. By 1866, exports had increased to 40,000 metric tons, representing 3.3% of China's tea exports.[11] Initially, Indian black tea, with its bitter taste and intense aroma, was not well-received in the London tea market.

Consequently, it was blended with Chinese tea, which was 25% cheaper than its Indian equivalent, and sold under the guise of Chinese Gongfu tea. Subsequently, Indian tea improved its roasting techniques, resulting in a robust, intensely aromatic brew that was difficult to sell. It was then blended with the more delicately flavored Chinese tea to enhance the taste of the Indian tea before being marketed. Indian tea was the first to adopt mechanization and had a shorter, more economical route to Britain than China. Moreover, it enjoyed tax exemptions. This resulted in Chinese tea being disadvantaged, gradually, Indian tea became popular in the British market. Thus, the Chinese tea market was cunningly encroached upon.

[10] Shen Fuwei, *Western Culture and China (1793–2000)*, Shanghai Education Press, 2003, pp. 542–544.

[11] Chen Chuan, *A History of Chinese Tea Export*, p. 76.

In 1885, the market share of Chinese tea in Britain fell to 61%, while that of Indian tea rose to 39%.[12] Following this, the annual export of Chinese tea to Britain, mainly black tea, continued to decline. In 1889, the volume of Indian tea shipped to Britain surpassed that of China. That year, China exported 58,161,531 pounds of Gongfu tea to Britain, compared to 80,509,995 pounds of Indian tea and 26,099,840 pounds of tea from Sri Lanka, a country that only began exporting tea in 1877, yet experienced rapid growth. Over 20 years, from 1897 to 1917, tea's position in the total export value from China fell from first to second place, surpassed by raw silk.

### *Tea Exports and Maritime Transport to America*

The nineteenth century saw a significant increase in America's imports of Chinese tea. Between 1805 and 1806, 42 American ships arrived in Guangzhou, carrying 9,830,480 pounds of tea. Although the annual figures fluctuated, by 1818–1819, the amount of tea transported by 47 American ships reached 12,035,280 pounds, breaking the 10 million pound mark. After 1823, American ships became involved in the opium smuggling trade, transporting even larger quantities of tea. Between 1832 and 1838, according to reports from the U.S. Department of the Treasury on trade with China, an average of over 14 million pounds of tea was transported annually, peaking at 16,942,122 pounds in the 1837–1838 fiscal year. Unlike Britain, the United States mainly imported green tea from China, accounting for approximately 80% of imports. The proportion of Chinese tea in the total value of Chinese goods imported by American merchant ships continued to increase. In 1821, Chinese tea made up 42.5% of the Chinese goods imported by the United States, rising to 78.6% by 1845. Between 1845 and 1860, tea accounted for 60% to 80% of the total value of Chinese goods exported to the United States, still holding a significant 61.6% share by 1885.[13] In 1844, the *Treaty of Peace, amity, and commerce between the United States of America and the Chinese Empire* was signed between China and the United States. The total value of goods shipped from China to the United States, Mexico, Peru, and the Hawaiian Islands amounted to 6,686,171 yuan, of which

[12] Chen Chuan, *A History of Chinese Tea Export*, p. 285.

[13] Chen Chuan, *A History of Chinese Tea Export*, p. 165.

tea alone accounted for 5,064,926 yuan. That year, the price of tea was high, and green tea became a hot commodity in the international market. As a result, the 43 American ships heading for the Chinese coast (with a cargo capacity of 19,269 tons) reaped substantial profits.

The opium smuggling trade presented business opportunities for British and American merchant ships, enhancing their interests in dealing with China and instilling confidence in American merchants to rush the transportation of the year's new tea. Innovative fast sailing ships were first developed by the United States and launched on the China–America route. These ships, known as clipper ships, could sail against the wind, traveling rapidly day and night. In the history of sailing ships, the clipper ship era spanned from 1840 to 1860, representing the final glory of wooden sailing ships and an era overlapping with the advent of steamships. From 1831 onward, the British also had small three-masted schooners like the "Jamesina," which transported opium from India to Fuzhou, Xiamen, Ningbo, and other places for sale, returning with smuggled tea. In 1839, East Boston and New York, respectively, built three-masted fast ships and dispatched them to Hong Kong for operations. The largest was the 650-ton *Akbar*, which set a record of 109 days on its maiden voyage from New York to Guangzhou, sailing north against the northeastern winds in the South China Sea. New York shipbuilders supported John W. Griffiths' design for the highly refined clipper ship *Rainbow*, launched in January 1845. On its first voyage, it set a new record by sailing from New York to Guangzhou against the northeastern wind in 92 days and 88 days for the return trip against the wind, totaling six months and 14 days for a round trip. This included a two-week stopover in Guangzhou, setting sail in March and returning in September. The appearance of the *Rainbow* marked the establishment of the clipper ship era.[14]

A ship similar to the Rainbow was the *Sea Witch*, an 890-ton vessel with a dragon carved into its bow, gnashing its teeth and claws. It was the most impressive clipper ship made in New York. In December 1846, it braved the northwestern winds from New York to Hong Kong in 104 days. In July 1847, it took 81 days to return to New York from Guangzhou. Later, in 1848, it set a new record of 77 days on its return journey from Guangzhou. In 1850, the ship *Houqua* set a new record of

[14] Arthur H. Clark, *The Clipper Ship Era*, N.Y. and London: Putnam's Sons, 1911, pp. 65–67.

just 88 days from Shanghai to New York, establishing the shortest route between China and America. This record was later broken by the *Surprise* in 1857, which completed the journey in 81 days.

By the time the American clipper ship *Memnon* was built in 1848, it had broken the thousand-ton mark, reaching 1068 tons. Its maiden voyage was also bound for China. The discovery of gold in California in 1848 attracted American clipper ships from the East Coast to the West Coast. Clippers sailing north to San Francisco via Cape Horn typically passed through Valparaiso, with a voyage time of over 110 days. Only the *Flying Cloud* set a record of 89 days in 1851; no new record was set after that. The daily sailing record set by the *Flying Cloud* was 374 miles, which remains the highest record for a sailing ship's 24-hour journey.

The American green tea market, which was initially a strong point in the China–U.S. tea trade, was gradually taken over by Japan with its export of green tea after the 1860s. In 1860, Japan exported 35,000 pounds of green tea to the United States, while China exported 30,558,000 pounds of tea to the United States, accounting for 96.41%. By 1880, Japanese tea exports to the United States had risen to 33,688,000 pounds, accounting for 46.68%, while Chinese tea exports to the United States reached 36,187,000 pounds, accounting for only 50.25%. Between 1880 and 1915, over 36 years, the proportion of Japanese tea in U.S. imports consistently maintained between 40% and 45.23%, while the share of Chinese tea decreased from 50.25% to 23.82% in 1915. During the same period, the percentage of Indian and Sri Lankan tea rose from 3.17% to 30.95%.[15] In the nineteenth century, the United States transitioned from a country that drank Chinese green tea to one that consumed Indian and Chinese black tea. In 1889, the United States shifted from drinking Fujian Kung Fu tea to consuming Indian and Sri Lankan black tea. The share of Chinese tea in the U.S. market has been declining. In 1915, China exported 2,310,000 pounds of tea to the United States, accounting for 23.82%; Japanese tea accounted for 43,869,000 pounds or 45.23%; and Indian and Sri Lankan tea made up 30,018,000 pounds or 30.95%. By 1930, Chinese tea exports to the United States had fallen to 6,467,000 pounds, representing just 7.62%; Japanese tea accounted for 20,448,000 pounds or 24.67%; and Indian

[15] Chen Chuan, *A History of Chinese Tea Export*, p. 195.

and Sri Lankan tea rose to 57,510,000 pounds or 67.72%.[16] Since then, black tea produced in India and Sri Lanka has dominated the international tea market.

### *Export of Chinese Porcelain to America*

In the eighteenth century, most exported Chinese porcelain set sail from Guangzhou, reaching England before being transshipped to continental European countries, a route that ran through the Indian Ocean. This was the maritime route that Portugal and the Netherlands used to transport Chinese porcelain back to their countries. The porcelain excavated from Fort Jesus in Mombasa, a southern Kenyan port, is of typical significance. Excavator J.S. Kirkman divided the artifacts from three centuries into four periods: the Early Portuguese period (1593–1634), the Late Portuguese period (1634–1698), the Omani and Mazrui period (1698–1837), and the Zanzibar period (1837–1895). In the first two periods, there were numerous pieces of porcelain dating from the late Ming Dynasty to the Qing Dynasty's Kangxi era, as well as Persian blue and white ware in Chinese styles, and Portuguese blue and white ware. The first half of the third period was a time of scarcity, with porcelain being rare. The latter half saw a large number of rose purples, mainly in soft pastels. After Zanzibar took control of Mombasa in 1837, the amount of European porcelain increased, and Chinese porcelain was squeezed out by the mid-nineteenth century.

In the second half of the eighteenth century, European porcelain from Sèvres, Dresden, and Chelsea flooded the market. High taxes were levied on imported Chinese porcelain. Although Chinese porcelain remained a best-selling brand in Europe, it was relatively expensive, resulting in a gradual decrease in its export volume to Europe. However, the demand for Chinese porcelain in America remained high up until the nineteenth century, and the supply could not keep up with the demand. Chinese porcelain exported to the United States began to adapt its form and decoration to cater to American tastes.

Before independence, the Thirteen Colonies in North America had to have their porcelain transported via British ships, which resulted in high prices for Chinese porcelain. On July 4, 1777, an advertisement was

[16] Chen Chuan, *A History of Chinese Tea Export*, p. 195.

placed in the *New York Gazette and Weekly Mercury*: "Henan porcelain (referring to the Western-style porcelain made in the southern suburbs of Guangzhou) shipped from London. The quantity is large and the types are complete, including a set of six pieces of blue and white dinnerware, bowls and cups of various sizes, breakfast bowls, tea cups and saucers, and plates of various sizes and styles."[17] At that time, Chinese porcelain was transported to North America via London, with the profits all going to the British. After 1760, the residents of the Thirteen Colonies started to drink tea commonly, resulting in an increasing need for Chinese porcelain year by year. The upper class even pursued the Rococo fashion, which was already outdated in Europe, using Chinese porcelain to decorate their interior furnishings. This quickly became a trend, with dining utensils often made of Chinese porcelain. Blue and white porcelain became everyday tableware in the states of New England in the eighteenth and nineteenth centuries. The willow pattern, popular in eighteenth and nineteenth-century England, which often features a willow tree, pagoda, bridge, and boat in the center, surrounded by lotus flowers and mandarin ducks, was also quite popular in America. These plates often have a diameter of 20.3 centimeters.[18] Armorial porcelain, a type of porcelain specifically ordered and customized in Guangdong, was also a trend in America, imitating European royalty and nobility by engraving family crests onto porcelain. Before America gained independence, there is no need for further elaboration, this kind of armorial porcelain was also customized by the British in China. Therefore, after New York City passed the heraldic symbol in March 1778, the city's crest appeared frequently in Chinese export resources. The crest features two figures dressed in German attire, representing the goddess of freedom and the goddess of justice. In front of them, however, is a British crown, indicating they are still subjects of the British monarchy.

After independence, armorial porcelain in the United States took on new forms. They were no longer the crests of powerful families. Some porcelain pieces were engraved with the seals of American states, some were customized in China with city badges, and some were adorned with medals or commemorative badges; the variety was plentiful. In 1809, the

[17] J. M. Mudge, *Chinese Export Porcelain for the American Export Trade*, Delaware, 1962, p. 67.

[18] Duncan Macintosh, *Chinese Blue and White Porcelain*, Vermont, 1977, p. 108.

state seal of Pennsylvania appeared on cups and saucers made by Peter Hetert, accompanied by the English words "Brave, Liberation, Independence." Some ceramics shipped to the United States were also engraved with Western words. The U.S. Congress has a porcelain plate from the late eighteenth century in its collection. The plate bears the inscription "This place is both glorious and safe" in foreign characters, and around the rim of the plate are 15 interconnected rings, each containing the English name of one of the 15 states.[19] This reflects the united efforts of the 15 states during the struggle for independence, striving to establish a federal government under the republic. The national emblem of the United States is the bald eagle. After 1786, clouds and rays of sunlight piercing through the clouds were added above the eagle, signifying that light had spread across North America.[20] This emblem is often found on porcelain pieces such as cups, plates, jars, and dishes, serving as a common symbol among the public.

The introduction of political and military themes into porcelain artwork is a distinctive feature of porcelain exported to the Americas during that era. Such panoramic depictions date back to the reign of Louis XIV of France, representing imperial grandeur in the French art scene of the eighteenth century. Influenced by French political ideals and cultural background, the United States, after gaining independence, ordered batches of porcelain with such illustrations from the Guangzhou porcelain factories. Significant historical political events from around the time of American independence were depicted on porcelain plates, cups, jars, and dishes.[21]

On December 1, 1822, Pedro was crowned Emperor Pedro I of Brazil. In 1825, Portugal officially recognized Brazil's independence. After Brazil's independence, some patriotic Brazilians gifted Emperor Pedro I a set of Chinese porcelain dinnerware. The plates were engraved with the Brazilian royal emblem, composed of tobacco and coffee leaves, and surrounded by the English inscription "Long Live the Independence

[19] Peter Herbert and Nancy Schiffer, *China for America Export Porcelain of the 18th and 19th Centuries*, Schiffer, 1980, p. 95.

[20] Peter Herbert and Nancy Schiffer, *China for America Export Porcelain of the 18th and 19th Centuries*, Schiffer, 1980, p. 11 ff.

[21] Peter Herbert and Nancy Schiffer, *China for America Export Porcelain of the 18th and 19th Centuries*, Schiffer, 1980, pp. 143–152.

of Brazil." This set of dinnerware became a valuable memento of Brazil's independence and is treasured by collectors worldwide.

During their stopovers in Guangzhou, American ships sailing to China often ordered porcelain pieces with human portraits, requesting them to be ready for return voyages within a month. Captains frequently used photos of wealthy merchants, shipowners, and themselves as samples to be depicted on the porcelain, leading to portrait art becoming a popular theme for porcelain exported to America. In fact, from the late eighteenth century to the mid-nineteenth century, the volume of such orders was quite significant.[22] The image of George Washington, the first president of the United States, also became a motif on Chinese porcelain exported to the United States.

Chinese porcelain exported to the United States often featured American scenes or commemorative images, tailored to the request of the buyer. These scenic depictions were common on seventeenth-century Chinese porcelain exported to the Netherlands, typically showing Rotterdam landscapes, windmills, mills, and fishermen whaling. On the porcelain plates exported to America are images of the first ship to sail to China, the Empress of China. This plate is currently held in the New Jersey State Museum.[23]

In 1802, a porcelain painter in Guangzhou replicated the design of the Pennsylvania Hospital in Philadelphia onto a batch of large-sized porcelain bowls intended for serving drinks like wine and lemonade. The design was initially sketched by British artist Perkins and depicted the hospital's magnificent architecture, well-equipped facilities, and advanced medical practices. The hospital's board of directors selected this painting and requested British porcelain factories to print the design onto the porcelain. However, the British side was unable to fulfill this request, so the task was handed over to Guangzhou, where it was quickly accomplished. Similar examples abound, such as porcelain jars shipped from China to New York in 1785, which already displayed the American national emblem and flag, with popular commercial slogans of the time engraved around the emblem, proclaiming, "Brave resolution secured our country's freedom, established the foundations of our empire, and

[22] Zhu Peichu, *Ming and Qing Ceramics and the Exchange of World Cultures*, Beijing: Light Industry Press, 1984, p. 203.

[23] Peter Herbert and Nancy Schiffer, *China for America Export Porcelain of the 18th and 19th Centuries*, p. 154.

expanded commerce."[24] In the nineteenth-century American market, a famous type of pottery called the Morton tub emerged. These tubs had diameters ranging from 21 to 38 inches (53 to 80 centimeters) and stood 10 inches (25.4 centimeters) tall, with a capacity of 8 gallons. Resembling Chinese wine jars in size, these ceramic tubs were a new product specifically crafted by Guangzhou factories for their American clientele. It goes without saying that these products were deliberately designed by American ship captains in Guangzhou, who sought to emulate Chinese cylinder-shaped ceramics.

In the nineteenth century, the emergence of the new American market for Chinese export porcelain provided a timely boost to production, preventing a contraction in response to the increasing demand in the European market. However, the decline of the porcelain industry in Jingdezhen was ultimately unavoidable after the 1850s.

### *Sino-Russian Tea Trade by Merchant Caravans*

In the second half of the sixteenth century, Russia began to expand eastward. The Russian Cossack captain Petrov first tasted tea in the Kalmyk Khanate in 1616. It wasn't until after 1628 that tea began to be imported into Russia. In 1640, Russian envoy Vasily Starkov returned from the Kalmyk Khanate with 200 bags of tea, each weighing 3 Russian pounds (1 Russian pound is equivalent to 409.51 grams), as a tribute to the Tsar. It was at this time that Chinese tea officially entered Russia.[25]

After 1660, Russian merchants began to arrive in Beijing. In 1676 (the 15th year of Kangxi), Emperor Kangxi received Russian merchant Nikola, presenting him with four boxes of tea and entrusting him to gift eight boxes of tea to the Tsar.[26] In the late seventeenth century, a small amount of tea was available on the Tobolsk market. In 1689, China and Russia signed the *Treaty of Nerchinsk*, setting boundaries and allowing trade between the two nations. Tea began to be traded with Russian merchants in Kulun. Russian merchant Nikitin purchased 5 puds (1 pud

[24] Peter Herbert and Nancy Schiffer, *China for America Export Porcelain of the 18th and 19th Centuries*, p. 97.

[25] John F. Baddeley, *Russia, Mongolia and China*, vol. 2, London: Macmillan, 1919, p. 118.

[26] John F. Baddeley, *Russia, Mongolia and China*, vol. 2, London: Macmillan, 1919, p. 398.

equals 16.38 kilograms) and 7 Russian pounds of tea in 1698, which was valued at 20 to 25 rubles per pud based on Moscow market prices. In 1723, the trading market moved from Kulun to Kiakhta. In 1727, the *Sino-Russian Kiakhta Treaty* was signed in Beijing, making Kiakhta the main market for tea and fur trade. Tea was transported from Tianjin to Zhangjiakou by horse, then transported to Kiakhta by camel. Typically, a caravan would consist of two to three hundred camels, each carrying 4 boxes of tea, each box weighing 16 puds (600 pounds). Traveling an average of 40 kilometers per day (calculating with 10 hours per day.), it would take 16 months to reach Kiakhta, crossing the Gobi Desert. Initially, the tea transported to Kiakhta was exclusively for domestic transport by the Russian government's caravans. Another port of entry was Zuluhaitu, where Chinese and Russian merchants would trade tea for fur via the Nen River and Nerchinsk, but the market was only open for 6 weeks each year.

After the trading city of Kiakhta officially opened in 1728, tea imports increased year by year. In 1820, the export of Chinese tea to Russia reached 100,000 puds (equivalent to 4 million Russian pounds, or 1,678,000 kilograms). The price of tea in Saint Petersburg was 60% to 80% higher than in London. However, due to the exposure to the elements during overland transportation, the tea flavor was superior to that of tea transported by sea through the hot and humid South Sea, which would diminish the tea flavor.[27] As a result, the volume of tea transported by overland caravans saw a steady rise. By 1838, the tea transported to Kiakhta included 71,950 pieces of large brick tea, 60,430 pieces of small brick tea, and 43,070 boxes of loose tea (each box weighing 100 pounds). Russian merchants had monopolized this caravan tea, selling it to Mongolia, Siberia, and other nomadic peoples in western Asia. These peoples had become dependent on brick tea, often cooking it with milk, cheese, salt, and spices. Tea was not only their beverage, but also their food. In 1839, Russian merchants spent 7 million rubles on tea in Kiakhta, which they could sell for 18 million rubles at the Nizhny Novgorod Fair, yielding a profit of 145%.

Before 1861, the merchants transporting tea to Kiakhta were from Shanxi, commonly known as the "Xibang Tea Merchants." They transported tobacco, silk, and general merchandise from Zhangjiakou to

[27] Weiyuan, *Illustrated Treatise on the Maritime Kingdoms*, vol. 38.

Kiakhta in exchange for various furs and felt. In the late Ming and early Qing dynasties, the merchants who traded with foreign countries for the Imperial Household Department were all from Shanxi, including Fan Yongnian, Wang Dayu, and Liang Jiabin, among others. To transport tea, they ventured deep into the Wuyi tea region in the south, each with a capital ranging from two to three hundred thousand to one million. The tea purchased from Fujian was stockpiled in Zhangjiakou and then transferred to the trading city via the trading route (from Zhangjiakou to Kiakhta). This route was divided into east, west, and middle paths, totaling over two thousand kilometers. The terrain was sparse and sparsely populated, so they had to travel in convoys. Each group had over a hundred carts, with each cart carrying about 500 pounds. One ox could pull more than ten carts, and they traveled only thirty to forty miles a day. In winter, camels had to be used for transportation. The goods that arrived at Kiakhta were measured in units of camels and large carts. There were about 100 Chinese trading houses in Kiakhta, but only 37 of them did wholesale business with Russian merchants, and the rest were small traders. The Russian merchants often paid with "foreign merchant money notes," amounting to tens of thousands of taels.

In 1845, Russia opened a highway from Orenburg to Yili, adding a commercial port between China and Russia in Yili. Starting from 1850, Russian merchants began to buy tea in Hankou, initially ordering black tea, but soon switching to selling Mongolian red brick tea. By 1851, this type of brick tea accounted for 75% of total exports. Between 1853 and 1856, the armies of the Taiping Heavenly Kingdom occupied areas near the Fujian tea region, causing tea prices to rise by half. Russian merchants then switched to buying tea from Hunan and Hubei, and over the next 10 years, the area of tea plantations in the two lakes region significantly increased. The Sino-Russian *Treaty of Tianjin* in 1858 allowed trade between the two countries not only overland but also through seven seaports, including Shanghai, Ningbo, Fuzhou, Xiamen, Guangzhou, Taiwan, and Qiongzhou. As a result, Sino-Russian trade in Kiakhta declined dramatically between 1859 and 1871. After Hankou officially became a treaty port in 1862, Russian merchants immediately established a brick tea factory there using steam-powered machines to press tea leaves into brick tea. After the half-tax was abolished, it became cheaper to transport Wuyi tea by boat along the Yangtze River to Tianjin than by overland transport. From 1861, Russia lifted the ban on the maritime transport of tea, allowing tea to be imported by both land and sea, with a tariff of 35 kopecks per Russian pound of floral tea, green tea, and yellow tea (equivalent to 1 1/2 pence per pound). As a result, smugglers took advantage of the price difference to sell tea transported by sea as Kiakhta tea, or to infiltrate smuggled tea into Kiakhta tea. The trading city established next to Kiakhta gradually evolved into a market for tea and cloth. In 1861, the Russian government had to allow the export of gold and silver coins. Russian merchants seized the opportunity to establish tea trading enterprises in Hankou and Tianjin, causing Hubei tea to gradually replace Fujian tea in the trading city and compete with Shanxi merchants for the tea market.

In 1862, Russia moved its border customs from Kiakhta to Irkutsk. In the same year, Russian merchants transported 1,485,027 pounds of brick tea from Tianjin to the Siberian market via ship. Brick tea began to be shipped into Russia from Vladivostok by sea. In the following years, there was an increase in the transport of brick tea from Tianjin to Kiakhta. In 1864, Russian merchants learned to manufacture brick tea, and many Russian commercial firms relocated to Hankou, where they set up their own factories to produce brick tea for transport to Russia. After the treaty between China and Russia was signed in 1869, Russian merchants set up a brick tea factory in Fuzhou in 1870, and additional tea factories were established in Jiujiang and Fuzhou in 1875. They used foreign transport tax receipts to protect the transportation of tea. There were three types of brick tea. The largest was made from second-grade green tea, 1.75 feet long, 7 inches high, and weighed 3.5 pounds. Each basket contained 36 pieces, cost 4 taels and 5 qian, and 20,000 baskets were exported each year. It was sent from Jiujiang to Tianjin and then transported overland to Kiakhta. The second type was made from black tea and was 8.5 inches long, 6% inches wide, each piece weighed 2.5 pounds, and each basket contained 64 pieces, valued at 6 taels, for sale to farmers throughout Russia. The third type was a high-quality brick tea, 8.25 inches long, 5.25 inches wide, weighing 1.5 pounds, each basket contained 80–90 pieces, valued at 5 taels and 5 qian, for sale in western Siberia, Kazan, and the Amur region. Russian merchants in Hankou indeed used high-quality tea dust to press a type of small "Jing" brick tea, each piece weighing 2.5 pounds. After being wrapped in tin foil, these were packed in lead boxes and transported to Russia. Jiujiang also produced this small "Jing" brick tea. Because of its convenient size and packaging, this small "Jing" brick tea became popular not only in Russia but also found markets far and wide including countries like Germany and France. Its portability made it especially favored by armies, travelers, and nomadic peoples living in border regions. As a result, the export volume of this type of tea exceeded 1 million pounds. Together, the tea factories in Hankou, Jiujiang, and Fuzhou produced about 100,000 baskets of brick tea for export annually.

Despite the competition from maritime transport, the caravan tea trade from Zhangjiakou to Kyakhta remained prosperous from 1860 to 1880. In the mid-nineteenth century, the average annual shipment of Chinese tea to Russia was 60,000 metric tons, which increased to 300,000 metric tons from 1871 to 1880. With the eastward extension of the Siberian railway, the journey that used to take 16 months by caravan was reduced to 7 weeks. However, the volume of caravan transport still increased due to the expansion of markets in Central Asia and Siberia. In 1886, China exported 2.21 million tons of tea, 600,000 tons of which were destined for Russia. The amount of brick tea transported by Russian merchants reached 105,850 tons (14,113,300 pounds). This again underscores the significant role tea played in the trade between China and Russia during this period.[28]

In 1888, Russia imported 920,610 tons of Chinese tea, of which 463,325 tons were transported overland through Kyakhta and 245,433 tons through Fancheng.[29] By 1904, with the completion of the Trans-Siberian Railway, the era of the caravan tea trade came to an end. However, between 1894 and 1938, Russia remained the largest consumer of Chinese brick tea, accounting for 95.6% of China's brick tea exports. As for other types of tea, Russia consumed 32.7% of China's total black tea exports and 16.8% of green tea exports.

## 8.2 The Forced Open Doors: Gunboats, Port Opening, and Concessions in China

China, the great East Asian nation with rich business opportunities in the eyes of Western maritime powers, was an indispensable link for Britain, which had defeated Spain and the Netherlands, in establishing its global network of shipping stations. The maritime competition between Britain and France in the eighteenth century ended with Britain gaining vast territories in Canada and India. At that time, Britain had obtained Australia

[28] Chen Chuan, *A History of Chinese Tea Export*, pp. 183–184.

[29] *Sources on the History of Foreign Trade in Modern China*, vol. 2, pp. 1283–1284.

and New Zealand overseas and steadily became China's largest trading partner. However, Britain had always envisioned occupying one or several islands along the coast of China as permanent residences and trading stations. Therefore, in 1793, a mission headed by George Macartney was sent via the Cape of Good Hope in South Africa to Beijing, China, to pay homage to Emperor Qianlong in Chengde, Rehe. The ambassador's flagship was the *Lion*, armed with 64 cannons. The Hindostan of the East India Company was loaded with gifts, while the Jackal and the Clarence were supply ships. Of the 95 members of the delegation, there were 20 Royal Artillery, 10 Light Cavalry, and 20 Infantry. After participating in the Grand Birthday Celebration in Rehe, the delegation proposed six requests to the Chinese government. The most important were: permission for British merchants to trade in Zhoushan, Ningbo, and Tianjin, expanding the number of treaty ports from one in Guangzhou to four; granting an island near Zhoushan to British merchants for storage, trading, and living; and setting up a warehouse in Beijing to sell goods. These requests were rejected by the Qing government, and the mission left Guangzhou to return to Britain in January 1794.

George Macartney and Amherst's embassy to China did not achieve their anticipated goals. However, starting with Macartney, the mission was undoubtedly an investigation team dispatched by the British government. They collected social, business, resource, shipping, and military intelligence along the coast of China. The conclusion they drew was that the Chinese Empire was just an old and shaky ship, luckily steered by some clever military men, which could maintain the status quo without sinking. On their way home, they arrived in Macau, where 4000 Portuguese people lived. They were very envious of this piece of land, only 8 square miles in size, which had made the Portuguese wealthy.

The gifts brought to Beijing by the British Macartney Mission were like a scientific instrument and a craft technology exhibition. In addition to planetariums, celestial globes, earth globes, high-power telescopes, and self-striking clocks, there were also round lights, woolen goods, cotton textiles, and steel products, copper cannons, howitzers, Mauser guns, machine guns, a model of the largest British warship with 110 heavy cannons, and countless exquisite pictures reflecting British cities and natural landscapes, characters, sea battles, and land battles. The mission brought batches of astronomical instruments, mechanical parts, and guns, intending to showcase the high level of military and craft technology achieved by Britain. They intended to remind the Chinese, who were "far behind Europe in the field of science,"[30] hoping that they could make a comparison between the two sides.

The British mission to China aimed to seek further improvement in Sino-British relations. Britain, proud of its status as a first-rate maritime power, was determined to achieve its goals. Therefore, half a century later, the incident of Britain forcibly occupying Hong Kong eventually occurred, which was caused by the rampant smuggling of opium.

In 1839, the anti-opium faction led by Imperial Commissioner Lin Zexu was ordered to strictly ban the opium trade, angering British opium merchants and the British government. Lin Zexu arrived in Guangzhou on March 10, 1839, confiscated opium, and burned a total of 2,376,254 catties of opium at Humen Beach. He demanded that foreign merchant ships coming to Guangzhou "never carry opium" or face consequences. Britain dispatched a large number of warships to protect opium trade ships and conducted private transactions at sea. The British fleet headed north to Xiamen, Dinghai, and Tianjin in July 1840. After Dinghai's fall, the Daoguang Emperor was afraid of British warships heading north to Tianjin, so he sent Qi Shan to Guangdong to negotiate peace and dismissed Lin Zexu. On January 7, 1841, British troops occupied the Dajiao and Shajiao forts at the mouth of the Pearl River, forcing Qi Shan to agree to ask the emperor to cede Hong Kong to them for "anchoring and residing." During the negotiations, British troops forcibly landed in Hong Kong on January 25, 1841, and never left. Soon, China and Britain officially declared war, and Humen was immediately captured. Xiamen, Dinghai, Ningbo, and Shanghai successively fell into British hands, and British troops entered the Yangtze River and captured Zhenjiang, and the *Treaty of Nanking*, which ceded territory and indemnified money, was officially signed on August 29, 1842. The treaty stipulated: opening Guangzhou, Fuzhou, Xiamen, Ningbo, and Shanghai as trading ports; ceding Hong Kong forever; indemnifying 21 million silver dollars. For the next 60 years, China embarked on a humiliating path of frequent wars, successive defeats, ceding territory and indemnifying money, and losing rights and interests. It was in this context that China was incorporated into the world market and began its difficult journey toward modernization, marked by industrialization.

[30] George Leonard Staunton, *An Authentic Account of an Embassy from the King of Great Britain to the Emperor of China,* Chinese translation, Shanghai Bookstore, 1997, p. 34.

The first of the five ports to be opened was Guangzhou (July 27, 1843), followed by Xiamen (November 2, 1843) and Shanghai (November 17, 1843), then Ningbo (January 1, 1844), and Fuzhou (July 3, 1844). Among these newly opened ports, only Guangzhou was of old standing. Since the lifting of the sea ban in the early Qing Dynasty, it had been a commercial port that dominated one side and controlled the other four. In addition to China's coastal routes, there were the Straits of Malacca in the west, the Sunda Straits in the south, and the sea route to Manila and the southern Philippine Islands in the east, sufficient to cross the Pacific, Indian, and Atlantic Oceans and connect with the whole world. In 1846, a Chinese sailing ship named "Keying," under the command of the British Kellet, set sail from Guangzhou with a group of British people and 30 Chinese. They rounded the Cape of Good Hope, passed by Saint Helena Island, crossed the Atlantic Ocean, and arrived in New York. They then crossed the ocean from New York and arrived at Gravesend, United Kingdom, on March 28, 1848, taking a total of 17 months. The *Illustrated London News* praised, "The first junk that ever rounded the Cape of Good Hope, or appeared in British waters."[31] This was the first well-known Chinese sailing ship to cross the Atlantic Ocean and shuttle between Asia, America, and Europe after the Chinese ships entered the South African waters in the era of Zheng He. This voyage proved that Chinese wooden sailing ships indeed had the capacity to continuously cross two oceans, but the time consumed was four to five times that of the clipper ships.

After the First Opium War, the newly opened ports were limited to four southeastern coastal provinces. In 1857, Britain and France used the pretext of protecting missionaries and concentrated their troops in Hong Kong. In December, they captured Guangzhou and advanced northward. In April 1858, envoys from Britain, France, the United States, and Russia led their fleets to invade the outskirts of Tianjin. On May 26th, the Qing government signed the *Treaties of Tianjin* with the four nations, which established permanent foreign legations in Beijing, opened 10 additional ports, and allowed foreign merchants and warships to enter various ports along the Yangtze River. The Allied forces attempted to force their way into Baihe, leading to a fierce battle at Dagu. In September 1860, the Allied forces reached the outskirts of Beijing, looting and burning down

[31] *The Illustrated London News*, 1 April 1848, pp. 220–221.

the Old Summer Palace, the New Summer Palace, and the Yuanmingyuan Garden. The Qing government signed the *Convention of Peking* with Britain and France, acknowledging the validity of the *Treaties of Tianjin* and opening Tianjin, Dengzhou, Niuzhuang, Chaozhou, Qiongzhou, Taiwan (Tainan), Tamsui, Hankou, Jiujiang, Nanjing, and Zhenjiang as treaty ports. The five treaty ports expanded to 16 commercial ports along the coastal areas and the middle and lower reaches of the Yangtze River. Kowloon, opposite Hong Kong, was ceded to Britain and became part of Hong Kong. In the *Treaty of Shimonoseki* in 1895 between China and Japan, additional ports such as Shashi, Chongqing, Suzhou, and Hangzhou were opened. The encroachment of foreign powers on China gave them a territory where they could freely exert their influence. The concessions and leases that gradually appeared in these opened ports became sharp blades embedded by Western developed countries in the Chinese Empire. They utilized various privileges and turned them into advanced frontiers for Western countries on Chinese soil. In the process of the growth and expansion of these ports, the model of Western-style industrial cities and transit ports provided by the concessions directly laid the blueprint for the rise of modern cities in China.

Concessions often exhibited a trend of continuous expansion under the coercion of the foreign powers. Once established, concessions and leases completely detached from their home countries and became communities created by Western powers on Chinese soil. The majority of concessions were established near old cities. Tianjin in the north and Shanghai at the mouth of the Yangtze River, which were China's northern and southern gates, served as the residences of the Beiyang Minister and Nanyang Minister of the Qing government. The concessions in these two cities became the starting point of modern industrial cities.

In the twelfth century, Tianjin was known as Zhigu and became a port in northern China. Prior to 1900, concessions were already established by Britain, the United States, France, Germany, and Japan. The concessions of Britain, the United States, and France were located in the southeast of the county town, near Zizhulin, and expanded continuously along the west bank of the Hai River. In 1900, the Eight-Nation Alliance occupied Tianjin, and Russia, Italy, Belgium, and Austria forcefully acquired concessions on the east bank of the Hai River, seizing control of the territories. The American concession was transferred to the British concession in 1901, resulting in four-country concessions on both the east and west banks of the Hai River along the railway. In 1918, the Russian and Austrian concessions in the east and the German concession in the west were taken back by the authorities. Japan took the opportunity to expand westward. The five concessions operated independently, resulting in a disconnected road system and a lack of integration in urban public facilities. The circumference of the old city of Tianjin was only about 4.5 kilometers, but by the beginning of the twentieth century, the concession areas were seven times the size of the old city, forming the bustling sections of the city. It was not until 1945 that the concessions were wiped off the urban map of Tianjin.

In the twentieth century, Shanghai, which became the largest city in the country, was originally a small town located about 2 kilometers south of the confluence of the Huangpu River and the Suzhou River (formerly known as the Wusong River). In the Tang Dynasty, it was called Huating, and in the Yuan Dynasty, it belonged to the Songjiang Prefecture. In 1292, it was established as a county named Shanghai County. After the opening of Shanghai, British Consul Balfour leased a 130-acre wasteland on the west bank of the Huangpu River in the northeast of the county. By 1846, the boundary of the concession had been formally demarcated and expanded to 830 acres. In 1848, the British concession had expanded northward to the present-day Suzhou River and westward to the present-day West Tibet Road. In the same year, the United States, through the church, established a concession on the north bank of the Suzhou River, and in 1849, France established a concession south of the British concession. In 1862, the French concession established its own Municipal Council to govern the city, while the British and American concessions merged in September 1863 to form the International Settlement, continuing to use the name Municipal Council to govern the city. By 1849, the area of the concessions was roughly equal to that of the county town. In that year, Shanghai surpassed Guangzhou and became the largest seaport for foreign trade in the country, accounting for about half of the total import and export trade. The area of the concessions continued to expand, and several expansions in 1899 and 1914–1915 increased the area of the concessions to more than 30 times that of the county town. In 1943, the United States, Britain, and France were forced to abolish the unequal treaties with China, and the concessions finally disappeared from China.

## 8.3 New Knowledge and New Schools

### *Lin Zexu Found the Key to International Knowledge*

The rampant opium smuggling and the struggle against it eventually escalated into an inevitable conflict known as the First Opium War between the outdated Qing government and the emerging industrial power, the British government, which dominated the seas. At this time, China had already been involved in the global market for many years, yet the Qing government still considered itself self-sufficient and independent from Western powers. Even before the outbreak of the opium war, Lin Zexu, who had already sensed the brewing conflict, focused on military training and preparation upon his arrival in Guangzhou. He procured modern ships and cannons, recognizing the disparity in military strength between China and Britain in naval warfare.

In the face of Britain's modern artillery and powerful ships, it was necessary to have extensive knowledge in dealing with them. Lin Zexu was among the first in his circle to pay attention to international affairs and actively seek knowledge about foreign affairs. During that time, foreign affairs were still referred to as "barbarian affairs" due to the traditional concept of distinguishing between the "barbarians" and the Chinese. Upon his arrival in Guangzhou, Lin Zexu sought to meet with Liang Tingnan, who served as the superintendent of the Yuehua Academy in 1836. Liang Tingnan had previously served as the editor-in-chief of the *Guangdong haifang huilan* and was responsible for the compilation of the *Guangdong Customs Gazette* in 1837. They discussed national matters together. However, effective diplomacy required individuals who were proficient in foreign languages and well-informed about foreign affairs. Lin Zexu relied on individuals like Liang Jinde and Yuan Dehui. Liang Jinde was the son of Liang Fa, the earliest Jesuit pastor in Guangdong. He was baptized in 1823 by Robert Morrison, a London Missionary Society missionary, and studied English under American missionary Elijah Coleman Bridgman. He attended the Morrison School in Macau and worked as an interpreter for foreign firms in Guangzhou. In 1839, he became Lin Zexu's interpreter. Another assistant was Yuan Dehui, a native of Sichuan. He studied at a Catholic school in Penang and was proficient in Latin. After attending the Anglo-Chinese College in Malacca, he returned to China and worked at the Beijing Imperial College. In 1839, he became Lin Zexu's aide and assisted in handling diplomatic affairs.

Upon his arrival in Guangzhou, Lin Zexu immediately hired translators to translate English newspapers published and sold in Macau, such as *Canton Register, Canton Miscellany, Singapore Free Press,* and *Bombay Courier.* These newspapers compiled relevant reports and commentaries about China and were regularly assembled into volumes for easy reference. As these newspapers were obtained from the Macau market, they were collectively referred to as the *Macau Newspapers.* They provided crucial and timely information about British activities, and some of them were even copied and forwarded to colleagues or sent to Beijing for reporting purposes.

Lin Zexu actively collected English books for his studies and must have also read Chinese books and newspapers published by Christian missionaries in Singapore. During that time, there was frequent commercial interaction between Guangzhou and Singapore. The missionaries in Singapore published a periodical called the *Dong Xi Yang Kao Mei Yue Tong Ji Zhuan.* This Chinese-language periodical was first established in Guangzhou on August 1, 1833, then moved to Singapore in 1834, and ceased publication in September 1838. It was primarily edited by missionary Karl Friedrich August Gützlaff (1803–1851) under the pen name "Aihan Zhe" (爱汉者) and covered various topics including news, history, geography, customs, religion, nature, crafts, trade, and miscellaneous information. The narratives were concise, popular, and vivid, representing a new style of modern vernacular writing. Gützlaff's works such as *Gujin Wanguo Gangjian* and *Wanguo Dili Quanji* were serialized in this periodical. The circulation of the publication increased from 600 to 900 copies per issue. In 1834, during the imperial examinations, copies of this magazine were distributed in Beijing, Nanjing, and other places. Later, Wei Yuan cited more than 20 places in *Dong Xi Yang Kao Mei Yue Tong Ji Zhuan* for his work *Illustrated Treatise on the Maritime Kingdoms.* This magazine became an exemplary model for missionaries in the treaty ports who engaged in publishing and propagating their teachings. Prior to the Opium War, missionaries in Singapore compiled and published several books on world geography, trade, and history. For example, *Gujin Wanguo Gangjian* (compiled by Gützlaff, 1838, 244 pages, 20 volumes), *Brief Geographical History of the United States of America* (compiled by Elijah Coleman Bridgman, 1838, 125 pages, under the pseudonym Gao Liwen), and *Comprehensive Treatise on Trade* (compiled by Gützlaff, 1840, 63 pages). These books were referenced in Wei Yuan's *Illustrated Treatise on the Maritime Kingdoms,* which consisted of a hundred volumes. Interestingly, these materials were originally collected by Lin Zexu before his resignation and exile to Xinjiang and later handed over to Wei Yuan for use in his book compilation.

Lin Zexu personally reviewed a batch of important English books that served as valuable literature for understanding Western countries and handling foreign affairs. He organized a team to translate these books. There were three types of books. The first was *The Law of Nations* by Swiss international jurist Emerich de Vattel (1714–1767), originally published in French in 1758. The English translation was already available in 1759, and since then, there have been 10 editions in the United Kingdom and 7 editions in the United States. It became a popular international law classic. Lin Zexu instructed Yuan Dehui to translate the chapters on warfare and hostile measures between nations based on the translation by J. Chitty and sought the assistance of Peter Parker, an American missionary physician in Guangzhou, for proofreading. The second type of book was *The Chinese* by the British East India Company's resident in Guangzhou, John Francis Davis. It was later renamed *The Chinese: A General Description of the Empire of China and Its Inhabitants.* Originally published in London in 1836, it represented the views of the British upper class on China at the time and provided some basis for British policies toward China. The third type was an abridged translation of *An Encyclopaedia of Geography* published by Hugh Murray in the United Kingdom in 1834. It was titled *Records of the Four Continents* (Si Zhou Zhi) and condensed the massive work of over 1500 pages into a manuscript of less than 90,000 characters. Later, it was categorized and included in Wei Yuan's work. This comprehensive overview of Asia, Africa, Europe (Atlantic and Northern regions), and America (Outer Atlantic) reflected the major countries of the nineteenth century, with a particular focus on the important economic and trade relations between China and the two Western countries of England and America. The book introduced the democratic system of the United States, using transliterations for terms such as "United States" (Yunai Shi Die Guo), "President" (Bolie Xi Ling), "Senate" (Shensiyuan), "House of Representatives" (Zhongyiyuan), and "Congress." However, due to its limited circulation, it could only serve as a temporary reference for government departments. From this, it can be seen that the Qing government lacked international knowledge when initially handling foreign affairs. However, in Guangdong at that time, they were finally able to distinguish between the United States and England as distinct and separate nations.

In May 1840, along with the establishment of the newly equipped Guangdong Navy, Lin Zexu, through the arrangements he made, found the key to open the country's doors and connect with the world—international knowledge. He directly obtained firsthand knowledge from Western countries to understand their national conditions and customs and found a pathway for learning. Despite facing numerous obstacles, Lin Zexu managed to cover the ground that would have taken previous generations decades to traverse in just a short period of two years. The banner of new knowledge that he raised inspired intellectuals who were disillusioned with the corrupt Qing government and determined to break away from the shackles of national humiliation. They emerged as leaders of their time. This movement of acquiring new knowledge gradually spread across China's essential regions during the second half of the nineteenth century, passed down from generation to generation. The brave and innovative individuals, willing to break with tradition, fearlessly advanced to embrace the arrival of a vibrant new era.

### *Missionaries and the Exchange of Chinese and Western Scripts*

One important channel for communication and information exchange between China and Western countries was the active involvement of Protestant Christian missionary societies in overseas missions in the late eighteenth century. Due to the Qing government's prohibition of Catholicism in China, the London Missionary Society (LMS) of the Congregational Church from Protestant Christian nations, particularly Britain, initially had to secretly dispatch missionaries to China. The pioneer of this Protestant mission activity in China was Robert Morrison, a LMS missionary who arrived in Macau on September 4, 1807. After entering Guangzhou on September 8, Morrison resided in foreign factories for a year, diligently studying Mandarin Chinese and Cantonese. In 1809, he joined the British East India Company as an interpreter and worked missionary in Macau, Guangzhou, Malacca, Batavia, and Singapore. Preaching, translating the Bible, and publishing were Morrison's earliest tasks. Starting in 1810, he secretly commissioned the engraving and printing of his own authored missionary booklets. In 1813, the LMS sent a Scottish missionary named William Milne (1785–1822) to Macau to assist Morrison, who soon moved to Malacca and founded the Anglo-Chinese College. In 1815, he established a printing press in Penang and later used movable type printing in Malacca to publish books and periodicals. The first Chinese periodical he published was the *Cha shi su mei yue tong ji zhuan*, which debuted on August 5, 1815, using woodblock printing. Each issue consisted of 7 sheets, 14 pages, and was distributed in stitched form. The initial print run was 500 copies, later increased to 1000 copies, and it was distributed free of charge. It ceased publication in 1821 after 7 volumes. The majority of its content focused on promoting Christian teachings, with only a few sections dedicated to popular discussions on astronomy, geography, and other subjects. In 1828–1829, Malacca also published a monthly magazine called "World News" ( "天下新闻"), which covered not only the promotion of Christian teachings but also a considerable amount of domestic and international news, science, history, and general knowledge, making it a true magazine.

The most significant work carried out by Robert Morrison was twofold: first, the translation of *the Bible*, and second, the compilation of *A Dictionary of the Chinese Language*. Before Morrison embarked on the translation of *the Bible*, existing translations were still in draft form. One version was a partial translation of the *New Testament* by Father Jean-Baptiste Basile (1662–1707) of the Paris Foreign Missions Society, and another was a complete translation from Latin by a Jesuit priest, Louis Antoine de Poirot. Morrison had not read either of these versions. There was also a translation of the *New Testament* by William Mosley, a British missionary, which was kept at the British Museum. Morrison had carefully read this version before leaving for China, and it inspired him to make a commitment to translate the entire Bible into Chinese. After William Milne arrived in Macau, he assisted Morrison in the translation of the Bible. In November 1819, Morrison completed the translation of the *Old Testament*, with 26 volumes translated by himself and the remaining 13 volumes translated by Milne. In 1823, the complete Bible, translated by Morrison, was printed using movable type in Malacca, consisting of 21 stitched volumes. This complete translation of *the Bible* marked a significant milestone in both missionary work and cultural exchange, as it became available for Chinese believers and readers. When Morrison returned to his homeland in 1824, he received recognition from King George IV for this accomplishment. A few years earlier than Morrison, Joshua Marshman, a Baptist missionary in Serampore, India, had also completed a translation of the Old and New Testaments with the help of Chinese assistants. His translation was one of the earliest published comprehensive translations, while the Morrison Version had the greater impact.

Another important work by Robert Morrison was the compilation of *A Dictionary of the Chinese Language*. This dictionary consisted of three parts: the first part was a Chinese-English dictionary translated from the *Yiwen Beilan* (Comprehensive Collection of Literary and Artistic Works) published in 1807 and published in 1815; the second part was the *Wuche Yunfu*, arranged in alphabetical order using English letters, printed in 1819; the third part was the *English-Chinese Dictionary*, arranged in alphabetical order and included individual characters, vocabulary, idioms, and example sentences, engraved in 1822. By 1823, these three parts were combined into a massive work consisting of over 2500 pages. This dictionary eliminated the barriers between Chinese and English characters for the first time. In addition to compiling the dictionary, Morrison also authored a series of books and grammars that provided guidance to Protestant missionaries from Western countries in learning Chinese, including *A Grammar of the Chinese Language* (1815) and the *Vocabulary of the Canton Dialect: Chinese words and phrases* (1828). Subsequently, other works facilitating missionary work and communication between Chinese and English languages were published, such as *A Dictionary of the Hok-këèn Dialect of the Chinese Language* (1832) and the *Chinese and English Dictionary* (1842) by Walter Henry Medhurst, *A Selection of Three Thousand Characters Being the Most Important in the Chinese Language for the Purpose of Facilitating the Cutting of Punches and Casting Metal type in Chinese* (1834) and *Vocabulary of the Hokkien Dialect* (1838) by Samuel Dyer, *Notices on Chinese Gramma* (1842) by Karl Gützlaff, and *A Chinese chrestomathy in the Canton dialect* (1841) by Elijah Coleman Bridgman. Robert Morrison also translated the *San Zi Jing* and the *Great Learning* in 1812. Following Joshua Marshman's translation of the *Analects* in 1809, he advocated for translating Chinese classics and textbooks, enabling British and American missionaries to grasp the essence of Chinese culture. This laid the foundation for David Collie's translation of the *Four Books* (1828) and Bridgman's translation of the *San Zi Jing*, the *Thousand Character Classic*, and the *Classic of Filial Piety*.

While translating, Robert Morrison also wrote descriptive works about Chinese society, including *A view of China for philological purposes, containing a sketch of Chinese chronology, geography, government, religion & customs* (1817), *China, a dialogue, for the use of schools: being ten conversations between a father and his two children, concerning the history and present state of that country* (1824), and *Report on China and Canton* (1823). These works guided Karl Friedrich August Gützlaff in compiling *A Sketch of Chinese History, Ancient and Modern* (1834) and *China Opened* (1838), establishing the foundation for sinology in Protestant countries such as England, America, and the Netherlands.

The missionaries were the first to introduce the use of lead–tin movable type for printing Chinese and English books, shifting the Chinese printing industry from woodblock and wooden movable type to the use of machine-based relief letterpress printing technology. After compiling *A Dictionary of the Chinese Language,* Robert Morrison decided to adopt the widely used lead-type technology from Britain. In 1814, an English printer named Thomas brought relief letterpress equipment to the East India Company's office in Macau. With the assistance of the Cai brothers, it took several years to cast the first set of Chinese lead movable type, which was used to print Morrison's Dictionary. Later, Samuel Dyer, a missionary from the London Missionary Society, arrived in Penang in 1827 to study the casting of Chinese lead movable type. He settled in Malacca in 1835 and was in charge of the printing press. He created a steel typeface of Chinese characters, successfully cast two sets of aesthetically pleasing and durable typefaces, totaling 1845 characters. These typefaces were used for printing Chinese books and newspapers at the Anglo-Chinese College in Singapore. When the Anglo-Chinese College relocated to Hong Kong in 1843, this set of typefaces, which is equivalent to today's "Size 4" font, came to be known as the "Hong Kong typeface." The oldest English newspaper in China, the *Hongkong Daily Press*, was also printed using this typeface. In 1858, Wu Tingfang used this set of typefaces to publish the earliest Chinese evening newspaper, the *Chinese and Foreign Gazette,* which was also printed in Hong Kong. After the Treaty of Nanking opened up five treaty ports for trade, missionaries began to imitate this lead movable type and establish printing presses to print various religious pamphlets and scientific booklets, marking the beginning of the transition to machine printing in the Chinese printing industry.

The establishment of the Singapore Printing Press marked the beginning of machine-printed Chinese books and periodicals. This endeavor was initiated by the Society for the Diffusion of Useful Knowledge in China, which was founded on November 29, 1834. The society, led by Ewing Matheson, a British merchant, had Samuel Dyer, Karl Friedrich

August Gützlaff, and John Robert Morrison (son of Robert Morrison) as its secretaries. The purpose of the Society was to facilitate communication between China and the West. They aimed to publish a category of books that would enlighten the Chinese people and impart Western skills and sciences to them.[32] In 1833, American Presbyterian minister Ira Tracy arrived in Guangzhou and was later sent to oversee the printing press in Singapore. Publishing began in 1835, marking the shift of the overseas Chinese publishing center from Malacca to Singapore before the outbreak of the Opium War. From 1834 to 1842, a total of 42 Chinese books and periodicals were published in Singapore. In addition to religious propaganda books such as the *Old Testament* translated by Samuel Dyer based on Robert Morrison's work (1836) and the *New Testament* translated by David Collie based on Morrison's work (1837), several books on history, geography, science, and trade were also published.

After the opening of the five treaty ports, the first group of British and American diplomats set foot on Chinese soil, allowing them to directly experience and appreciate Chinese society while comparing the differences between Chinese and Western cultures. British diplomat T.T. Meadows (1815–1868) published *Desultory Notes on the Government and People of China and on the Chinese Language* in 1847. He took note of the fact that Chinese officials were selected through the imperial examination system and praised China's system of appointing and promoting officials. He suggested that Britain should establish a civil service examination and a regular promotion system for civil servants. The British government established a civil service system in 1855, which was later emulated by several European countries.[33]

American missionary Samuel Wells Williams (1812–1884) arrived in China in 1833 and lived in the country for half a century. He had been chargé d'affaires for the United States in Beijing. He wrote a monumental work titled *The Middle Kingdom* in 1848. In this book, he candidly discussed the backwardness of Chinese society and provided frank commentary on the achievements and shortcomings of Chinese culture. He declared that one of the purposes of the book was "to

[32] *Chinese Repository*, December 1834, p. 384. Gu Changsheng, *From Morrison to John Leighton Stuart*, Shanghai People's Publishing House, 1985, p. 59.

[33] Teng Ssu-yu, "Chinese Influence on The Western Examination System", in Huang Pei and Tao Jinsheng, eds., *Selected Academic Papers of Teng Ssu-yu*, Taipei, Shihuo Publishing House, 1980, pp. 125–181.

wipe away the almost inexpressible contemptuous impressions often made about the Chinese and their civilization." "The Middle Kingdom" was a comprehensive encyclopedia-style work that marked the first comprehensive exploration of Chinese society by Williams. In 1883, it underwent revisions and additions, supplementing new materials, and was translated into Chinese, totaling about 1.1 million Chinese characters.[34]

During the same period, important works on international knowledge written by Chinese authors were Wei Yuan's *Illustrated Treatise on the Maritime Kingdoms* (Hai Guo Tu Zhi) and Xu Jiyu's *A Short Account of the Maritime Circuit* (Ying Huan Zhi Lue).

Wei Yuan (1794–1857), a native of Shaoyang, Hunan, became acquainted with figures such as Lin Zexu and Gong Zizhen after passing the imperial examination in 1822. During the Opium War, he personally participated in the resistance against the British in Zhejiang and aspired to break away from the conservative authorities and pursue reform and self-strengthening. His work, *Illustrated Treatise on the Maritime Kingdoms,* was initially published in a 50-volume edition in early 1843, and an abridged version was later published in Japan from 1854 onward. Xu Jiyu (1795–1873), native of Wutai County in Shanxi, achieved the top score in the imperial examination in 1826. He wrote *A Short Account of the Maritime Circuit,* which included reproductions of world maps printed in the United States. It was officially published in 1848. *Illustrated Treatise on the Maritime Kingdoms* and *A Short Account of the Maritime Circuit* were the efforts of enlightened individuals in post-Opium War China to break free from the imprisonment imposed by the stubborn faction within the backward Qing government, raising the banner of national self-strengthening and independence. They sought to catch up with the tide of industrialization and responded to the call for progress by embracing the strategies proposed by Wei Yuan ("learning from the barbarians to subdue the barbarians") and developing the slogan of Lin Zexu ("learn the superior techniques of the barbarians to subdue the barbarians"). Xu Jiyu, in particular, gained praise from foreigners through his interactions with Westerners and his study of foreign politics and customs. Observing that industrialized or industrializing Western powers such as Britain and the United States not only possessed abundant resources and advanced modern industries but also had established democratic political

[34] Samuel Wells Williams, *The Middle Kingdom*, translated by Chen Ju, Shanghai Ancient Books Publishing House, 2005.

systems distinct from China, he was especially impressed by the democratic republic governed by a president in the United States. Taking the words of the first U.S. president, George Washington, he declared, "to establish a country and pass it on to future generations is a sign of selfishness."[35] Praising Washington, he said, "Expanding the territory for thousands of miles without usurping the throne or passing it on to his descendants, he established the system of popular election, which was almost like making the world a public domain, overturning the traditional notions of the three dynasties."[36]

In Volume Nine of *A Short Account of the Maritime Circuit,* it praised that after the founding of the United States of America, "they did not establish noble titles, nor did they follow the rules of the old regimes. Public affairs were entrusted to public discussions, creating a unique system never seen before. How remarkable it is! Among the outstanding figures of the East and West, who can surpass Washington?" Washington opposed lifelong rule and advocated for a maximum of two terms as president through elections held every four years. This earned him the admiration of the Chinese people, who had lived under despotic feudal systems for thousands of years. Xu Jiyu's statements, however, offended his political enemies, and he was stripped of his position. Nevertheless, his works had already been appreciated by individuals in Japan, Europe, and America, with Japan even having printed editions of his book as early as 1861. In 1865, Xu Jiyu was in charge of the language school Tongwenguan, and in 1867, Tongwenguan adopted his book as a geography textbook. During that year, American envoy Anson Burlingame brought a statue of Washington, and a presentation ceremony was held. In his response, Xu Jiyu declared that Washington had displayed astounding abilities in laying the foundation of the United States, making him a model and mentor for all of humanity.

[35] Xu Jiyu: *Yinghuan kaolue*, in *The Hundred Types of Qing Dynasty Manuscripts*, 40, Taipei, Wenhai Publishing House, pp. 209–210.

[36] Ibid.

### *Establishing Western-Style Schools and Updating the Structure of Knowledge*

Unlike the emerging Western-style schools in nineteenth-century China, traditional Chinese education for over a thousand years was closely tied to the imperial examination system and focused on teaching Confucianism and traditional cultural and historical knowledge. Confucian education was provided in government-sponsored schools at the prefectural, county, and local levels, with the addition of academies (shuyuan) promoted in 1733 to supplement the official schools and expand the number of educated students, although the overall number remained limited. The curriculum mainly revolved around the Four Books and Five Classics, with a particular emphasis on the Four Books and the Cheng-Zhu school of Neo-Confucianism. The prevalence of private schools and village schools, especially in prosperous coastal regions, increased the chances of basic education for urban and rural students. However, by the nineteenth century, all these educational institutions were isolated from the global currents and had become outdated, unable to meet the needs of the changing times.

London Missionary Society sent missionaries to China who engaged in preaching, publishing, teaching, and providing medical services. The first free school established to educate Chinese children was the Anglo-Chinese College, funded by Robert Morrison and overseen by the London Missionary Society. It officially opened in December 1818 in Malacca. Following the British model, the school offered a curriculum spanning primary to secondary education, using both Chinese and English as the medium of instruction. The primary classes were taught in Hokkien (Fujian dialect), while the higher grades used English. Malacca, where the school was located, had a Chinese printing press, facilitating the production of teaching materials. The faculty mainly consisted of London Missionary Society missionaries, with the aim of promoting Christian doctrines and fostering the study of Chinese and Western literature. Prominent individuals who served at the school included William Milne, James Humphreys, Samuel Kidd David, David Collie, Jacob Tomlin, John Evans, and James Legge. During James Legge's tenure as the seventh principal (1840–1843), the college relocated to Hong Kong, after having operated in Malacca for 25 years. By 1831, the college had established eight Chinese primary schools with 200 students and four Malay-medium primary schools. In 1835, the secondary school had

70 students. Additionally, the college later opened schools for Chinese girls. The curriculum included the Chinese *San Zi Jing* as a primer, and subjects covered Chinese, English, mathematics, astronomy, geography, and ethics. Students were placed in different classes based on their knowledge level; thus, textbooks varied. After relocating to Hong Kong, the college ceased operations in 1856, and the original site was converted into the Central School.

The Morrison School in Macau was established earlier than the Anglo-Chinese College in Hong Kong. After Robert Morrison's passing in Macau, foreign businessmen and missionaries in Guangzhou and Macau formed the Morrison Education Society to support the primary students recruited by the missionaries. One of these schools was a girls' school founded by Mrs. Karl Friedrich August Gützlaff in 1834, which later expanded to include a boys' school. On November 4, 1839, the Morrison School officially opened with five students under the leadership of Samuel Robbins Brown, a missionary from the American Presbyterian Mission recommended by Yale University. In 1842, the school relocated to Hong Kong, and in 1847, Brown returned to the United States, taking three Chinese students, Rong Hong, Wong Koon, and Wong Shing, to pursue their high school education in the United States. These were among the first group of students from a Western-style school to study abroad. The school ceased operations in 1850. The curriculum offered at the time included Chinese, English, arithmetic, algebra, geometry, astronomy, geography, physiology, chemistry, music, and physical education, reflecting an educational model with the three pillars of intellectual, physical, and aesthetic development. Moral education was imparted through religious courses as well as Chinese and English language classes. The Morrison School was the first school on Chinese soil established according to the teaching and management model provided by American churches, and its curriculum was equivalent to primary and junior high school levels in later progressive schools. The number of individuals who received this type of education was very limited.

The Western-style new schools were first established in the treaty ports starting from 1843. They were initially initiated by Protestant Christian churches, followed by Catholic churches, gradually increasing in number. As they were run by foreigners in a Western manner, they were referred to as "yang xue tang" by the Chinese. These schools were established along with the footsteps of missionaries, first appearing in the treaty ports and later penetrating into the inland regions, becoming a means of missionary work. During the period before 1860, there were more girls' schools than boys' schools in the early stages of the "yang xue tang" establishment. This was due to the fact that foreign missionaries established their presence in the treaty ports through the use of gunboats, and the local people inevitably had biases and suspicions, making it difficult to gain their trust. Boys already had educational opportunities, and people were not willing to easily send their sons to receive education from foreigners, even if it was free or included board and lodging. As for girls, they had no access to education, but under the efforts of foreign missionaries, several girls could often be recruited. However, they were also subject to public criticism and often faced forced withdrawals or hesitations to attend school in places like Guangzhou, Ningbo, and Xiamen. The earliest girls' school was established in Ningbo in 1844 by Mrs. Aldersay of the Oriental Female Education Society of Britain. Despite widespread rumors, Mrs. Aldersay visited the households near the school, not only waiving the boarding fees but also providing assistance in the daily lives of the enrolled girls. Only a few students were recruited at first, but by 1845, there were 15 students. Initially, the duration of girls' education was calculated on a day-to-day basis. It wasn't until the Second Christian Education Convention held in Shanghai in 1896 that a maximum limit of 8 years was established. Mrs. Aldersay's girls' school merged with the Presbyterian Girls' School in Ningbo in 1857. After the establishment of the Fuzhou Boys' School in 1848, the first girls' school in Fuzhou was founded in 1850. The most famous school was the boys' boarding school founded by American Board missionary Justus Doolittle in 1853, which later developed into Foochow College. In 1854, Mrs. Doolittle established a girls' boarding school, which later became the renowned Wenshan Girls' School. Presbyterian missionary Andrew Patton Happer first established the Guangzhou Boys' Day School in 1850 (formerly known as the Macau Boys' Boarding School established in 1845), followed by the establishment of the Guangzhou Girls' Day School in 1853.

After the establishment of Huai'en Elementary School in Shanghai in 1844, various Christian missions also started schools in Shanghai. In 1850, the Anglican Church established Yinghua School, and the American Baptist Churches founded a girls' school. In 1851, the Presbyterian Mission established Biwen Girls' School (later renamed Shanghai Ninth High School), and American female missionary Emma Jones opened Wenji Girls' School (later known as St. Mary's School). John M.W. Farnham of the American Northern Presbyterian Mission founded a boys' school in 1860 and a girls' school the following year, which later became the renowned Lowrie Institute. Various Catholic religious orders from France also established schools in Shanghai, including St. Joseph School in 1847, St. Louis School in 1850, and Xuhui Girls' School (later renamed Shanghai Fourth High School) in 1855. St. Ignatius College for Chinese Christian students, founded by Angelo Zottoli in late 1849 in Zikawei, Shanghai, was the most famous foreign-style school at that time. In terms of establishing girls' schools, Shanghai, like Hong Kong, took the lead on the mainland.

The establishment of foreign-style schools, recognized by society as a progressive educational model, occurred after the opening of China in 1860, when foreign legations were stationed in Beijing. These schools were capable of meeting the new social demands and nurturing talent for customs, foreign trade, banking, industry, road construction, telegraphy, ports, and missionary work. They demonstrated their developmental potential during the period from 1860 to 1902, making significant progress in terms of school size, educational levels, and management approaches. Successive batches of these schools, reaching the level of secondary education, were established along the coast and riverside, with some even claiming to be equivalent to Western universities.

Among the leading schools were:

Bridgeman Girls' College in 1864, renamed Peking Bridgman Girls' High School in 1895, was the earliest women's school in North China. In the autumn of 1864, Calvin Wilson Mateer, an American Presbyterian missionary, opened an elementary school for boys (Mengyang Educational Society) in Dengzhou, Shandong, later renamed Tengchow College in 1872. In 1865, the first fully fledged secondary school under the Christian Church, the Chongshi School, was established by William A.P. Martin of the American Presbyterian Mission and renamed Chongshi High School in 1885. In 1879, St. John's College was founded by combining two preexisting Anglican colleges in Shanghai by Joseph Schereschewsky.

The foreign-style schools not only provided education for girls and individuals with disabilities (establishing schools for the deaf and blind) but also imparted scientific and international knowledge. They emphasized social practice, including participation in social surveys, engaging in mechanical operations, and observing photoelectric effects. These schools focused on providing comprehensive education in morality, intellect, and physicality, which allowed them to grow and eventually replace the old-style private schools and government schools.

### *The Birth and Development of Textbooks*

Educational textbooks emerged following the rise of Western-style schools. The earliest of these schools, the Morrison Education Society, offered English classes in subjects such as geography, arithmetic, algebra, chemistry, music, and composition, using textbooks from England. Chinese classes utilized the *Four Books* and *Five Classics* and were taught by Chinese teachers. In 1856, James Legge of the Anglo-Chinese College published a textbook called *Graduated Reading: Comprising a Circle of Knowledge in 200 Lessons,* which spanned 55 pages, and covered topics related to humanity, governance, trade, and more. The book was presented in a bilingual format, with English at the top and Chinese at the bottom. In 1859, only the Chinese text was included when it was reprinted in Guangzhou. In 1857, Wilhelm Lobscheid compiled the *Thousand Character Classic* in Hong Kong. Later, missionaries in Guangzhou, Shanghai, Ningbo, and other places created several scientific books, which served as early textbooks introducing Western scientific and technological knowledge. These textbooks were widely used in various church-run schools. These publications differed significantly from traditional Chinese children's readings. The new books introduced contemporary scientific and international knowledge, connecting education with social realities and the natural environment. Teachers employed effective teaching methods, guiding students from the shallow to the deep, in contrast to the rote memorization common in private schools, where students recited texts without understanding their meaning.

After more than 20 years of establishing Western-style schools in the treaty ports, Christian missionary schools had grown to around 350 by 1876. However, there was no unified arrangement for curriculum and textbooks among these schools, making it difficult to establish a common standard of education. In May 1877, during a national conference held in Shanghai, Christian missionaries formed a "School and Text-books Series Committee," chaired by William A.P. Martin, with committee members including William Cooper, Calvin Wilson Mateer, Young John Allen, Rudolf Christian Friedrich Lechler, and John Fryer. The Committee embarked on developing a set of primary textbooks, with John Fryer serving as the chief editor, and a set of advanced books, with Young John Allen as the chief editor. The goal was to combine religious and scientific principles and connect with Chinese society, ensuring that students from mission and government schools could use these textbooks. In May

1890, during the second committee conference held in Shanghai, it was decided to establish the Educational Association of China. John Fryer was appointed as the general editor and secretary, responsible for compiling, reviewing, and publishing textbooks for various subjects used in schools.

John Fryer (1839–1928) served as the general editor from 1879 until his resignation in 1896. He dedicated nearly 20 years to the compilation and revision of educational textbooks. By 1890, the School and Textbooks Series Committee had approved and printed over 98 different kinds of textbooks, with more than 30,000 copies in circulation. It had also developed over 80 new titles, of which John Fryer had compiled and translated 42 textbooks for various subjects.

In 1895, a new trend emerged in textbook editing, where Chinese individuals began to compile their own textbooks and use them for teaching in modern schools. That year, Chen Qiu (1851–1904) from Rui'an in Zhejiang Province introduced his self-compiled textbook called *Liji Jiaojing* at his privately run Liji School. Modeled after the *San Zi Jing*, the book introduced new knowledge from various subjects, covering 36 topics, including Chinese and Western studies, machinery, military training, current affairs, concessions, and religions. In 1897, Chen Maozhi and others at the Nanyang Public School in Shanghai edited the enlightenment textbooks, which students in the primary school division used. They also published books such as *Calculation Textbook* and *Object and Calculation Textbook*, following the style of British and American textbooks, printed with movable type but without illustrations. The texts were written in parallel couplets, marking the beginning of a new era in compiling Chinese self-authored primary school textbooks.

The turning point in the production of new-style textbooks began in 1901. That year, Nanyang Public School published Zhu Shuren's compilation of enlightenment textbooks in three volumes, following the format of English and American textbooks but without illustrations. In 1902, Yu Fu established the Wenming Publishing House in Shanghai and lithographed the seven-volume enlightenment textbooks in regular script, accompanied by illustrations. This set of textbooks presented a completely new appearance. It became the most comprehensive set of primary school textbooks at the time, gaining popularity in the field of primary education for five to six years. In conjunction with the publication of school regulations, Wenming Publishing House introduced various compiled textbooks under the general title of "Scientific Encyclopedia." This series of books reflected the spirit of self-compiled textbooks by

Chinese educators and also included a few translated textbooks from Japan. The range of subjects was extensive, and these textbooks were widely adopted, meeting the needs of new-style schools established in various regions. As a result, Wenming Publishing House quickly became the most comprehensive publisher of textbooks at the time. In 1902, the Commercial Press, established in 1897, established a compilation office with Zhang Yuanji from Nanyang Public School Translators' Institute as its director. Zhang, along with Jiang Weiqiao, Gao Fengqian, and Zhuang Yu, compiled textbooks following the style of Japanese textbooks and created the "Latest Textbooks." This series was divided into two sets: elementary and higher elementary school textbooks. Later, they also compiled the "Latest Secondary School Textbooks," comprising seven volumes for elementary school, nine volumes for higher elementary school, and eleven volumes for secondary school. This series of textbooks also included "Botany" by American author Winfield Scott Sweet.

With the increase in students studying in Japan and the presence of a considerable number of Japanese educators teaching in China, the trend of translating Japanese textbooks emerged in Shanghai. Three notable figures played significant roles in this trend. Firstly, Fan Bingqing, a member of the East Asian Literature Society, established the Education World Publishing House in 1901. He translated various secondary school textbooks such as *Modern Natural History Textbook* and *Modern Chemistry Textbook*. Secondly, Wu Qisun of Wenming Publishing House translated numerous Japanese textbooks and also translated works in the natural sciences such as *General History of the World* and *Revised World Geography*. Thirdly, the Huiwen Study Society, established in 1903 with Fan Diji as the main compiler, published 100 editions of secondary school and college textbooks in lithographed paperback. These textbooks covered a wide range of subjects and were collectively referred to as the "General Encyclopedia." Among them, there were 42 titles in the social sciences, 37 titles in the natural sciences, and 21 titles in applied technology. Books on politics, law, geography, biology, chemistry, agriculture, and forestry were also included. Besides Shanghai, publishing institutions such as the Tianjin Beiyang Official Press and the Wuchang Hubei Official Press also introduced various Japanese-translated textbooks. These textbooks, along with teaching reference books, filled the new-style schools in coastal provinces, surpassing the 110 Western-style textbooks approved and adopted by the School and Text-books Series Committee in previous years. In the past decade, secondary school textbooks reached a saturation point. The dominance of church schools in secondary education had been completely disrupted by the wave of compiling and printing textbooks brought about by educational system reforms.

## 8.4 The Westernization Movement and China's Modernization

### *The Bureau of Manufacture and the Translation Institute*

Lin Zexu proposed learning new knowledge and technologies from the West after 1839. However, the conservative forces aligned with the ruling Qing dynasty responded with criticism, resistance, and even suppression. It was only after the two Opium Wars, particularly the Taiping Rebellion in 1850, which struck at the heart of the Qing government's control in the Yangtze River region, that the high-level decision-makers were forced to lean toward adopting advanced naval and artillery technologies to consolidate their power.

Zeng Guofan, the Viceroy of Guangdong and Guangxi province, supported the Hunan Army and the Huai Army in using Western firearms and cannons to combat the Taiping armies, who had also acquired Western firearms and cannons from foreigners in Shanghai, Ningbo, and other places. Starting from 1861, Zeng Guofan began to purchase American steamships and converted them into training vessels. He established an arsenal in Anqing and laid the foundation for a modern arms industry. He objectively recognized that Western technology far surpassed China's and that shipbuilding and military training required knowledge in disciplines such as cartography, mathematics, astronomy, hydrography, manufacturing, and machinery.[37] With his support, in April 1862, Li Hongzhang, the leader of the Huai Army, swiftly expanded his forces to a size of 10,000 Western firearms. This was done in order to confront the forces of the Taiping Heavenly Kingdom and advance up the Yangtze River to capture its capital, Nanjing.

In 1865, Li Hongzhang purchased the largest foreign machinery factory in Yangjingbang, Shanghai, capable of building both large and small ships, cannons, and firearm components. He established the Jiangnan Arsenal and also had machinery ordered from the United States transported to Shanghai and incorporated into the arsenal. In 1866,

[37] *The Complete Works of Tseng Wenzheng*, vol. 4.

Shen Baozhen established the Fujian Shipbuilding Bureau in Fuzhou, followed by the official opening of the Tianjin Machinery Manufacturing Bureau the following year. Together with the Jinling Machinery Bureau, they formed the four pillars of the emerging modern machinery industry. The ships and gunboats produced by the Jiangnan Arsenal and the Fujian Shipbuilding Bureau were the earliest examples of modern ironclad warships in Chinese history. The firearms they manufactured were among the first mass-produced and large-scale military products. Although China started relatively late compared to the highly developed shipbuilding industry in Britain, China's progress was not considered too late.

In addition to shipbuilding and artillery production, the Jiangnan Arsenal had the foresight to prioritize talent training and the dissemination of technology. Recognizing the importance of talent development and further advancement of mechanical products, they established the Translation Bureau in May 1868 based on the recommendations of Xu Shou, Hua Hengfang, and others. In 1870, they also established the Jiangnan Ship Training Bureau, followed by the establishment of the Artillery Training School in 1874. This comprehensive approach encompassed the entire process, from manufacturing and development to utilization and research, demonstrating their visionary planning.

Establishing a translation bureau promptly translating books on advanced scientific and technological knowledge, encompassing both technical applications and theoretical research, ensured that the contributions of the Jiangnan Arsenal to the movement of new knowledge were not restricted by the development and manufacturing of products. More importantly, the pioneers of that time had already recognized the groundbreaking significance of disciplinary construction and were dedicated to translating academic works in various natural sciences, with particular emphasis on acoustics, optics, chemistry, and electricity. Xu Shou (1818–1884), one of the initiators of the translation center, was a native of Wuxi, Jiangsu Province. He and his younger son Xu Jianyin (1845–1901) were involved in manufacturing the wooden ship "Huanghu" at the Anqing Arsenal. In March 1867, they joined the Jiangnan Arsenal and assisted in shipbuilding. Xu Shou once presented proposals to the authorities regarding coal smelting, self-manufacturing of cannons, training of navies, and translation of Western books. The translation of Western books involved collaboration between Chinese and Western individuals, with English-speaking missionaries proficient in Chinese serving as translators and Chinese scientists assisting in writing. At that time, there were no specialized English-Chinese dictionaries, so translating scientific and technological works not only involved specialized knowledge but also required the establishment of terminology. Without the participation of Chinese professionals, it would have been difficult to carry out this work. Therefore, the translation of Western books was not only initiated by the first group of Chinese scientists who had a grasp of Western scientific progress but also a task that could only be accomplished under their impetus. This work was initially undertaken by mathematician Li Shanlan (1811–1882) in Shanghai. He collaborated with Alexander Wylie, Joseph Edkins, and others to translate many Western mathematical works into Chinese, including *Elements of Analytical Geometry and the Differential and Integral Calculus* by Elias Loomis, Augustus De Morgan's *Elements of Algebra*, and the last nine volumes of *Euclid's Elements*, the first six volumes of which having been rendered into Chinese by Matteo Ricci and Xu Guangqi in 1607. However, in 1865, Li Shanlan accepted a position as a mathematics instructor in Tongwen Guan, which led Xu Shou and Hua Hengfang (1833–1902) to take over. They collaborated with Alexander Wylie, John Fryer, and Daniel J. Mac Gowan to translate Western books.

The translated books include *Manual of the Steam Engine* (interpreted by Alexander Wylie, translated by Xu Shou, 9 volumes), *A Catechism of the Steam Engine* (interpreted by John Fryer, translated by Xu Jianyin, 12 volumes with an appendix), *Practical Geometry* (interpreted by John Fryer, translated by Xu Jianyin), and *Coal and Coal-Mining*(interpreted by John Fryer, translated by Wang Dejun, 12 volumes). These books were officially printed and published in woodblock format in 1871, Xu Shou collaborated with John Fryer in translating specialized books on chemistry, technology, metallurgy, and forensic medicine. They translated 25 books, with a particular emphasis on chemistry books, which had the widest circulation. In 1871, Xu Shou translated David Ames Wells' textbook *Principles and Applications of Chemistry*. Once the Chinese translation of this book was released, it became popular for three to four decades.

Xu Jianyin, the second son of Xu Shou, collaborated with John Fryer in translating 12 technical and military books, including *An Introduction to Practical Chemistry* by John Eddowes Bowman the Younger in 1871 and *Sound* by British acoustics expert John Tyndall.

Wang Dejun was an engineering and technical expert proficient in modern ship cannons. His expertise was highly respected even by foreign experts. He translated *Coal and Coal-Mining* by Warington Wilkinson Smyth with John Fryer and *The China Sea Directory* by J.W. King with Karl Traugott Kreyer.

Mathematician Hua Hengfang collaborated with Daniel J. Mac Gowan in translating the *Manual of Mineralogy* by James Dwight Dana (1813–1895) and *Principle of Geology* by Charles Lyell (1797–1875), the founder of modern geology. The book explores the formation and evolution of the Earth and introduces the concept of biological evolution to China for the first time. Hua Hengfang left the Jiangnan Manufacturing Bureau in 1876. Between 1873 and 1888, the translation of six works on mathematics, physics, and coastal defense collaborated by Hua and Fryer was published by the Jiangnan Arsenal. In 1897, Hua Hengfang published a 10-volume work titled *Jueyi Shuxue* based on the research of Thomas Galloway(1796–1851) and R.E. Anderson, introducing probability theory to China for the first time.

The works of the Translation Bureau of Jiangnan Arsenal covered coastal defense, naval forces, metallurgy, astronomy, mathematics, chemistry, electricity, optics, military systems, manufacturing techniques, military engineering, medicine, public law, finance, and agriculture. The breadth of subjects covered placed the bureau in a leading position in the publishing and scientific communities of the time. It published 238 books over the years, with the majority published before 1900. By the end of the nineteenth century, the introduction of Western knowledge primarily focused on applied technology and natural sciences, accounting for 65% of the total number of translated books. Among the 360 books on science and technology, the books translated and published by Jiangnan Arsenal's Translation Bureau accounted for 66%.[38] This demonstrates the significant role of the bureau in introducing Western knowledge during the nineteenth century.

## *The Earliest Scientific Center in China*

On June 22, 1876, the Shanghai Polytechnic (格致书院) opened in Shanghai. The board of directors consisted of five members, four Westerners. They included Walter Henry Medhurst, the British Consul in Shanghai, Francis Blackwell Forbes of the American trading firm Forbes &

[38] Shen Fuwei, *Western Culture and China (1793–2000)*, pp. 87–91.

Co., as well as Alexander Wylie and John Fryer. The Chinese representative on the board was Tang Tingshu, the General Manager of the newly formed China Merchants' Steam Navigation Company in 1872. By June 1874, the board of directors expanded to eight members, with Tang Tingshu, Xu Shou, Wang Ronghe, and Xu Jianyin joining as Chinese directors alongside the four Western directors. The Academy officially raised funds and collected books and equipment. Eight directors in London also contributed to the establishment of the Shanghai Polytechnic. Machinery and equipment intended for replacement by the British Museum of Science arrived in Shanghai in 1875, laying the foundation for the academy, and a complete set of chemical instruments was donated by the King of Belgium. The academy established a museum room to exhibit collections and built a library and reading room that were open to the public.

The modern scientific experimentation and training center initiated by the Shanghai Polytechnic was born in China's largest foreign trade base and port city. It was an important step for China to catch up with industrialized nations in the field of science. The Shanghai Polytechnic's attached museum was the first free-of-charge museum in China to be open to the public. It was also the earliest science exhibition hall in China. The library building was also the earliest public library in Shanghai.

At its inception, the Shanghai Polytechnic aimed to emulate Western countries and establish a higher-level technical institute in China. It had John Fryer as the main lecturer. Its enrollment did not reach a significant scale, and it functioned mainly as a venue for periodic scientific lectures. Starting in 1876, John Fryer compiled and printed a monthly scientific magazine called *Gezhi Huibian,* which became very popular. The initial print run was 1000 copies, gradually increasing to 3000 copies per issue, and it received widespread acclaim from various sectors of society.

In 1885, Wang Tao (1828–1897), who had traveled to Western Europe and Japan to give lectures, was appointed as the head of the Shanghai Polytechnic, overseeing its daily operations. Under his advocacy, starting, quarterly courses were held for nine consecutive years, from 1886 to 1894. For each course, the examiners included high-ranking officials of the Self-Strengthening Movement and John Fryer. The special courses held from 1889 to 1893 had even higher-ranking examiners, with the spring course being assigned by the Beiyang Minister and the autumn course by the Nanyang Minister. With the support of the Nanyang Minister and the Beiyang Minister, the Shanghai Polytechnic exclusively organized regular courses and examinations focused on current affairs and policy discussions. These examinations were different from the traditional imperial examinations in terms of the topics, candidates, and channels for recommending talents.

Exam candidates were required to provide appropriate commentary on the introduction and assimilation of Western learning, economic development, political reform, strengthening national defense, and handling diplomatic affairs. They needed to carefully weigh the pros and cons, present insightful solutions, and provide guidance on the path to resolution. Through these examinations, the Polytechnic awarded the top performers with superexcellent, special, or first-class rankings, along with material rewards and support. Wang Tao compiled the examination papers and evaluators' comments each year, making them public. Many exceptional essays gained widespread acclaim, and publishers even compiled these reform-oriented essays into books, such as *Strategies for National Prosperity,* which circulated widely.

The examination system of the Polytechnic openly challenged the long-standing problems of the traditional imperial examination system. Wang Tao boldly advocated for a new approach to talent selection, blending Eastern and Western influences. He established a new form of education with a grassroots approach, using Shanghai as its base. This approach aimed at reforming the imperial examination system and laid the groundwork for the subsequent reforms during the Hundred Days' Reform in 1898. It included abolishing the formal writing style based on the eight-legged essay and introducing policy discussions as a means of selecting candidates, as well as opening special economic courses and promoting practical learning. Wang Tao's efforts provided a precedent for these reforms.

### *Tongwen Guan: The Precursor to the University*

After Beijing was forced to open up to the outside world in 1860, in order to meet the demand for diplomatic interpreters, Prince Yixin proposed the selection and training of talented children under the age of thirteen or fourteen from the Manchurian sons of the Eight Banners in each province to learn the languages of England, France, and the United States. On July 11, 1862, Tongwen Guan was officially established. The first department to be opened was the English department, enrolling ten Manchu children.

In the following year, the French and Russian departments were established, each with ten students. Soon after, similar colleges were established in Shanghai and Guangzhou for teaching Western languages.

At the beginning of the establishment of the Beijing Tongwen Guan, students were recruited through recommendation methods. In 1867, the recruitment method was changed to examination-based. From 1870 onward, batches of students were admitted every seven to eight years, and the selection was based on academic performance regardless of ethnicity. By 1885, a total of 108 individuals had been admitted through the examination process. From 1868 to 1899, a total of five batches of 74 students were recommended by the Shanghai and Guangzhou Tongwen Guan. Initially, the Beijing Tongwen Guan focused on the study of Western languages and had a very narrow curriculum. In January 1867, Prince Yixin proposed the establishment of an astronomy and mathematics institution to introduce Western astronomy and to train officials for the Imperial Academy. However, this proposal faced opposition. It wasn't until 1869 when Li Shanlan was hired and appointed as the mathematics instructor, that the Astronomy and Mathematics institution could officially enroll students and commence teaching. In Beijing at that time, the idea of hiring foreigners as teachers was not widely accepted. However, the reform of teaching content and knowledge structure was the trend. In comparison to the various Western-style schools along the southeast coast, the mere language training offered by the Beijing Tongwen Guan was no longer sufficient to meet the societal demands. It neither supplemented officials at various levels like the imperial examinations nor cultivated individuals with new knowledge like the Western-style schools in other regions. Consequently, even though they appointed Willian Alexander Parsons Martin, as the chief instructor of the Tongwen Guan in November 1869, the number of students at the Tongwen Guan did not increase but instead declined.

One distinctive aspect of the Beijing Tongwen Guan's work was the collaborative translation efforts of teachers and students during the teaching process. The first publication of the Tongwen Guan was Martin's translation of *Elements of International Law*. This book had been printed in 1864, before Martin's appointment. In that same year, the Ministry of Foreign Affairs, based on the principles of maritime autonomy for each country outlined in *Elements of International Law*, successfully negotiated the transfer of a Danish merchant ship seized by Prussia within Chinese territorial waters. This greatly impressed Prince Yixin. As a result, 300 copies of *Elements of International Law* were printed and distributed to various provinces. Soon after, Japanese versions and translations appeared, and in 1870, it was included as a textbook in Japanese universities. *Elements of International Law* became one of the most important legal texts during the Meiji Emperor's reign and played a significant role in implementing reform by Emperor Meiji. After Martin arrived at the Tongwen Guan, he collaborated with Wang Fengzao and Feng Yi to translate Theodore Dwight Woolsey's *Introduction to the Study of International Law* (1878) and Johann Kaspar Bluntschli 's *Das moderne Völkerrecht*. Additionally, he had Wang Fengzao to translate his own English work, *Traces of international law in ancient China* into Chinese. Another influential translated book by the Tongwen Guan was Fawcett's *Political Economy*. This work, advocating the accumulation of wealth through free trade, was the most insightful book on commercial and economic principles before the translation of Adam Smith's *The Wealth of Nations* in 1902. It remained a textbook in many modern schools until the end of the Qing dynasty. The Tongwen Guan played a significant role in translating and disseminating Western knowledge, contributing to the development of new schools and the spread of modern ideas in China.

Compared to the Translation Bureau of the Jiangnan Arsenal, the translated works produced by the Beijing Tongwen Guan were primarily collaborative efforts of students and instructors. These works focused more on political, legal, and literary texts, and their quantity could not match the translated publications produced in Shanghai. However, *Elements of International Law* and *English Grammar* were indeed influential publications from the Tongwen Guan. These works played a significant role in introducing legal education to China and nurturing the first generation of Chinese diplomats, leaving an indelible contribution.

### *The Cradle of Engineering and Technical Talent: National Peiyang University*

The New Knowledge Movement advocated by Lin Zexu, together with the talent development plans of the Translation Bureau of the Jiangnan Arsenal and the Shanghai Polytechnic during the 1870s, along with the continuous improvement of talent cultivation goals at the Beijing Tongwen Guan, inevitably led to the establishment of a comprehensive university with multiple disciplines. After the First Sino-Japanese War in 1895, the urgent need for nurturing engineering and technical talent became even more pressing. On September 19 of that year, Sheng Xuanhuai, the director of the Tianjin Customs, submitted a proposal to establish the Tianjin Peiyang Western School, which was approved by the Guangxu Emperor on October 2. It was renamed National Peiyang University in 1896 and was the first government-run university in modern China where Western science and technology was its main focus.

From its establishment, Peiyang University closely followed the curriculum and teaching content of Harvard University and Yale University in the United States. Charles D. Tenney, an American missionary, was appointed as the president and was responsible for the management of the university. After arriving in China in 1882, Tenney initially worked as a missionary in Shanxi and later founded the Tianjin Anglo-Chinese College and acted as principal in 1886. Upon receiving the appointment as the president of Peiyang University, he closed the Anglo-Chinese College and devoted himself wholeheartedly to the establishment of the university. He recruited a large number of American professors to teach in Tianjin, with classes conducted in English and assignments completed in English. Initially, students were recruited from Tianjin, Shanghai, and Hong Kong, and they enjoyed government funding for their education. Monthly exams were held, and scholarships were awarded to outstanding students. In addition to civil engineering, mining and metallurgy, and mechanical engineering, laws were included to meet the urgent demand for legal professionals by the government. This allowed the engineering-focused university to preliminarily qualify as a comprehensive institution with disciplines in both engineering and law. The curriculum in the first-class section spanned four years, with the first year covering geometry, trigonometry, and physics. The second-year curriculum included driving skills, measuring, mechanics, calculus, physics, and chemistry. The third year covered astronomical engineering, chemistry, and botany. The fourth year covered lithology (geology), geography, zoology, international law, and political economy. Zoology and botany were placed alongside mathematics, physics, and chemistry. Upon graduation, students were able to enter prestigious research institutes at universities such as Harvard and Yale without further examinations. The first graduating class of Peiyang University in 1899 was led by Wang Chonghui, who later became a renowned diplomat. These engineering and law experts nurtured on Chinese soil were already on par with the leading higher education institutions in the United States, which had the world's largest GDP at the time. This demonstrated the extraordinary vision and foresight of Peiyang University. This crucial step marked the beginning of a reform that would ultimately overthrow the traditional examination-based educational system, setting the stage for its true transformation.

Peiyang University did not have a peer institution until Sheng Xuanhuai established Nanyang Public School (predecessor of Shanghai Jiaotong University) in Shanghai in April 1897. Although The Imperial University of Peking was approved as early as July 1898, it was officially opened until 1902.

After sending off its first graduating class, Beiyang University was ravaged by the Eight-Nation Alliance. On July 13, 1900, Tianjin fell, and the German troops occupied the campus, incorporating the area into the German concession. In response, Charles D. Tenney traveled to Berlin to negotiate and received the compensation of 50,000 taels of customs silver. The university was temporarily forced to suspend its operations. In 1902, the university reopened, using the compensation funds to construct a new teaching building. It was officially renamed Peiyang University and offered programs in civil engineering, mining and metallurgy, and law. French and Russian language programs, as well as a normal school, were added. The duration of the law program remained four years, while civil engineering and mining and metallurgy were reduced to three years. It was required that applicants have completed their studies at provincial higher schools (equivalent to high school level) before being eligible for admission. Leveraging his role as the supervisor of Western learning in Zhili, he gradually aligned the curricula and courses of primary schools and middle schools in Zhili, establishing a seamless educational system from primary school to university. This enabled the new schools in Zhili to draw from the educational system and facilities of the United States, putting them on a path of modernization. From 1907 onward, students from Peiyang University became an important reserve force for sending students to study in the United States. In 1913, the university was renamed National Peiyang University. In 1920, based on a proposal by Cai Yuanpei, the president of Peking University, the engineering program of Peking University was merged into Peiyang University, while the law program of Peiyang University was merged into Peking University. As an engineering university, Peiyang University nurtured a successive wave of highly skilled technical experts and managers, contributing to China's modernization efforts.

### *Abolish the Imperial Examination System, Promote the Establishment of Schools*

The introduction of Western learning, by the late nineteenth century, had increasingly taken on a political dimension. In November 1887, foreign residents in Shanghai established the "Society for Diffusion of Christian and General Knowledge among the Chinese," to westernize the thinking of the officials, gentry, and intellectuals who dominated the current state of affairs in China.

Timothy Richard, an English Baptist missionary who arrived in China in 1870 and was recommended by the Commissioner of Customs, Horace G. G. Hurd, joined the Society in October 1891. He soon became the superintendent (later known as the General Secretary). In 1894, the society was renamed the Christian Literature Society for China to further promote Western learning. Through the society, Timothy Richard used the newspaper *The Review of the Times* as their official publication to promote their reform ideas. After the First Sino-Japanese War, it became a popular publication with a circulation of over 20,000 to 30,000 copies.

Following the defeat in the First Sino-Japanese War, China experienced territorial losses and indemnity payments. Timothy Richard actively established connections with influential figures to drive political reforms. He emphasized the need to educate the people, support their livelihoods, ensure their safety, and create a new citizenry. In essence, he called for the government's restructuring and implementing of new policies with the direct involvement of foreigners. Andrew Young John William Allen, published *Zhongdong Zhanji Benmo* in eight volumes in April 1896, exposing the inside story of the war. He also published a two-volume book called *Education in Japan* by Japanese scholar Mori Arinori, serving as a model for advocating educational advancement for national prosperity. The reform proposals of Richard and Allen influenced the policies implemented by Kang Youwei and Liang Qichao during the Hundred Days' Reform. However, the reform movement ultimately failed to achieve its goals and ended.

The abolition of the civil service examination system and the establishment of a new educational system were two sides of the same issue. Amid a nationwide fervor for reform, even the Qing government, which had been reluctant to let go of the examination system, ultimately had to abolish it more thoroughly and promote the establishment of schools. As a prelude to this, Minister of Education Zhang Baixi compiled the

*Approved Regulations of Educational Institutions* in August 1902. The blueprint for these regulations was drawn from "the successful practices of Europe, America, and Japan." The education system followed a unified structure, with students progressing from elementary schools to university, adopting the 20-year continuous education system. The basic educational system was modeled after the Japanese system established in the 33rd year of the Meiji era (1900), while the administrative management system drew inspiration from the French university system. Revised by Zhang Zhidong, this education system, known as the "School System in 1904," was officially promulgated. The new system stipulated that students enter elementary school at the age of 6 (for 5 years), then advance to higher elementary school (for 4 years), secondary school (for 5 years, or secondary industrial school), higher school and university preparatory division (for 3 years), followed by specialized universities and choosing majors (with 8 disciplines, studying for 3–4 years), with a total study period of up to 21 years. After the implementation of this system, new-style schools sprouted up across the country, surpassing the earlier Christian and Catholic schools in terms of growth. However, the schools under this new system still emphasized the study of Confucian classics and regarded Confucianism as the "religion of China." These schools rejected Christianity while venerating Confucianism and simultaneously incorporated Western learning. They were schools caught between tradition and new knowledge, but once such schools emerged, they were bound to deviate from the original goals set by those who believed they could steer and determine their direction. Despite the Qing government's persistence, on September 2, 1905, an edict was issued to abolish the ancient examination system and promote the expansion of schools.

## 8.5 The Theory of Evolution by Natural Selection and the Ideological Emancipation

### *Evolutionary Theory and the Reform Movement*

In the late nineteenth century, China, facing border crises, experienced a series of territorial losses and humiliations through unequal treaties. It not only realized that its goal of self-strengthening was hindered by its technological backwardness but also recognized that the constraints of traditional ideologies prevented groundbreaking reforms. As a result, a group of forward-thinking individuals turned to evolutionary theory, borrowed from the realm of Western learning, as a powerful tool for reform and innovation.

The earliest introduction of evolutionary theory to China was through the joint translation efforts of Mac Gowan and Hua Hengfang, who translated *Principle of Geology* in 1873. The explicit mention of Darwin's theory was made by Zhong Tianwei. In the 1880s, he introduced Darwin's theory of biological evolution and Herbert Spencer's theory of social evolution in his *Gezhishuo*. The text mentioned that Darwin's book published in 1859 discussed the origin of species and the principle of survival of the fittest, asserting that plant and animal species undergo changes and are not fixed since their creation. Unsuitable species gradually disappear, while well-adapted ones persist, following the natural law of the survival of the fittest. This passage represents one of the earliest introductions to the essence of Darwin's theory of evolution, encapsulated in the phrase "survival of the fittest."

The systematic introducer of Darwin's theory of evolution was Yan Fu (1854–1921), a native of Houguan, Fujian province. In 1866, Yan Fu entered the Foochow Arsenal Academy and in 1876, he was sent to the United Kingdom to study naval science. He returned to China in 1879 and taught at the Beiyang Naval Officers' School in Tianjin from 1880. It was not until after the Chinese defeat in the First Sino-Japanese War that Yan Fu became famous. He is celebrated for his translation of *Evolution and Ethics* by the British biologist Thomas Henry Huxley (1825–1895), which was a compilation of lectures Huxley delivered as the Romanes Lecture at Oxford University in 1893. The book was published in 1894. The fervent and captivating style of the lectures, combined with Yan Fu's translation, invoked the principle of "survival of the fittest" in biological evolution to awaken the Chinese people to "struggle against fate" and strive for national salvation. The translation adopted the popular classical style of the Tongcheng School, making it sonorous and powerful. When *Evolution and Ethics* was translated in 1896, it was vigorously promoted by progressive individuals. In December 1897, Yan Fu serialized the translation in his Tianjin newspaper *Guowenbao* for three months, causing a sensation nationwide. Even conservative literati and bureaucratic elites became readers. In April 1898, the translation was published in woodblock print, creating a nationwide sensation.

The introduction and dissemination of *Evolution and Ethics* in China was like a miraculous remedy for a nation desperately seeking survival. Just as Darwin's theory was destined to bury the Christian doctrine of

God's creation, the introduction of *Evolution and Ethics* in China would inevitably serve as a powerful blow to the dominance of Confucianism and traditional thought, becoming a scientific weapon to dismantle the dogmas of Confucius. Thus, Cai Yuanpei ranked Yan Fu as the foremost figure in introducing Western philosophy during the past fifty years (1872–1922).[39] Hu Shi, a leading advocate of the New Literature Movement, found the latest scientific methods in the "skepticism" put forth by Darwin and Huxley. Once *Evolution and Ethics* took root in China, it sparked a revolutionary shift in thinking during the initial decades of the twentieth century, guiding China toward progress amidst tremendous social transformations. Terms such as "competition," "natural selection," "survival of the fittest," and "struggle for existence" gradually integrated into Chinese society, becoming catchphrases for those who had received a modern education and embraced new ideas.

In 1905, Hu Shih recalled in his *Autobiography at Forty* that, "After the publication of *Evolution and Ethics*, it didn't take long to become popular nationwide and even became reading material for secondary school students... What they could understand was merely the significance of the formula 'survival of the fittest' in international politics. After repeated military defeats and the national humiliation of the Boxer Rebellion and the *Treaty of Shimonoseki*, this formula of 'survival of the fittest' was indeed a powerful wake-up call, providing a tremendous stimulus to countless individuals. In just a few years, this idea spread like wildfire, igniting the hearts and blood of many young people... Terms such as 'evolution,' 'competition,' 'elimination,' and 'natural selection' gradually became familiar phrases in newspaper articles and the common parlance of patriotic individuals."

Sun Yat-sen utilized evolution theory as a weapon for revolution, aiming to accelerate the pace of civilization's progress. Li Dazhao, guided by historical materialism, believed that Confucianism was not an eternal and unchanging truth. Evolution theory became the most powerful ideological weapon in overthrowing the Confucian orthodoxy during the May Fourth Movement.

[39] The Last Fifty Years, *Shenbao*, Shanghai, 1923.

### *The Establishment of Western Medicine*

Traditional Chinese medicine, based on texts such as the *Yellow Emperor's Classic of Internal Medicine* and *Treatise on Cold Damage Diseases* from over 2000 years ago, emphasizes pulse diagnosis and treatment based on the theories of the six channels and eight principles. This differs significantly from Western medicine, which analyzes the spatial position and structural morphology of objects from a materialistic perspective. While China has had the Imperial Academy of Medicine, which provided healthcare exclusively to the imperial family, they were distinct from the hospitals that emerged in the West during the Middle Ages. Macau was the first place in China to establish Western-style hospitals, with the creation of the St. Raphael Hospital in 1569, which had 70 beds and provided medical services to the local population as a Catholic hospital. In 1770, the East India Company decided to form a permanent council in Guangzhou, and British doctors began to gradually arrive in Guangdong. However, the Chinese had little trust in Western medicine, and it was only when Chinese individuals were injured in clashes with Europeans that British missionary doctors had the opportunity to treat Chinese patients. Dr. A. Pearson played a key role in changing this situation. He arrived in Macau in April 1805 and taught the smallpox vaccination technique. Over a period of 12 months, thousands of Chinese children were vaccinated by Pearson. The book on smallpox vaccination written by Pearson was translated into Chinese by G.T. Staunton. Several Chinese were specifically trained in the technique of smallpox vaccination. However, it took another century for Western medicine to be promoted and gradually become mainstream in Chinese medicine, which occurred during the First World War when Western medicine found more advocates and successors in China.

After the London Missionary Society sent Robert Morrison to Guangdong for missionary work, Morrison and his assistant opened a clinic in Macau in 1820 to establish closer contact with the Chinese people. In 1827, the LMS sent medical missionary Thomas Richardson Colledge to Macau to join Morrison in opening an ophthalmic hospital, providing eye treatments for the Chinese. Colledge later set up a clinic in Guangzhou as well. Through providing free medical consultations, Colledge connected with several Chinese individuals and appealed to the British and American Christian missions to send more doctors to China, emphasizing the use of medical relief and other methods as alternatives to formal evangelism.

Subsequently, missionary doctors began arriving in Guangzhou and also established clinics in cities like Shanghai and Ningbo to promote Western medical knowledge. The most renowned missionary physicians were Peter Parker (1804–1889), Benjamin Hobson (1816–1873), and John Glasgow Kerr (1824–1901).

Peter Parker was the first American missionary physician to come to China. In November 1835, Parker officially opened the Ophthalmic Hospital in Guangzhou. Parker was highly skilled in medicine and had a reputation for successfully treating patients. Within the first three months of its opening, the hospital received nearly 1200 patients, including 270 women, which was uncommon at the time due to the prevailing customs of gender separation. Parker diligently recorded the diagnoses of various diseases and categorized them, documenting a total of 24 different diseases. In the spring of 1836, the clinic expanded and was renamed "Hospital of Universal Love" (Boji Hospital). When Lin Zexu arrived in Guangzhou for his official post, he wrote to Parker seeking a remedy for opium addiction and also sought treatment for a hernia. Parker provided him with a hernia belt, which was sent to Lin by messenger.

In February 1838, Thomas Richardson Colledge, Peter Parker, Elijah Coleman Bridgman, and John Robert Morrison established the "Medical Missionary Society of China" in Guangzhou. They conducted fundraising activities to support medical clinics and aimed to counteract the negative impression created among the Chinese people by the involvement of foreigners in the opium trade. During the First Opium War, the hospital suspended its operations for over two years and resumed its activities in November 1842. In 1846, he successfully performed surgical procedures using the newly invented method of ether anesthesia. In 1855, Parker transitioned to a diplomatic role, and John Glasgow Kerr took over the management of the Hospital of Universal Love.

Benjamin Hobson followed in Peter Parker's footsteps as a medical missionary in Guangzhou. After arriving in Macau in December 1839, he joined the local Medical Missionary Society. He opened Wo Ai Clinic in 1848 in Guangzhou. In addition to his medical work, Hobson translated various scientific books and published them in Guangzhou. His works include *Summary of Astronomy* (1849), *Treatise on Physiology* (1851), and *Natural Philosophy* (1855). These three books introduced new knowledge in astronomy, anatomy, and natural history to the Chinese intellectual community. *Treatise on Physiology* was the first printed book to explain physiology and medical knowledge based on anatomical principles. Eight anatomical illustrations from the book were displayed outside the office of the Governor-General of Guangdong and Guangxi by Ye Mingchen, which was regarded as a significant event in the scientific community.

After the publication of *Natural Philosophy*, Shanghai Mohai Shuguan reprinted this book along with the *Treatise on Physiology*. Hobson's reputation spread from Guangzhou to Shanghai, influencing the study of Western knowledge by scholars like Xu Shou and Hua Hengfang. He had briefly visited Shanghai, and in February 1857, Hobson returned to Shanghai and took over medical work at the Renji Hospital from William Lockhart. With the assistance of Guan Maocai, he published three Western medical works: *First Lines on the Practice of Surgery in the West* (1857), *Treatise on Midwifery and Diseases of Children* (1858), and *Practice of Medicine and Materia Medica* (1858). These books were printed in a quantity of a thousand copies by the Renji Hospital. The introduction of Western scientific knowledge and medical theories by Benjamin Hobson stimulated the intellectual curiosity of a new generation of Chinese scholars.

John Glasgow Kerr was an American medical missionary. He was born in Ohio and graduated from Jefferson Medical College in Philadelphia in 1847. He arrived in Guangzhou on May 15, 1854, and ran Huiji Dispensary of the Presbyterian Mission and the Canton Hospital. In 1859, Kerr opened Boji Hospital, located on Zhengsha Street in a southern suburb of Guangzhou. In 1860, Kerr opened the second dispensary in Foshan, later becoming the largest hospital in the city. In 1899, Kerr opened the first mental hospital in China. Kerr kept working until his death on August 10, 1901. During his 47-year tenure, approximately 750,000 patients sought medical treatment at his clinic, with 40,000 inpatients and 49,000 surgical operations performed by him. Among these surgeries, 1300 cases involved successful treatment of previously deemed incurable diseases. He was highly acclaimed for his exceptional surgical skills and relentless pursuit of excellence in internal medicine.

Boji Hospital was the most significant Western medical publishing house in the nineteenth century. Starting from 1859, most of its publications were compiled by John Glasgow Kerr. His Chinese colleagues Lin Xiangdong, Kong Qinggao, and He Liaoran assisted him in completing 34 medical, pharmaceutical, and nursing works. Among foreign physicians in China, John Glasgow Kerr was renowned for having translated and authored the most works and possessing extensive knowledge in the field.

John Glasgow Kerr took a significant step in cultivating Western medical talent. In 1866, he established the South China Medical School at Boji Hospital. It introduced the system of independent medical schools dedicated to training medical professionals, which was prevalent in Western countries, to China. With the establishment of the South China Medical School, China witnessed the first medical school attached to a hospital. Assisting John Glasgow Kerr in running the school was Huang Kuan (1828–1879), the first Chinese physician to receive formal medical education in the United Kingdom. Starting in 1862, he joined Boji Hospital and participated in teaching anatomy and surgery. In 1879, the South China Medical School admitted its first female student, marking the beginning of training female doctors in China.

Since the establishment of the South China Medical School, various church hospitals in Shanghai, Dengzhou, Beijing, Tianjin, Xiamen, Suzhou, Hangzhou, and other cities have gradually opened medical schools or medical classes of varying sizes, offering courses and training qualified medical personnel.[40]

The modernization of medical education gradually took shape in the early twentieth century, but before that, there was an important initiative. With the support of Li Hongzhang, the Viceroy's Hospital Medical School was officially established in Tianjin on December 15, 1881. In April 1888, the Medical School was taken over by the Qing government and became the first official medical school in China. In 1894, it expanded and was renamed as the Beiyang Medical School.

In the twentieth century, as Christian universities were established one after another, the medical schools operated by the Church also gained independence and opened up this attractive pathway for training professionals to various sectors of society to expand their enrollment.

From 1908 to 1917, various medical schools were established across the country. These included Hubei Medical College in Wuchang (1908), Hankou Medical College (1908), Tongji German Medical School in Shanghai (1908), North China Union Women's Medical College in Beijing (1908), Shanghai Aurora Medical School (1908), Shandong Qilu Medical College (1909), West China Union Medical School in Chengdu (1909), Fujian Union Medical School (1910), German Medical School in Qingdao (1910), Beijing Medical School (1912), Jiangsu Medical

[40] *Wong ji-ming & Wu Lien Teh,History of Chinese Medicine*, The Tientsin Press, 1932, pp. 219–310.

School (1913), and Mukden Medical College (1917). These schools were established in provincial capitals, major cities, or port cities to expand professional talent training in Western medicine further.

In 1913, the Rockefeller Foundation was established in the United States and immediately sent people to investigate the medical field in China. At that time, 500 individuals in China had received Western medical education. There were 244 hospitals, 446 Western doctors, 112 Western nurses, and 94 Chinese doctors trained in Western medicine. Western medicine in China represented Western countries' medical knowledge and practices, and even the faculties were of Western nationality. However, a small group of qualified Chinese doctors already practiced Western medicine. Although their numbers were limited, they had established a foothold in the Chinese medical community and represented a promising direction for treatment. Despite the need for more effective measures to rapidly expand this group, they held great potential.

### *The Innovation of Drama and Stage Arts*

The history of drama in Western countries spans over two thousand years, and it was introduced to China in the mid-nineteenth century. Shanghai was the first place where Western-style drama was performed. Within the Shanghai International Settlement were numerous tea houses and theaters owned by Chinese individuals, where traditional Chinese operas like Kunqu and Peking opera were performed. Besides organizing dances, sports events, and competitions, foreign residents in Shanghai also began staging plays from 1850 onward. In 1867, the Lánxīn Grand Theater was completed near the Suzhou River and staged 19 Western plays. However, it was destroyed by a fire and later rebuilt. After its official opening in January 1874, the Lánxīn Theater began rehearsing Western classics. The audience was mainly Westerners, with few Chinese.

The rise of drama in China was facilitated by a group of new-minded individuals who studied in Japan and were determined to save the country. In 1903, Chinese students studying in Japan were inspired by Japanese Shinpa plays of social tragedy. They recognized the significant role that French and Japanese dramas played in inspiring patriotic spirit and advocating for national independence movements. In September 1904, Chen Peiren (1874–1933) and Liu Yazi (1887–1958) founded China's first theater magazine, *The Twentieth Century Grand Stage* in Shanghai. The magazine advocated the use of drama to promote social reform and nationalism. However, it was banned after two issues were published.

In 1904, Chen Duxiu (1880–1942) published an article titled *On Opera* under the pen name San'ai in the magazine *New Fiction*. He advocated for "more new plays that promote moral values" and "the adoption of Western methods" while refraining from performing plays involving gods, ghosts, or explicit content. Students from Shanghai's church schools were the first to take the initiative and perform short, politically charged dramas using a Western-style format and incorporating songs, lyrics, and dialogue. Wang Youyou (1888–1937) organized the Wenyou Society, which ventured beyond the confines of schools and publicly performed revolutionary contemporary plays. These new plays departed from the uncivilized aspects of traditional operas and advocated for reform, earning the nickname Wenming xi (civilized drama). Wang Youyou and others, such as Zhu Shuangyun, formed the Enlightenment Amateurs' Theater, which further promoted contemporary dramas as a new genre in Shanghai alongside Peking opera and Kunqu opera.

In the drama movement, two groups left a historical imprint on theatrical reform: the Spring Willow Society, established by Chinese students in Tokyo and the Spring Sun Society, initiated by Chinese students in Shanghai. The Spring Willow Society primarily performed plays with foreign themes. In February 1907, members of the Spring Willow Society, including Li Shutong, Zeng Xiaogu, Wu Wozun, and Xie Hangbai, staged a one-act play of *The Lady of the Camellias* by Alexandre Dumas fils. The second production was *Uncle Tom's Cabin* by Harriet Beecher Stowe. The play celebrated the freeing of black people during the American Civil War, and the members of the Spring Willow Society rehearsed passionately and performed it to critical acclaim. This marked an inspiring beginning for the Chinese to perform Western plays on Western-style stages officially. The performances of the Spring Willow Society encouraged sympathizers of revolution among the students studying abroad. The actions of the Spring Willow Society quickly resonated in Shanghai, and Wang Zhongsheng (approximately 1874–1911) and Ren Tianzhi, established the Spring Sun Society in 1907. They also staged *Uncle Tom's Cabin* in Shanghai theaters and achieved success.

The emergence of Wenming Xi in the lower Yangtze River Delta provided a foundation for the development of drama as a new genre. In 1910, Ren Tianzhi gathered Wang Youyou, Chen Dabei, and others to organize the first professional drama troupe. They toured cities along the Yangtze River, such as Shanghai, Nanjing, and Hankou, gaining a considerable audience.

In 1909, Nankai School in Tianjin also began staging contemporary plays. It established the student drama troupe, with Henrik Ibsen's *An Enemy of the People* being a regular play performed during school anniversaries and graduation ceremonies. Ibsen's social dramas had already gained favor among the emerging drama community before the Xinhai Revolution. In 1914, Lu Jingruo Wu Wozun and Xie Hangbai, who had returned from Japan, established the New Drama Comrades Association in Shanghai and formed the Spring Willow Theater. They performed plays such as *The Lady of the Camellias, East Lynne, Joan Haste, Resurrection,* and *A Doll's House.*

After the May Fourth Movement, drama underwent reforms, breaking away from its previous focus on political propaganda or the lack of scripts. It became a literary genre, adopting Western theatrical styles for artistic creation.

In 1923, Chen Dabei, together with 48 organizations from both domestic and international spheres, established the New Chinese Drama Society in Beijing. They produced *Heroes and Beauties,* a model for school drama groups nationwide. Simultaneously, they abolished the outdated practice of men playing female roles for the first time on stage, promoting male and female actors performing together. Furthermore, Hong Shen, the first Chinese student to study drama abroad, returned after completing his studies in the United States. In 1923, he adapted American playwright O'Neill's famous play *The Emperor Jones* into *Yama Zhao.* This paved the way for subsequent modern plays focusing on social issues, such as *The Puppet Family* and *The Haunted.* Hong Shen also adapted Oscar Wilde's *Lady Windermere's Fan* into *The Young Mistress's Fan.* The play was tremendously popular and highly influential in modern drama development in China.

### *Translation in Literature and the Literary Revolution*

In the early nineteenth century, translation in literature emerged and quickly formed a close bond with vernacular writing close to spoken language and characterized by concise wording. The initial translation of *the Bible* was rendered in Classical Chinese by British missionaries. However, due to traditional constraints and the inadequate linguistic proficiency of the missionaries and their assistants, this version did not capture the attention of scholars and the bureaucratic class. Despite its pioneering role, the scholarly and elaborate translation of the Bible could not be circulated for long without modifications—consequently, new versions with revisions kept appearing.

In church schools, there was a growing demand for vernacular writing that suited the needs of the general public. Thus, alongside the repeated revisions of the scholarly translations, early Bible translations emerged in vernacular Mandarin Chinese. The version using the Southern Mandarin dialect (Nanjing dialect) was published in 1857, with the assistance of a literatus from Nanjing, translated by Medhurst and Schereschewsky. The *New Testament,* based on Beijing dialect (Mandarin), was officially released in 1867. Joseph Schereschewsky, an American Episcopal missionary who participated in the translation of this Mandarin (Beijing dialect) version, independently completed the translation of the *Old Testament* in the Beijing dialect, which was published in 1868. In 1878, the British and Foreign Bible Society published a combined New and Old Testament edition titled *The Holy Bible*. This Mandarin translation of the Bible was most suitable for spreading Christianity in the northern regions. Thus, the first vernacular Mandarin Chinese version of the *Bible* was born.

Through the translation of the Bible, Christian churches continuously expanded their religious propaganda frontiers. Simultaneously, they quietly utilized vernacular or semi-literate styles in textbooks, religious propaganda materials, newspapers, and specialized works to disseminate news and scientific knowledge, promote the Gospel, express political opinions, and incite the masses. As the educated public grew in numbers, a significant amount of written propaganda emerged, gradually forming a grand and sweeping wave in the literary revolution by the early twentieth century.

A group of new-style literati who engaged in translation literature through magazine publications contributed to this trend. Zhou Guisheng and Xu Nianci are two early rising stars in the Shanghai translation scene. They pioneered the practice of translating Western novels directly into vernacular modern Chinese.

In the early history of Western literature translation in China, some translated works that reflected Western countries' social life during the capitalist growth period were successfully spread. These works garnered sympathy, understanding, and anger from the Chinese public, laying the foundation for future translation reform. Among these notable works were two French novels (*La Dame aux Camélias* and *Les Misérables*), one British novel (*David Copperfield*), one American novel (*Uncle Tom's Cabin*), and one Russian novel (*Resurrection*). They were translated by Su Manshu, Lin Shu, and Ma Junwu and marked the beginning of a new era in translation.

### *Vernacular Literature and the Literary Revolution*

The literary revolution was initially proposed under the banner of literary reform. Using vernacular language to translate Western novels became a new phenomenon in China. These Western novels broke away from the traditional episodic structure of Chinese novels and vividly portrayed real life. This was entirely due to the influence of the new literary style. The new style was best suited for expressing new things, ideas, and trends, leading to the proposal for literary reform.

In January 1917, Hu Shih (1892–1963), studying in the United States, published an article titled *A Preliminary Discussion on Literary Reform* in the magazine *New Youth*. He put forward eight proposals for literary reform and advocated for using colloquial language, which became known as vernacular literature. After proposing literary reform, Hu Shih began creating vernacular fiction and poetry and translating Western novels and poems into vernacular Chinese.

At the same time as Hu Shih's "Preliminary Discussion," Chen Duxiu published *On the Literary Revolution* in the magazine *New Youth*. He advocated for three principles: overthrowing the refined and flattery-laden aristocratic literature and constructing a plain and lyrical national literature; defeating the stale and extravagant classical literature and building fresh and sincere realistic literature; overthrowing the obscure and esoteric literati literature; and establishing a clear and accessible social literature. These propositions unified the literary revolution in form and content, putting forward popular and national literature ideas.

New literature emerged alongside the retranslation of European literature. After the May Fourth Movement, various literary groups inspired by French literary salons were established. In Shanghai, there were the Literary Research Society, Creation Society, Crescent Moon Society, and Sun Society, while in Beijing, there was the Unnamed Society. These groups became the backbone of the literary revolution. Their members engaged in creative writing and translating, introducing, and studying European and American literature. The new literature, created in vernacular language, adopted European models in terms of themes, language, and artistic techniques. Initially, it was referred to as "Europeanized literature." The new literature revolutionized the literary scene through vernacular novels, vernacular poetry, dramas and scripts, literary critiques, and author introductions. Western myths, allusions, and many English vocabulary words frequently appeared in vernacular novels, prose, and literary critiques through transliteration, becoming popular terms known to the public. Promoting the national language literature encompassed the standardization of dialects and the reform of Chinese characters, leading to the movement for Chinese character romanization. Learning and studying Western literature became a fundamental subject in many schools. In coastal cities, English-version works such as *The Arabian Nights, Treasure Island, Aesop's Fables,* and *A Tale of Two Cities* by Dickens were directly used as textbooks for reading and examination in elementary and secondary schools.

The new vernacular translations of Western literary works swept away the traces of previous translations where the translators had to speculate on the original meaning and engage in creative reinterpretation. As a result, the translation works of Lin Shu, Bao Tianxiao, and Zhou Shoujuan had new versions. The literary pioneers of the May Fourth Movement, such as Lu Xun, Mao Dun, Zheng Zhenduo, Guo Moruo, and Xu Zhimo, contributed to the creation and translation of literature. In the 1920s, many translators began translating directly from the original versions. Geng Jizhi translated Russian literature, producing the first vernacular translation of Leo Tolstoy's *Resurrection* and notable works by Ivan Turgenev, such as *Fathers and Sons*. Liang Shih-chiu translated Shakespeare's plays, Guo Moruo translated Johann Wolfgang von Goethe and John Galsworthy's works, Li Qingyai translated Guy de Maupassant and Anatole France's works, Fu Lei translated Honoré de Balzac and Romain Rolland's novels, and Pan Jiaxun translated Henrik Ibsen's works. These translators were prominent figures in the literary world after the literary revolution. They played a crucial role in bridging the gap between Chinese and Western literature, serving as a communication bridge.

CHAPTER 9

# Cultural Interactions in China's Reform and Modernization

## 9.1 China's Slow Economic Development Under the Grip of the Western Powers

### *China's Shipping Industry Controlled by the Western Powers*

In the late Qing Dynasty, China's shipping industry was controlled by the Western powers which had a catastrophic influence on China's foreign trade and economic development. The world shipping industry underwent revolutionary changes after the introduction of steamships in ocean shipping. As an indispensable part of the Pacific shipping line, China was coveted by the British Empire from the very beginning, which was the leading maritime power at that time. In 1835 British company Jardine & J. Matheson Company attempted to launch its first steamer passenger service between Macao and Canton with the steamer *Jardine,* but in vain. The *Jardine* was blocked from entering the port. After the Opium War, China had to open five treaty ports for foreign trade and residence. Then, foreign ships began to appear in these port cities, among which most were British ships. In 1827, the British Peninsular and Oriental Navigation Company was established, and the company opened the Chinese-British shipping service in 1834. In 1845, the company's London to Hong Kong shipping line was extended to Shanghai. In 1853, the *Confucius* steamship of the American Russell, Sturgis and Company made its maiden voyage to Shanghai. In 1860, the Blue Funnel Navigation Company



F. Shen, *Cultural Communications Between China and The Outside World Throughout History,*
https://doi.org/10.1007/978-981-97-4696-5_9

opened its shipping line to Shanghai, and the French Mail Company began its voyage to China.

Since the 1860s, Western countries have adopted protective and incentive policies for their shipping industry, which helped reduce ship manufacturing and operating costs year by year. Ocean ships of high speed and high-tonnage gradually came to the forefront of the global ocean shipping industry. At first, the Chinese called this type of ship steamship and later paddle ship. Through the unequal treaties and the most-favored-nation treatment, Britain, France, and the United States seized the privilege of navigating in Chinese coastal and inland waters and operating the shipping industry. Countries like Denmark, the Netherlands, Spain, Belgium, Italy, Austria, Japan, Sweden, Russia, Germany, Peru, Brazil, and Portugal followed suit and achieved the same privilege.

After the establishment of the British-controlled Chinese Maritime Customs Service in 1855, the administration of China's coastal ports fell into the hands of the Western consulates and customs, as did China's rights in the shipping industry embodied by the customs duties, tonnage taxes, and inland taxes. Taking the pilotage as an example, the consular groups of Western countries issued their regulations on pilots and set up their pilot committees. The British, American, and French consular groups elected three to five captains and one naval officer as the pilots. According to the 1858 *British Treaty of Tian Jin* and the 1843 *General Regulations for British Trades at China's Five Ports*, the British Empire was empowered to manage China's taxation and navigation issues. Under the leadership of the British Inspectorate-general of Customs, foreign customs commissioners administered China's waterways and treaty ports. Meanwhile, the Qing government was forced to allow foreigners to serve as pilots in Chinese waters. In 1861 the Qing government issued the *Regulations Governing Trade on the Yangtze River*. These regulations allowed foreign merchants to navigate between Shanghai and the inland ports along the Yangtze River. The steamer *Fire Dart* of the American Augustine Heard Company made the maiden voyage along the Yangtze River, which marked the beginning of foreign ships' entry into China's inland waters. In 1867 Inspector General Robert Hart issued *the General Regulations of Pilotage* which shifted the pilotage rights from China to the Western powers. In his September report to the Tsungli Yamen (Office for General Management) in 1877, Hart alleged there were 203 pilots in the treaty ports, including 103 Chinese, 40 British, 35 American, 10 Portuguese, 6 Danish, 6 Swedes, 2 Dutch, and 1 Austrian. Foreign pilots accounted for almost half. Gradually foreign pilots got the upper hand in numbers. In Shanghai's port, there were 35 foreign pilots and only 2 Chinese pilots. There were all foreign pilots in ports like Tian Jin, Qinhuangdao, Xiamen, Shantou, etc. Foreign pilots from eleven countries worked around Chinese ports, including those along the Yangtze River. Among the 175 Yangtze pilots, 38 foreign pilots worked in the middle and lower reaches of the Yangtze River and the Yangtze Delta. However, in the dangerous upper Yangtze, there were all Chinese pilots. This foreign-dominated situation did not change until 1929 when China resumed its pilotage rights. Even in 1948, there were still 20 foreign pilots in Shanghai, 3 in Tian Jin, 1 in Huludao, and 1 in Yingkou.

From their first appearance in Chinese coastal waters in 1835 to their withdrawal from China in 1949, foreign ships witnessed the change and turbulence of modern China's shipping history which covered four periods: the Beginning (1835–1914), the Development (1915–1937), the Collapse (1937–1945), and the Recovery (1945–1949). During this process, Chinese private shipping companies endured untold hardships and seized opportunities to survive, which laid the foundation for New China's shipping industry.

In the Beginning period, China's shipping industry underwent political upheavals in the late Qing and early Republican China. Facing the intrusion of foreign ships and losing control of navigation management, China began its tough journey in developing the shipping industry. Marked by the establishment of the government-supervised and merchant-managed China Merchants' Steam Navigation Company in 1872, the Beginning period could be divided into two stages. In the first stage, the British Empire dominated the Chinese shipping industry and controlled over 60% of its business. China did not have its own shipping company until 1872 when the Qing government established China Merchants' Steam Navigation Company. In the second stage, Japan began to participate in the furious foreign competition for the Chinese shipping industry. As a latecomer, Japan was dwarfed by the British shipping magnates. By the end of World War I, ships sailing on the Pacific routes accounted for less than 10% of the world's total ships. British companies like the Peninsular and Oriental Navigation Co. and Blue Funnel Navigation Co. pioneered the sea routes to the Far East and opened shipping services to Hong Kong and Shanghai. Likewise, British merchants were the forerunners in establishing shipping companies in China. In 1865, the British Hong Kong Canton & Macao Steamboat Company was established to support the market for regional ferry transport in the Canton area, including Hong Kong, Macao, and the West River. In 1862, Russell & Co. built up the first foreign steamboat enterprise in China Shanghai Union Steam Navigation Co. This Anglo-Chinese joint-stock enterprise plied the waters between the treaty ports and inland ports. Though established earlier than the Hong Kong Canton & Macao Steamboat Company, the Shanghai Union Steam Navigation Co. in essence was not the first foreign shipping company in China. In 1867 the British Butterfield & Swire Co. opened the China Navigation Co. to trade up the Yangtze River from their Shanghai base. By 1875, the China Navigation Co. owned one million pounds and about 80 riverboats and seaboats whose total tonnages reached 158,000 tons. In 1877 Jardine Matheson founded the Indo-China Navigation Co. for trade in Chinese coastal and inland ports. These four British shipping companies dominated China's shipping industry from the very beginning. When the Qing government established the China Merchants Steam Navigation Company in 1872, Japan also established its largest shipping company, the Nippon Yusen Kaisha. In 1875 Nippon Yusen Kaisha brought the Shanghai–Yokohama liner from the American Pacific Mail Steamship Navigation Company. Subsidized by the Japanese Government, Nippon Yusen Kaisha established a route between Nagasaki and Tianjin.

In accordance with *the Chefoo Convention* of 1876, China opened the ports of Wuhu, Yichang, Wenzhou, and Beihai to English trade, and Britain could use Hukou, Wuxue, Luxikou, and Shashi (likin-free ports along the Yangtze River) for cargo transportation. According to the 1897 Supplementary Treaty to the Convention between Great Britain and China relative to Burmah and Thibet, Wuzhou, Sanshui, and Tengyue (today Tengchong), inland ports were opened as new trade ports. British ships could sail up the West River from Hong Kong and Canton to Wuzhou (Guangxi Province). After the Qing government issued the Regulations Governing Trade on the Yangtze River and the Inland Waters Steam Navigation Rules in 1898, registered ships of both Chinese and foreign shipping companies could trade up the Yangtze River and West River from their coastal bases. Gradually all the coastal and inland waters fell into the hands of foreign companies.

By the end of the nineteenth century, Japan, Germany, and France were also competing for shipping rights on the Yangtze River. Subsidized by the Japanese government, Osaka Shosen Kaisha began to open the Yangtze River routes from Shanghai. Founded in 1884, Osaka Shosen

Kaisha focused on the operation of Japanese domestic routes. As a later entrant to the market, Osaka Shosen Kaisha had to compete with British companies including Butterfield & Swire Co., Jardine & J. Matheson Co., George McBain & Co., Hong'an Steamship Company, and so on. In 1903, Nippon Yusen Kaisha bought the Wayside Wharf (Shanghai) and some steamships from George McBain & Co. with 2.5 million yen and began its navigation up the Yangtze River. In 1904, Japanese merchants established the Hunan Kisen Kaisha (Hunan Steamship Company) to navigate between Changsha and Hankou. The Hamburg-Amerika and North German Lloyd Bremen initiated a joint service on the Yangzi River in 1900 that competed on the Shanghai–Hankou and Hankou–Yichang routes. In 1906, the French Compagnie Asiatisque de Navigation began its Yangtze River shipping service. Upon the Yangtze River, foreign shipping companies from Britain, Japan, Germany, and France competed against each other for cargo and passenger transportation. The British and Japanese shipping competition made the situation worse. After the merging of Nippon Yusen Kaisha, Osaka Shosen Kaisha, Hunan Kisen Kaisha, and Daito Kisen Kaisha in 1907, the Nisshin Kisen Kaisha (Japan–China Steamship Company) was established to manage all Japanese shipping businesses in China. With the establishment of Nisshin Kisen Kaisha, Japanese companies became a formidable force on the Yangtze and opened the Changsha–Changde and Jiujiang–Nanchang routes. Osaka Shosen Kaisha withdrew from the fierce competition on the Yangtze River and focused on the Chinese coastal routes. During the world economic recession in the early 1900s, Hamburg-Amerika, Compagnie Asiatisque de Navigation, and Hong'an Steamship Company also withdrew from the Yangtze shipping competition. With the help of Butterfield & Swire Co., Jardine & J. Matheson Co., China Merchants' Steam Navigation Company bought the properties of Compagnie Asiatisque de Navigation. Meanwhile, the American Hong'an Steamship Company was sold to three Chinese compradors. By the outbreak of World War I in 1914, North German Lloyd Bremen closed its businesses in China. After this point, the Yangtze River shipping was dominated by four companies: Jardine & J. Matheson Co., Butterfield & Swire Co., Nisshin Kisen Kaisha, and China Merchants' Steam Navigation Company. Though British shipping companies were in control of the market, still Japanese companies were catching up. The Fengyang (3977 tons) riverboat of Nisshin Kisen Kaisha shared a similar tonnage with the Longhe (3922 tons) riverboat of Jardine & J. Matheson Company.

By 1914, the market shares of Britain, Japan, Germany, and China in Yangtze River Shipping were as follows:

| | *Number of Ships* | *Percentage* | *Tonnage* | *Percentage* |
|---|---|---|---|---|
| Britain | 16 | 38 | 37,733 | 40 |
| Japan | 13 | 31 | 29,026 | 41 |
| Germany | 2 | 5 | 3364 | 3 |
| China | 11 | 26 | 24,580 | 26 |
| Total | 42 | 100 | 94,703 | 100 |

Source: *The Eastern Miscellany*, Volume 15, Issue 3, March 1928, p. 170

Meanwhile, the shipping industry in the three northeastern provinces of China was dominated by Japan and Russia. The Japanese East China Railway Shipping Department operated over 50 ships on the Songhua River and managed the regular Harbin–Changchun routes from April to mid-November. Russia established the Jiangsheng Steamship Company to control the shipping industry of Heilongjiang Province. In 1871, the Russian company only had more than 60 ships in operation. By the outbreak of World War I, it had more than 100 ships. As for Chinese shipping companies in Northeast China, the ships operated by the Heilongjiang and Songhua River Post Office were no more than twenty. The number of ships operated by Chinese merchants in Jilin Province was similar.

According to the second volume of *The International Relations of the Chinese Empire*, written by Hosea Ballou Morse, British companies controlled over half of China's shipping market before 1895. Though its number of ships doubled by 1912, Britain's market share in China's shipping industry fell to 44%. Japan's market share rose from 2% in 1897 to 23% in 1912. Germany's market share rose from 5% in 1899 to 13% in 1903 and maintained that level for almost ten years with only 14% in 1912. Companies from Britain, Japan, and Germany controlled over 2/3 of China's shipping market. According to the Inspectorate-General of Customs, Britain ranked top (with 38,106,730 tons) in terms of the total tonnage of ships registered and cleared by the Chinese Customs in 1912. With 19,913,385 tons, Japan came next. With 17,277,407 tons, to which Chinese private shipping companies contributed 4,404,309 tons, China ranked third. In the two decades after 1895, the total tonnage of foreign ships in China doubled, while the market share of Chinese companies was squeezed to less than 20%.

China Merchants' Steam Navigation Company was China's first and largest shipping company. In the company's founding years, Li Hongzhang entrusted Zhu Qiang and Zhu Qiyi to purchase foreign ships and set up offices in Shanghai and Tianjin ports. In 1872, heads of the China Merchants Steam Navigation Company, including Tang Tingshu, the Governor-General of Zhili, and Sheng Xuanhuai, raised 1 million taels to explore and navigate coastal routes and develop the Chinese port system. Chinese ships could sail north to Niuzhuang and south to Hong Kong, connecting the whole country. By 1873 the early raised funds ran out; another 500,000 taels was raised in 1876 to purchase properties from U.S. Russell & Co., including 7 seagoing vessels, 9 riverboats, and all kinds of pontoons, barges, harbors, warehouses in Shanghai, Tianjin, Hankou, Zhenjiang and other ports. During the Sino-French War in 1885, China Merchants' Steam Navigation Company shifted its properties to U.S. Russell & Co. in case of being taken over by France. After the Sino-French war, the company's properties were taken back by Sheng Xuanhuai who made the company a government-supervised merchant enterprise. During the Sino-Japanese War of 1894–1895, China Merchants' Steam Navigation Company authorized the British and American companies to operate its ships and resumed the operation after the war. In 1898, the company's total capital increased to 4 million taels. In 1914, the company's assets were all converted into operating capital which was worth 8.4 million taels of silver, among which industrial capital was 4.4 million taels. (*China Merchants' Steam Navigation Company 75th Anniversary Edition*).

Facing fierce foreign competition, Chinese private shipping companies began their long day's journey into night. Chinese companies, like Dada (1905), Zhengji (1905), Ningshao (1909), Zhaoxing (1910), Zhidong (1911), Sanbei (1914), and so on, spared no efforts to earn their market share. China Merchants, Dada, and Ningshao even participated in the inland waters shipping competition. Smaller companies such as Anhe Shipping Bureau, Liyi Shipping Company, Shenji Shipping Company, Qianjiang Company, and so on, were also trying to catch up.

In the early years of the Republic of China, shipping companies and routes were as follows:

| *Country* | *Company* | *Number of ships* | *Major Routes* |
|---|---|---|---|
| China | China Merchants' Steam Navigation Company | 30 | Between ports in China |
| China | Kaiping Coal Mines | 6 | Tianjin—Qinhuangdao—Yingkou |
| China | Ning-Shao Steamship Company | 5 | Shanghai—Ningbo—Hankou |
| China | Ching Kee Steam Navigation Company | 5 | Yantai—Beibu—Yangtze River |
| Foreign | Butterfield & Swire Co. | 47 | Shanghai—Tianjin—Inchon<br>Shanghai—Guangdong<br>Shanghai—Hankou<br>Shanghai—Ningbo |
| Foreign | Jardine & J. Matheson Co. | 42 | Shanghai—Tianjin<br>Shanghai—Hankou<br>Hankou—Xiangtan<br>Shanghai—Tianjin—Niuchuang |
| Foreign | Nippon Yusen Kaisha | 12 | Osaka—Shanghai<br>Yokohama—Shanghai<br>Kobe—Niuchuang<br>Yokohama—Niuchuang<br>Osaka—Tianjin<br>Osaka—Tsingtao |
| Foreign | Osaka Shosen Kaisha | 13 | Osaka—Dalian<br>Keelung—Hong Kong<br>Dakou (today Kaohsiung)—Guangdong<br>Dakou—Tianjin<br>Osaka—Tianjin<br>Osaka—Tsingtao |
| Foreign | Nisshin Kisen Kaisha | 11 | Shanghai—Hankou<br>Hankou—Yichang<br>Hankou—Xiangtan<br>Hankou—Changde |
| Foreign | Melchers & Co. | 6 | Shanghai—Hankou<br>Shanghai—Tianjin<br>Shanghai—Jiaozhou |
| Foreign | Siemssen & Co. | 4 | Shanghai—Guangzhou |
| Foreign | South Manchurian Railway Company | 2 | Dalian—Tsingtao—Shanghai |
| Foreign | Harada & Co. | 2 | Osaka—Tsingtao |

From 1915 to 1937, China's shipping industry underwent great development. With the outbreak of World War I, Britain, the United States,

Germany, Austria, France, Norway, and other countries had to withdraw from the Far East shipping industry. Dozens of Chinese shipping companies made their debut on the market and seized the opportunity to develop. However, these Chinese companies were of limited capital and small scale. If the Western powers returned, they would easily go bankrupt. By 1915 there were 718 Chinese ships registered on the marker, whose total tonnage was 40,472 tons. During WWI, ships of German and Austrian companies were taken over by the Qing government; and Japanese shipping companies scrambled to fill the business vacuum left by the withdrawal of Western ships. In 1918, Japanese companies had the largest market share in China's shipping industry.

According to the statistics of *China Year Book*, the number and tonnage of ships operating in Chinese ports in 1920 were as follows:

| *Country* | *Number of ships* | *Tonnage* |
|---|---|---|
| Britain | 29,543 | 40,313,707 |
| Japan | 25,152 | 28,191,592 |
| China (State) | 54,791 | 23,632,198 |
| China (Civil) | 85,586 | 4,021,111 |
| U.S | 5547 | 4,718,251 |
| France | 603 | 852,979 |
| Others | 1911 | 1,603,299 |
| Total | 214,601 | 104,266,695 |

The Development Period (1915–1937) of the shipping history in modern China could be divided into three stages. The first stage (1915–1924) was the golden era of the Chinese shipping industry. China Merchants, Sanbei, Hong'an, and Ching Kee formed the pillar of the Chinese shipping industry. China Merchants had sufficient funds and government support. Sanbei came next. Founded by Yu Hede in 1914, the company had only one boat and navigated just between Shanghai and Ningbo. By 1919, the company had developed into a shipping magnate with a capital of 2 million yuan, navigating in Chinese coastal and inland waters with 17 ships (total tonnage of 37,973 tons). With Hong'an and NingXing, the company competed with foreign companies on the Yangtze River. Sanbei also had ocean lines connecting Southeast Asia with East Asia. Aboard ships of Sanbei, passengers could sail to Vladivostok, Yangon, Nagasaki, and so on from Shanghai. In 1919, Yu Hede bought the control of Hong'an (founded in 1915) from British shareholders and turned it into a Chinese shipping company. Like Sanbei, Hong'an's headquarters was in Shanghai, from where all its ships sailed up the Yangtze River and traded between inland ports. The total tonnage of the company's ships, including five river tugboats, was 11,528 tons. Ching Kee was founded in 1905 and reorganized into a limited company in April 1920, with its capital increased to 2.5 million yuan. The headquarters of Ching Kee was in Yantai, with branches in Tianjin, Shanghai, Guangzhou, and Hong Kong. The company had 26 ships (total tonnage of 37,990 tons) operating on routes to Southeast Asia and Northeast Asia and along the Yangtze River. In June 1924, Datong Steamship Company was established in Shanghai, which marked the peak of the Chinese private shipping industry. The company only had four ships with a total tonnage of 5630 tons, but it gained momentum in the private sector.

The second stage (1924–1931) witnessed the return of foreign companies. In terms of their market share, Britain still held onto the top spot, and Japan came next. Italy, Portugal, Belgium, Spain, Sweden, and Norway resumed their coastal trade in China. The total tonnage of global ships reached 60 million tons, of which Britain accounted for about 30%. To enhance their competitiveness in the market, the shipping companies built new, larger ships and increased the speed of their fleet. On China's ocean routes and coastal routes, the average tonnage of cargo boats and passenger boats surpassed 4000 tons. Meanwhile, Chinese companies began to take the lead in China's shipping industry and even gained more profits than the British companies. The following table shows the market share of Chinese, British, and Japanese companies in China's shipping industry.

| *Year* | *Country* | *Tonnage (thousand tons)* | | | *Sum (thousand taels)* | | |
|---|---|---|---|---|---|---|---|
| | | *Export* | *Import* | *Percentage (%)* | *Export* | *Import* | *Percentage (%)* |
| 1925 | China | 14,846 | 14,820 | 33.51 | 545,198 | 514,431 | 45.6 |
| | Britain | 15,284 | 15,105 | 24.23 | 352,111 | 343,057 | 29.92 |
| | Japan | 10,286 | 10,161 | 23.04 | 203,959 | 206,727 | 17.67 |
| 1926 | China | 13,324 | 13,360 | 27.84 | 494,667 | 476,263 | 37.8 |
| | Britain | 18,183 | 18,278 | 38.03 | 433,025 | 399,955 | 32.51 |
| | Japan | 11,866 | 11,728 | 24.59 | 296,226 | 290,670 | 22.84 |

(continued)

(continued)

| *Year* | *Country* | *Tonnage (thousand tons)* | | | *Sum (thousand taels)* | | |
|---|---|---|---|---|---|---|---|
| | | *Export* | *Import* | *Percentage (%)* | *Export* | *Import* | *Percentage (%)* |
| 1927 | China | 8596 | 8561 | 23.29 | 260,273 | 344,283 | 32.44 |
| | Britain | 12,639 | 12,655 | 35.27 | 396,732 | 370,391 | 35.32 |
| | Japan | 10,647 | 10,428 | 29.39 | 275,856 | 263,581 | 24.84 |

*Source* Cai Kecheng, The Development of Shipping and the Necessity of Restoring Navigation Rights, Transportation Journal, Volume 1, Issue 5, February 1933, p. 116.

Although China was catching up in the shipping competition, still its market share was below one-third of the total. In 1925, foreign ships accounted for more than 65% of the total tonnage in the Chinese market, while Chinese ships made up the remaining 34%. In terms of market share revenue, foreign companies accounted for 54%, and Chinese companies for the remaining 46%. In 1926, foreign ships accounted for 72% of the total tonnage in the Chinese market, while Chinese for the remaining 28%. In terms of market share revenue, foreign companies accounted for 62%, and Chinese companies for the remaining 38%. In 1927, foreign ships accounted for 77% of the total tonnage in the Chinese market, while Chinese for the remaining 23%. In terms of market share revenue, foreign companies accounted for 68%, and Chinese companies for the remaining 32%. With less tonnage, Chinese ships made relatively more profits which made them more competitive on the market. If China could implement protective policies in shipping, Chinese companies could have gained a larger market. As for shipping market revenue in China, Britain ranked top, followed by China, Japan, the United States, France, Russia, the Netherlands, and other countries.

For Chinese shipping companies, the third stage (1931–1937) was the calm before the storm. Japan began its occupation of Northeast China in 1931 and wanted to launch a full-scale invasion against China. Coastal trade was frustrated by the looming war. Under the fierce competition with British and Japanese companies, Chinese companies spared no efforts to keep their market share in coastal and inland shipping. In 1932, China Merchants Steam Navigation Company was nationalized by the government to coordinate its cooperation with private companies and defend the domestic market. In the global market, Chinese companies were unable to compete with those of the Western maritime powers. While the

Western powers were building more ships to dominate the global market, China was still in the first stage of exploring ocean routes to Europe and the trans-Pacific routes. Chinese companies fell behind in the shipping competition with foreign companies in China. Before the outbreak of the Anti-Japanese War (1937–1945), the total tonnage of Chinese ships was 576,000 tons, while that of foreign ships was 710,000 tons.

The Collapse Period of China's shipping industry overlapped with the Anti-Japanese War when Japan controlled almost all Chinese coastal ports and forced Chinese shipping companies to the verge of collapse. Restricted by war, only eight large companies survived in the Upper Yangtze River, including China Merchants, Minsheng, Sanbei, and Qianghua. With government support, Chinese shipping companies survived the shipping fiasco until the surrender of Japan in 1945. When the war was over, China only had 457 riverboats (with a total tonnage of 80,680 tons), including 17 state-owned boats (23,942 tons) and 440 private boats (56,738 tons).[1]

China's shipping industry entered the Recovery Period three years before the founding of New China. Through buying ships from the United States, Britain, Panama, Norway, Canada, and other countries as well as salvaging Japanese wrecks (237 ships, 487,323 tons) in coastal waters, China increased the number of its ships to 4032 (1,092,217 tons) by June 1948, including 1455 sea boats (823,963 tons) and 2,577 riverboats (268,254 tons).[2] Meanwhile, China also ordered some new ships from foreign countries. The founding of New China opened a new chapter in China's shipping industry. After years of development and hard work, China has developed into one of the greatest maritime powers in the world. However, we still have a long way to go.

### *Under the Three Mountains of Foreign Debts, Trade Deficit, and Great Depression*

On January 11, 1943, China and the United States signed new treaties in Washington, and China and the United Kingdom signed new treaties in Chongqing. The United States and the United Kingdom formally

[1] Zeng Baiguang: *Chinese Merchant Ships and Shipping Industry*, *World Transportation Monthly*, Volume 2, Issue 2, August 1948, p.38.

[2] Ibid., p. 40. Li Jinglu, *Annual Review of China's Shipping Industry*, *World Transportation Monthly*, Volume 2, Issue 1, July 1948, p. 24.

announced the termination of their unequal treaties concerning China. On February 23, 1943, the French government in exile in London announced the termination of their extraterritoriality and leased territory in China. Since the signing of the first unequal treaty in 1842, these unequal treaties have haunted China for over a century. Though the British government under the leadership of Winston Churchill did not return Hong Kong to China, and Japan did not return to Taiwan until its surrender in September 1945, still China was making progress in terms of sovereignty and territorial integrity.

The century before the founding of New China witnessed China's financial ruin and its people endured destitute poverty under the three mountains of foreign debts, the trade deficit, and the Great Depression of the 1930s.

After the Taiping Rebellion (1850–64), the Qing government fell to the verge of bankruptcy and turned to foreign loans for help in 1865. The foreign debts began to snowball and dragged China down. China's foreign debts consisted of two parts: governmental debts and foreign direct investment in China. To construct railways and develop industries, the Qing government turned to foreign banks for help and incurred huge debts. Meanwhile, China opened its industry, finance, transportation, and other sectors for foreign investments. China's market was gradually controlled by foreign governments and companies. In 1865, the Qing government borrowed 1,433,664 pounds from the Bank of London to pay its debts to Russia. To suppress rebellions in Shaanxi and Gansu provinces, the Qing government borrowed more money from the United Kingdom in the 1870s. In the three decades before the First Sino-Japanese War, China's British debts had accumulated to 4,468,891 pounds, which equaled 15,550,700 Chinese kuping taels. One kuping (treasury) tael is about 37.3 grams (1.32 oz) in weight. The British loans were managed by two British banks: the HSBC (Hong Kong and Shang Hai Banking Corporation) and the Bank of East Asia. After the First Sino-Japanese War, China fell deeper into the quagmire of foreign debts. The war added 40 million yuan in new foreign debts to the Qing government. The Treaty of Shimonoseki forced China to pay the war indemnity of 200 million taels to Japan. Later, Britain, Russia, and France intervened and Japan was forced to hand over control of the Liaodong peninsula to Russia (for Port Arthur), Japan asked for an additional indemnity of million kuping (1.12 million kg) taels of silver from China. Japan asked China to pay off the war indemnity of 230 million taels in three installments. China had to pay 80 million taels, the estimated amount of China's annual total revenue, in the first six months after the war and 50 million taels in the next six months. The remaining 100 million taels were asked to be paid off within three years. Cornered by the war indemnity, China had to increase its foreign debts. Meanwhile, Russian and French banks competed with British and German banks for loans (with an annual interest rate of 8%) to the Qing government. After deliberation and comparison, the Qing government borrowed 100 million taels (with an annual interest rate of 4%) from Russian and French banks at the mortgage of its customs revenue. Later, the Qing government borrowed another 100 million taels (with an annual interest rate of 5%) from British and German banks at the mortgage of its customs revenue. These 200 million taels of government loans had to be paid off within 36 years and deprived China of its customs administration.

Within two years, the 200 million taels loans ran out. In 1897, the Qing government proposed taking another 100 million taels loan. British and Russian banks competed for the bid. Then the British HSBC and the German Deutsche Asiatische Bank won the bid and provided the 100 million taels loan to the Qing government with an annual interest rate of 4.5%. However, China only got 83% of the loan and had to pay off the debts within 45 years. Besides, China had to promise not to cede the Yangtze River region to any other country and mortgaged its custom revenue, *likin*, and salt taxes.

During the Boxer Rebellion (1899–1901), the Eight-Power Allied Forces invaded China, and the Qing government signed the Treaty of Xinchou in 1901 with 11 countries, including Germany, Britain, the United States, France, Austria, Belgium, Spain, Italy, the Netherlands, Japan, and Russia. According to the treaty, China had to pay the war indemnity of 450 million taels which meant every Chinese person (with a population of 450 million) had to pay one tael to the invaders. Even the annual exchange profits reached 3 million taels. China was forced to pay off the war indemnity within 39 years (with an annual interest rate of over 4%). Thus, China had to pay over 980 million taels to the invaders before 1940. China's customs revenues could not balance these new debts; the Qing government had to levy more taxes on its people. From 1894 to 1901, the Qing government incurred 40,935,000 pounds of new debts for industrial development. With the war indemnity, 100 million taels of new debts were added to China's annual debt payment. China's creditors included the United Kingdom, Germany, France, Russia, Belgium,

Japan, the United States, and other countries. The Qing government made several long-term (with a maximum of 60 years) loans from these countries' banks with a maximum loan of 2.5 million pounds and mortgaged all its fiscal revenues, including the salt taxes, land taxes, and railway assets. Besides, these loans created compound interests for foreign banks, which in turn competed to control China's economy.

By the time the Republic of China was founded, China had reduced some of its foreign debts, but the debts it inherited from the Qing government were far from being paid off, in particular the Xinchou war indemnity, which was extended to 1941. To bolster his position as the president, in 1913 Yuan Shikai negotiated a so-called "Queryreorganization loan" of £25 million from the Five-Power Banking Consortium (Britain, France, Germany, Russia, and Japan). After Japan occupied Shandong in 1918, the Duan Qirui government colluded with the Japanese government and signed the Nishihara loans of 460 million taels, making Japan China's largest creditor. In September 1920, the new Four-Power Banking Consortium (Britain, France, America, and Japan) was established to tighten foreign control over China's economy. The founding of the new Four-Power Banking Consortium enraged the Chinese people who refused to follow the same old disastrous road of the Qing government and mortgage out their freedom and sovereignty.

According to Remer's statistics on *Foreign investment in China*, China's government loan was 248 million U.S. dollars in 1902, $323.2 million (141.10% of that in 1902) in 1913, $427.6 million (172.28% of that in 1902) in 1931. The percentage of government loans in foreign debts was respectively 87.2% (1902), 62.8% (1913), and 61.4% (1931). Foreign debts and military expenditures have become the Chinese government's major fiscal expenditure since 1900. From 1928 to 1934, the average percentage of China's military expenditure was 45%, while that of the foreign debts was 35%. Together these two items accounted for 80% of total government expenditure. Foreign debts and military expenditure could not be balanced by government revenues from customs and salt taxes. During the same period, China's annual debt payment exceeded 200 million yuan, except for 1928, and the percentage of foreign debts in government expenditure was respectively: 1928–36.82%, 1929–37.15%, 1930–40.52%, 1931–39.58%, 1932–28.43%, 1933–29.19%, and 1934–33.13%.

Overwhelmed by foreign debts, China lost control of its economy and industry, and the Chinese people suffered destitute poverty. Taking the railway industry as an example, it was dominated by foreign capital. Foreign investment in Chinese railways ($640 million) accounted for over 20% of the total foreign investment ($3242.5 million) in China.[3] The foreign debts in China's state-owned railways reached 247.96 million U.S. dollars with the Chinese Silver Yuan to U.S. Dollar exchange rate of 25%.[4] The French loans for the construction of the Zhengtai Railway were not paid off until 1932. With mounting foreign debts, China constructed its Railway system with a limited coverage of 15,232 kilometers. However, this left the distance and length of railway in China per capita well below the world average.

The world average rail mileage per thousand square kilometers was 10 kilometers at that time, but that of China was 1.4 kilometers. The rail mileage per 10,000 people was 6.7 kilometers, but that of China was 0.3 kilometers. Relevant figures of the world's top five railway countries were as follows:

| *Country* | *Rail Mileage (km)* | *Rail mileage per thousand square kilometres (km)* | *Rail mileage per 10,000 people (km)* |
|---|---|---|---|
| U.S.(including Alaska) | 402,859 | 43.0 | 33.9 |
| Soviet Union | 77,035 | 4.0 | 5.4 |
| Germany | 58,619 | 125.0 | 9.3 |
| France | 53,561 | 97.0 | 12.0 |
| United Kingdom | 39,291 | 126.0 | 8.1 |

Source: *The Eastern Miscellany*, Volume 32, Issue 17, September 1935, p. 17

Foreign banks competed to offer loans to China which brought them several benefits. Firstly, the interest rate of the foreign loans was usually between 7 and 8%, much higher than the foreign banks could not get in their domestic market. In 1877, the interest rate of HSBC's loans to China was as high as 15%. Secondly, the foreign loans were made with the mortgage of China's customs revenues, salt taxes, railway assets, and so on. Thus, the control over China's finances, railways, mines, etc., was shifted to foreign countries. Thirdly, foreign banks could take rebates from their loans. Taking the government loan from the Banque

[3] C.F. Remer, *Foreign investment in China*, New York, 1933, pp. 70, 88.

[4] *Railway Yearbook*, Volume 1, pp. 299–301.

Industrielle de Chine in 1914 as an example, the bank took 19% of the 100 million francs loan as its rebates. Fourthly, China's railroad loans and government loans brought foreign banks huge dividends and some political privileges. Fifthly, railroad loans or other industrial loans were provided with some imparity clauses. For example, when foreign countries provided railway loans, they gained the privilege of appointing the chief engineer and accounting director to the railway construction, selling relevant equipment to China, and supervising and monitoring the whole project. Sixthly, when China takes a loan from foreign banks, the banks usually charge China 3% of the whole loan as their service fees. Meanwhile, foreign banks speculated on the exchange rates of the loan. With the same amount of loans, China gained less but repaid more. Taking Yuan Shikai's "reorganization loan" of £25 million as an example, China only got £20.97 million. During the period from 1921 to 1925, when the Chinese government repaid the loan, HSBC deliberately increased the exchange rate and made China pay 1.89 million taels more. Both the old Four-Power Banking Consortium before the Revolution of 1911 and the new Four-Power Banking Consortium after WWI attempted to dominate China's foreign loans and take full advantage of the loans. However, the member states of the banking consortium were involved in intriguing against each other and dying for the largest piece of the pie.

Foreign countries took full advantage of China's debts, especially the war indemnity, and benefited a lot from it. Taking Japan as an example, Japan owned more than 300 million taels of China's debts according to the Treaty of Shimonoseki (230 million taels) and the Treaty of Xinchou (75 million taels), which was more than six times Japan's annual financial revenue in 1895. With China's debt payments and the assets and resources it plundered from China, Japan began its military expansion and industrial revitalization. Furthermore, Japan deposited part of the debt payments in the Bank of London as its foreign currency reserve, introducing a new financial system among the gold-standard countries.[5] Based on its currency reserve, the Japanese government issued public bonds to support industrial production and exports, which greatly enhanced its national strength and laid the foundation for its military expansion. Benefiting from the war indemnity, Japan kept on launching new wars. In 1905, Japan won the Russo-Japanese War and accumulated funds for

[5] Yuzo Kato, *Modern History of East Asia*, Chinese Edition, China Social Science Press, 1992.

the next war. After its occupation of the Korean Peninsula, Japan began to invade Manchuria and Mongolia. Japan fell deeper and deeper into the abyss of militarism.

From 1876 (the second year of Guangxu's reign), the Chinese Maritime Customs began to issue the *Foreign Trade of China*, providing reliable statistics of China's annual imports and exports. The *Foreign Trade of China, 1876* provided reliable data on China's imports and exports from 1869 (the eighth year of Tongzhi's reign) to 1875 (the first year of Guangxu's reign). In the 67 years from 1864 to 1930, China's exports increased by more than 16 times; imports increased by more than 25 times. The growth of imports was much faster than that of exports. In its trade with foreign countries, China only had a trade surplus for six years (1864–1875). During the other 61 years, China witnessed worsening trade deficits. The total amount of China's trade deficits reached 6 billion kuping taels. Though Germany also had years of trade deficits before WWI and Britain had years of trade deficits after WWI, their deficits occurred in visible trades and could be balanced by surpluses in invisible trade. However, China's deficit in visible trade could not be balanced by its surplus in invisible trade, and it kept on growing.

Considering China's economic structure at that time, China exported semi-produced goods and raw materials and imported foreign manufactured and semi-produced products. This economic structure defined China as an underdeveloped agricultural country. Its industrial imports mainly came from developed countries such as Britain, the United States, Germany, France, and Japan. In their *Nature and Trends of China's Foreign Trade in the Twenty Years Since the Founding of the Republic of China* (Peking Institute of Social Survey 1933), Chen Junhui and Cai Qian analyzed the average value of China's imported and exported goods from 1912 to 1931 based on their respective economic attributes. Taking the goods in 1912, 1921, and 1931 as example, relevant data were as follows:

| *Item* | *Category* | *Year* | | | *The increase or decrease rate of 1930 compared with that of 1921* |
|---|---|---|---|---|---|
| | | *1912* | *1921* | *1931* | |
| Export | Raw materials | 34.11 | 32.67 | 42.77 | +8.66 |
| | Semi-produced products | 37.72 | 38.43 | 32.48 | −5.24 |
| | Manufactured products | 25.59 | 25.15 | 24.27 | −1.32 |
| Import | Raw materials | 8.48 | 14.52 | 32.52 | +24.04 |
| | Semi-produced products | 26.34 | 25.54 | 19.71 | −6.63 |
| | Manufactured products | 51.77 | 58.27 | 46.18 | −5.59 |

First of all, it can be seen that foreign manufactured products accounted for almost half of China's imports, while the Chinese manufactured products, including those produced by foreign companies in China, accounted for less than 25% of its exports.

Meanwhile, China had to import more and more grain instead of exporting it. The percentage of grain in China's exports decreased from 17.66% in 1912 to 14.97% in 1931. However, the percentage of grain in China's imports grew from 16.88% in 1912 to 22.62% in 1931. Within 20 years, China's import of foreign food grew 8%, showing China relied more on the global food market. With the increase in raw material exports and the growth in manufactured product imports, China was chained to the global trade system, becoming a raw material producer and a market for dumping goods. Even the rural areas of China could not escape from the vicious cycle.

Furthermore, during the period from 1912 to 1934, the average percentage of luxury goods in import surplus was 22.35%, and it even rose to 403.15% in 1919, casting more negative impacts on the national economy.[6]

China's political situation improved after 1926 when the Northern Expedition (1926–1927) began. From 1925 to 1934, China's foreign trade took a heavy blow from the Great Depression. What made the situation worse was Japan's occupation of Northeast China after the *September 18th Incident* in 1931. The three northeastern provinces accounted for 1/10 of China's imports and 1/3 of its exports. The loss of the three provinces made China fall deeper into the financial quagmire. Taking the

[6] Gu Yuantian, *The Urgency of Reducing International Trade Surplus*, *Ta Kung Pao Economic Weekly*, No.119, June 26, 1935.

total value of imports and exports in 1926 (100.0%) as the benchmark, the total value of imports and exports and the trade deficit indexes from 1925 to 1934 were as follows:

| *Year* | *Total value of imports and exports (thousand yuan)* | *Index* | *Excess value of imports over exports (thousand yuan)* | *Index* |
|---|---|---|---|---|
| 1925 | 2,686,332 | 86.7 | 267,216 | 66.0 |
| 1926 | 3,098,108 | 100.0 | 404,966 | 100.0 |
| 1927 | 3,009,356 | 97.1 | 146,938 | 36.3 |
| 1928 | 3,407,851 | 110.0 | 318,789 | 78.7 |
| 1929 | 3,554,524 | 114.7 | 389,642 | 96.2 |
| 1930 | 3,434,765 | 110.9 | 646,433 | 159.6 |
| 1931 | 3,650,339 | 117.8 | 816,413 | 201.6 |
| 1932 | 2,402,261 | 77.5 | 867,191 | 214.1 |
| 1933 | 1,957,395 | 63.2 | 733,739 | 181.2 |
| 1934 | 1,564,879 | 50.5 | 494,451 | 122.1 |

Source: *The Eastern Miscellany*, Volume 32, Issue 12, June 1935, p. 6

In 1931, the industrialized countries dumped their products on the Chinese market to ease their own financial crises, which pushed up China's imports to 2.233376 billion yuan (the highest in a decade). The total import and export value in 1931 reached 3.650339 billion yuan, the highest in a decade. Besides, the export value of 1931 reached 1.416963 billion yuan, second only to that of 1929, 1928, and 1927. Due to war, China Customs' statistics of Northeast China ended in the later six months of 1932. For two consecutive years from 1931 to 1932, China became the foreign countries' dumping ground and its trade deficit soared. Judging from the price index, China's economy was devastated by the Great Depression. Taking the price index of 1929 as the benchmark, the price index of China's foreign trade from 1930 to 1934 was as follows:

| *Year* | *Import* | | | *Export* | | |
|---|---|---|---|---|---|---|
| | *Price* | *Amount* | *Value* | *Price* | *Amount* | *Value* |
| 1930 | 117 | 88 | 103 | 103 | 86 | 88 |
| 1931 | 139 | 81 | 113 | 102 | 88 | 90 |
| 1932 | 130 | 74 | 95 | 86 | 63 | 54 |
| 1933 | 123 | 69 | 84 | 78 | 73 | 57 |
| 1934 | 123 | 52 | 64 | 68 | 74 | 50 |

Source: *The China Economy Yearbook*, Volume 1, Shanghai Life Bookstore, p. 79

The first five years of the 1930s witnessed China's soaring export price index with the peak in 1931 and 1932. The export price index began to fall after 1932. During the process, foreign countries speculated in foreign exchange trade and benefited from the widening of the price scissors. To recover from the financial crisis, the U.S. implemented new silver policies as part of President Roosevelt's New Deal, which made silver-standard China suffer more.

In July 1933, eight countries, including the United States, Mexico, Canada, Peru, Australia, China, India, and Spain, signed a silver agreement in London which stipulated that the Governments of the five silver-producing countries (the United States, Australia, Canada, Mexico, and Peru) would each absorb from the mines in their respective countries a certain amount of silver, the total amount to be absorbed by said producing countries being 35,000,000 ounces per annum during the four years commencing the first day of January 1934; that such silver so absorbed would be retained in each of said respective countries for the said period of four years, to be used for coinage purposes or as reserves for currency, or otherwise to be retained and kept off the world market during such period of time. The U.S. Congress ratified the London Agreement on Silver in December 1933. Later, the U.S. Congress passed the Silver Purchase Act on June 19, 1934, to increase its purchase of silver from foreign countries and change its gold standard. The Silver Purchase Act of 1934 declared "that the proportion of silver to gold in the monetary stocks of the United States should be increased, with the ultimate objective of having and maintaining, one-fourth of the monetary value of such stocks in silver." In 1934, the United States purchased 185 million ounces of silver from abroad, which was more than 8 times the silver it absorbed from its domestic market, causing the price of silver to soar in the global market. From 1914 to 1931, China's annual outflow of silver reached 62.9 million yuan. In 1932, the world economy fell into the Great Depression and China's outflow of silver exceeded 13,905,000 yuan; in 1933, the outflow of silver exceeded 14,422,000 yuan. After the implementation of the Silver Purchase Act in July 1934 and Executive Order 6814 Requiring the Delivery of All Silver in August 1934, China's outflow of silver soared to 259,941,000 yuan that year, more than ten times that of the previous year.[7] To compete with the American banks, foreign banks, especially the British and Japanese banks, began to concentrate China's silver in Shanghai and shipped tons of silver out through Hong Kong and Dalian. Though the Chinese government raised the export tax on silver on October 15, 1934, the outflow of silver could not be stopped.

Facing America's predatory buying of silver and worsening trade deficit, China's economy fell into turmoil. With tons of silver shipped out to London, New York and Hong Kong, China's urban economy began to shrink; Shanghai's financial market was crushed; the rural economy almost went bankrupt. During the economic turmoil, national companies suffered catastrophic losses from the low market prices, shrinking imports and exports, and decreasing purchasing power. Meanwhile, foreign banks took advantage of China's financial turmoil and profited from the outflow of silver and floating foreign exchanges. Besides, the government over-issued currency and bonds, devastating the credit system and accelerating the closure of domestic banks and money shops. When the Anti-Japanese War broke out, China's economy was on the verge of collapse.

## 9.2 The Enthusiastic Trend of Studying Abroad and the Surging Tide of Social Revolution

### *Thoughts on Social Revolution Advocated by Students Studying in Japan*

After ten years invested in studying abroad at official expense, the Westernizationist officials realized the fact that without directly sending people to the West, it was impossible to make any progress in harnessing Western technologies in a short time. Rong Hong, the first Chinese graduate from Yale, after returning to China in 1854, proposed a plan of selecting children aged 12–14 to study in the United States for 15 years. Starting from elementary school, these children were expected to enter U.S. colleges and acquire military and scientific knowledge. 120 children were sent to the U.S. in four batches between 1872 and 1875. 86 of these children came from Guangdong Province, the hometown of Rong Hong. However, facing strong opposition from the conservative hardliners, the government had to recall all the students from the U.S. in July 1881.

[7] *Trade Monthly*, December 1934, Society and Economy Monthly, Volume 2, Issue 4.

Obviously, 15 years was too long for the students to stay rooted in traditional Chinese culture.

There existed another way of studying abroad—sending a group of carefully selected professionals overseas for further study and having them return home as backbones in their respective fields in four to five years. This plan was put forward by Shen Baozhen, Minister of Vessel Administration, but it didn't kick off until May 1875. In March 1876, Li Hongzhang, one of the main leaders of the Westernization Movement, seized the first opportunity and sent seven officers to the German Military Academy. Two of them returned home without any academic progress and the rest studied there for three to five years. Foochow Shipbuilding Institution sent 30 students to Britain and France for further study. In May 1877, 12 of the students arrived in London to study naval navigation, and 18 of them arrived in France to study manufacturing, mining, and metallurgy. The first batch returned to China in 1881 as the country's earliest professionals in mining, metallurgy, and mechanical engineering. Based on France's prototype invented in 1885, they managed to build China's first steel warship—Ping Yuan, in 1888.

During the 1880s, a total number of 89 students were sent overseas by official institutions such as Foochow Arsenal, the Office of the Governor-General, and the Imperial Naval College. These people made up the first batch of military experts and engineers educated in the West.

China's defeat in the Sino-Japanese War of 1894–1895 made it imperative for the Chinese government to understand the truth of Japan's vitalization and prosperity. As a result, 13 students were sent to the Higher Normal School of Tokyo in 1896. The Japanese government saw this as an opportunity to foster pro-Japanese forces in China and offered to fund 200 Chinese students to study in Japan. The Japanese government promulgated a decree allowing foreign students to enter schools under the direct administration of the Ministry of Education, established a Chinese Overseas Students Department in the Imperial Japanese Army Academy, founded Tokyo Shinbu Gakko, a military preparatory school providing basic military training to Chinese students, and set up all sorts of short-term courses for Chinese students in Hosei University and Tokyo Police School. In an effort to better prepare Chinese students in Japanese, a bunch of specialized schools carved out preparatory courses lasting one to two years. The number of Chinese students studying in Japan reached over 1000 in 1903 and peaked at 8600 in 1905. However, the number dropped to 1400 during the Revolution of 1911. By the end of the Qing Dynasty, the number of Chinese students in Japan totaled 50,000, with government-sponsored ones accounting for 30%, and non-government-sponsored ones accounting for 70%. However, most of them only managed to finish middle school courses. Those who finished college courses simply took up 3% of them all.[8]

Sending students to Japan was a convenient and cost-efficient way in the eyes of the incumbent Japanese whose main concern was to prevent revolution from happening, Zhang Zhidong, for example. However, it was for the purpose of finding a way to break free from the shackles put by the Western powers that the ambitious Chinese students decided to go to Japan. After all, the success of another oriental country could be a better example. The earliest revolutionary organization of Chinese students in Japan was established in 1901, with a mission to build a stronger China. The short-term goal was to largely and rapidly introduce into China Western books on science, technology, politics, law, and liberal arts. The long-term goal was to overthrow the crown, find a republic, and revitalize the Chinese nation. Such a goal was also set by the Revive China Society in their revolutionary propositions.

A large number of enthusiastic Chinese overseas students in Japan were involved in the campaign of introducing Western books, eager and anxious to save the country from subjugation and ensure its survival. In 1899, Chinese students in Yokohama launched a semimonthly publication, under the banner of "natural rights," which was a further step from the theory of natural selection. In 1900, members of the Reed Association established the Translation and Compilation Society in Tokyo, and on December 6, *A Monthly Magazine of Translated Political Works* was launched. It published classics advocating natural rights, freedom, and democracy, including *Du Contrat Social*, and *De l'esprit des lois*, serving as a trail blazer in spreading and promoting new thinking from the West and Japan in China. *A Monthly Magazine of Translated Political Works* continued to be published until 1903, when it was renamed the *Journal of Political Science*. *Guo Min Bao* (Nationals' Newspaper), launched in May 1901 and funded by Sun Yat-sen, went on for only four issues before it was forced out of business by the authorities for its pro-revolution stance. Since 1902, the surging number of new periodicals launched has been coupled with the increasing number of Chinese students in Japan.

[8] *Compilation of Materials on the History of Modern Education in China—Education for Studying Abroad*, Shanghai Education Press, 1991, 370 pp.

1907 solely witnessed the launch of 18 periodicals and the total number of periodicals ever launched reached over 40 by 1911.[9] In 1907, 18 new magazines were created in one year. Following the *Kai Zhi Lu* and *A Monthly Magazine of Translated Political Works*, Hunan students ran the *Hunan Travelling Translation and Editorial* (1902–1903), Jiangsu students had *Jiangsu*, and Zhejiang students ran *Zhejiang Tide*. Later, there was *Yunnan*, founded in Tokyo in 1906 by Huang Xing, Yang Qiufan, and Lv Zhiyi, which had already issued a thousand copies, and after it was shipped back to China for sale, the number of printings rocketed, and it was run until 1912 before it came to an end. The six magazines, *Compendium of Translated Books, Travelling and Translation, Hubei Students, Jiangsu, Zhejiang Tide,* and *Yunnan*, were the six major periodicals of the late Qing student community calling for reforms and propaganda of Western social trends.

Regarding the compilation of magazines, students who studied in Japan established a large number of translation agencies, including the Translation and Editorial Society (1900), the Hunan Translation and Editorial Society (1902), the Textbook Translation and Editorial Society (1902), the Hubei Law and Politics Editorial Society, the Fujian Society (1904), the East Asia Newspaper, and more than 20 other organizations. The number of translation agencies organized in China by students returning from Japan was even greater. Most of these agencies were situated in Shanghai, such as the Quang Jee Sue Kouk, the Kaiming Book Company, Wenming Publishing House, etc., which were all springing up at once. There were more compilation societies organized by students returning from Japan in China, mostly concentrated in Shanghai, such as Guangzhi Shusha, Zuoxinsha, Wenming Shusha, Kaiming Shusha, Kai Ming Shusha, etc., which were all springing up at once. These organizations built a new bridge between Tokyo and Shanghai to convey Western learning, and the translation of Western books shifted from Western art to Western politics, focusing on politics, law, history, geography, and international relations, with natural science and engineering books taking a back seat. According to statistics, from 1896 to 1911, nearly 1100 Japanese

[9] The third collection of Feng Ziyou's *A History of the Revolution has a list of revolutionary books and newspapers at home and abroad before the founding of the nation*, but it omits to list the *Journal of Political Science, the Compendium of National Newspapers*, and the *Journal of Learning* (compiled by a Japanese student and edited by the Shanghai Newspaper Society 1907).

books were translated in China,[10] and most of the books translated from Japanese came out only after 1905. By way of comparison, the number of translated Japanese books published in 1850–1899 was 86, second to the 286 translated from the United Kingdom during the same period, and slightly higher than the 82 translated from the United States,[11] which ranked third. Of the total number of 567 translated books, Britain accounted for 50.5%, Japan 15.1%, the United States 14.5%, and countries other than Germany, France, and Russia 12.2%. Although these statistics cannot fully demonstrate the actual situation of the translation of books at that time, the translation of Japanese books, from 1896 onwards took a big step forward, in less than 10 years, surpassing the United Kingdom and the United States, ranked first, which is indeed a fact. During the period 1902–1904 alone, there were 321 Chinese translations into Japanese, accounting for 60.2% of the total number of 533 translations during the same period; 57 British translations followed, accounting for 10.7% of the total; and 32 American translations accounted for 6.1% of the total, taking the third place.[12] In terms of the number of translated books, neither Britain nor the United States has been able to compete with Japan since then.

The first encounter with Japan, a transformed and fast-developing country due to the introduction of a Western political system, and science and technology, inspired and motivated the Chinese students studying there, who, by advocating Western thoughts and cultivating the spirit of democracy and science, later grew to be the main force in China's democratic revolution. Around the founding of Tung Meng Hui, the Chinese Revolutionary League, in 1905, Western revolutionary theories advocated by Chinese students ranged from freedom, equality, philanthropism, and natural rights, to Darwinism, populism, anarchism, and so on.

The Enlightenment Movement triggered the French Revolution. Rousseau's *Social Contract Theory* was the pioneering work in the promotion of "man's natural rights." The book was translated by Yang

[10] *The Comprehensive Catalogue of Chinese Translations from Japanese Books*, edited by Tan Ruqian (The Chinese University of Hong Kong Press 1908), lists 958 titles, and it has been pointed out by many in the academic community that at least dozens of titles have been omitted.

[11] Huang Fu-ching, *Students Staying in Japan at the End of the Qing Dynasty*, Taipei, 1976, pp. 183–184.

[12] Id. at 184–185.

Tingdong, a Chinese student in Japan, and then published in Shanghai. Important pieces such as *The Declaration of Independence* and *Declaration of the Rights of Man and of the Citizen* were translated by Ms. Xiaopin and published in 1903, playing a critical role in advocating democratic revolution, political freedom, and social equality. Ma Junwu, a Japanese student, was a member of the Guoxue Society and devoted himself to the introduction of Social Darwinism and the doctrine of civil rights. He published books such as *Darwin's Chapter on Divine Choice* (Juvenile New China Society 1902), *Spencer's Chapter on Women's Rights Darwin's Chapter on Competition for Goods* (Juvenile New China Society 1902), *Spencer's Principles of Sociology* (Juvenile New China Society 1903), and *The Principles of Freedom of Maitreya John* (The Translation and Repertory of Books) (1903), which were followed by Yan Fuxing's The Principle of Freedom (1903), and Principles of *Mile John Freedom* (Translation and Compilation Society 1903), the achievements were so great that they succeeded Yan Fu. When Yan Fu published John Muller's (1806–1872) *The Theory of the Boundaries of Rights of the Group and Self* (originally called The Theory of Freedom) in 1903, a Chinese translation had already been published in Tokyo. At this time, Yan Fu had already had competition from the translations of students who had stayed in Japan. Ma Junwu wrote *A Modern History of France* to promote the history of the French Revolution. Japanese scholars wrote a number of history books on the social changes in Western Europe, such as Nakae Shomin's *The Second Century of France before the Revolution* (Japan's Foreign Students' Compilation and Translation Institute 1901), Shibue Tamotsu's *History of the Revolutionary War in France* (Oratorio 1903), Shibue Tamotsu's *History of the War of Poland's Decline and Fall* (Chiang Zhe Lung's Translation 1904), and Hinata Takeshimatsu's *History of the Revolutionary War in France* (Young People's Association's Translation 1903) were translated and released one after another at this time. *The History of the French Revolution* (compiled by the YMCA 1903) by Takamatsu Hoda was also translated and released. Zhang Xiangwen's translation of Montesquieu's *The Essence of All Laws* by Zhang Xiangwen was one of the earliest translations of *The Spirit of the Laws*, which advocated constitutional monarchy and the separation of powers, and was published by the Translation and Compilation Society as early as 1903, long before the translation of *The Meaning of the Laws* by Yan Fu.

The revolutionary students Zou Rong, Chen Tianhua, and Yang Dusheng all held the French Revolution in high regard, believing that after paying a huge price, France had finally become a great power. Lu Xun also read Kato Hiroyuki's *Theory of Competition* and Shibue Tamotsu's *History of the War of Poland's Decline and Fall* during his youthful stay in Japan.[13] Kato Hiroyuki (1836–1916) was a representative of the German school of thought in Japan, and believed in Social Darwinism. The Chinese students in Japan translated seven of his works, six of which were published in Tokyo and Shanghai in 1901–1903, including *The Theory of the Race of Things* (translated by Yang Yim-hang and translated by Tokyo Translation and Repertory Society in 1901), and *The Collected Lectures of Kato Hiroyuki* (translated by Sakushinsha and translated by Sakushinsha in 1902). *A Brief Overview of the Great Situation in the Nineteenth Century* (translated by Master Yohozhai, Quang Jee Sue Kouk 1902), *Heavenly Law* (translated by Wu Jianchang, Quang Jee Sue Kouk 1902), *A New Theory of Human Rights* (translated by Chen Shanshu, Kaiming Book Company 1903), *The Rationale for the Evolution of Moral and Legal Laws* (translated by Jin Shoukang et al. Quang Jee Sue Kouk, 1903), and *The Theory of Evolution of Politics and Religion* (translated by Yang Tingtong, Quang Jee Sue Kouk). The original work, published in 1882, advocated constitutional monarchy and statism, opposed the liberal civil rights movement that advocated elected members, and converted the Western idea of "natural human rights" into the idea of legislation under the control of the monarch (the emperor) in line with the spirit of "peace and spirit." *The Theory of Moral and Legal Evolution* was a supplement to *The Competition for the Rights of the Strong* published in 1893 and written in 1894, and it is also a representative work of Kato Hiroyuki's social Darwinism in his later years. Influenced by the popularity of the theory of heavenly evolution in China at that time, Kato Hiroyuki's idea of feedback from the emerging countries under the dictatorship in the East soon resonated with the students who stayed in Japan. Liang Qichao, who published a fragment of *Heavenly Law* in *Xin Min Cong Bao* (1902), had already criticized Kato, who advocated egoism, for "specialising in the theory of evolution, and taking love of self as the standard of morality and law, and for the radicalism of his words, which are aberrant and have a flow of evils" in his commentary. Liang Qichao thought that it would be better to expand the principle of superiority and inferiority "from one's

[13] *Lu Xun Research Materials*, 4th series, p. 368.

own self to a family, a township, a country, and all beings in the world," and then this egoism would be able to promote the progress of society.[14]

Among all the surging social revolutionary thoughts at the turn of the nineteenth and twentieth centuries, Russian populism attracted the most attention of Chinese revolutionary parties and democratic radicals. Populism was the political proposition of a group of radicals among the petty-bourgeois reformers, who were trying to oppose feudal serfdom and realize socialism through the development of the rural economy. Since China was a feudal agricultural society, its urban economy was founded on small production as well. Therefore, the slogan of "going to the people" and the thought of "mobilizing the peasants to transform the society" won the sympathy of some people. In 1902, Ma Junwu wrote the first book that systematically introduced the revolutionary activities of Russian populists. Since 1903, the political activities of Russian populists had been reported under the name of "Nihilist" in the newspapers of the revolutionary parties in Shanghai and Tokyo.

The name Nihilist was first used by the Christian church publication, *the Bulletin of the Nations*, in an article published in August 1878, which was phonetically translated from English as Nihilist, meaning "nothingness." In the five years from 1879 to 1883, *the Bulletin of the Nations* published ten articles on the assassination of the Tsar by the Russian Nihilists, using the translations of Nihilist, Nihilist, or Nihilist, respectively. There were two reports in the 10th issue (November 5, 1896) of the "Zeitung" (Times), which was run by the reformists Kang and Liang, and a similar message in the 12th issue (January 15, 1897), which was published in the same issue. The Nihilist Party was originally a socialist group influenced by the Paris Commune movement and adopting the ideals of the First International. The activities of the Proudhonian anarchists in the First International influenced Russian parties, including Russian populists, and thus the Nihilist Party was confused with anarchism in some Japanese political writings, such as the book "*Anarchism in the Modern Age*" by Shotaro Smokyama. The book was translated by Kanichi in 1904, re-titled *Free Blood*, and published in Shanghai, focusing on the history of the Nihilist Party in Russia. 1903 saw the publication of

[14] Chen Yingnian, "*The Introduction and Dissemination of the Writings of Modern Japanese Thinkers in China at the End of the Qing Dynasty*," in Beijing Municipal Society for the *Study of the History of Sino-Japanese Cultural Exchanges*, People's Publishing House, 1982, pp. 262–282.

an article entitled *The Nihilist Party* in the Shanghai-based *Sojourner* (19 June), which was run by the revolutionaries and reported on the political activities of the Russian populists. In the same year, the publications of students in Japan also published articles on the activities of the Russian Populists, including *Lucia's Void Party* in the 4th issue of *JIANGSU* (July 1903) and *The Theory of a New Society* signed by Dai-me in the 8th issue of *ZHEJIANG TIDE* (August 1903), which also referred to the Russian Void Party as a party that pursued the Proudhonist doctrine. Liang Qichao's *New People's Series*, No. 40–41 (1903) (November 1903), in which Liang Qichao wrote *On the Russian Nihilist Party*, and the 1904 Shanghai edition of *The Newest Novels*, in which Coldblood translated *The Curious Story of the Russian Nihilist Party,* were published. Specialized books included *Heroine of the Nihilist Party* written by a Qingmin in Jiangxi, *The Nihilist Party* translated by Coldblood in 1904 (Shanghai Kaiming Bookstore), and "*The Truth about the Nihilist Party*" written by Mohasun and translated by the owner of Bone Sex Hall in 1907, printed by Guangzhi Bookstore. At that time, the Japanese and Chinese newspapers and books referred to the Russian revolutionary parties collectively as the Void Party, and on most occasions, even more so in reference to the Populists.

The revolutionaries decided to follow suit with the Russian populists and assassinate Qing court officials after the murder of Yang Quyun, leader of the Revive China Society in early 1901. Cai Yuanpei, a revolutionary and one of the founding members of the Nationalist Party, set up a patriotic women's school in Shanghai, in an attempt to cultivate "women of the Nihilist Party." A picture of the Russian Populist woman, Sophia, appeared in No. 2 of the League's journal, Min Bao, and *a biography of Sophia*, written by WuShou, appeared in No. 15. Under such political encouragement, Qiu Jin, a female revolutionary, threw herself into the tide of women studying abroad and women saving the country, sacrificed her life and made her name on the scroll of honor in history.

Early socialist ideas spread in China, ranging from *The Learning of Cosmos*[15] and cosmopolitan sayings of idealistic socialism, to anarchism and Marx's scientific socialism. Ideal socialism did not develop during the evolution of political and social thought in the late Qing Dynasty, and only Kang Youwei borrowed the prophecies of American science fiction

[15] Benjamin Kidd, Richard Timothy and Erkang Cai, *Da Tong Xue*, The Christian Literature Society for China, 1899.

about the future of society to attach to the Tatung theory of the Xia, Shang, and Zhou dynasties. The driving force was Social Darwinism, a progressive view of social evolution, which was recognized by most of the students who stayed in Japan. The writings of the Japanese thinker Hiroyuki Kato received more attention, as did Spencer's *New Meaning of Sociology* (1902), which was translated into Japanese by Shibue Haozhi and retranslated into Chinese by Han Tan.(1902), *The Evolution of Civilisation in the Nineteenth Century* (edited by the Minyosha and translated by Chen Guoyong), *Socialism* (Quong Jee Sue Kouk, 1903) by Murai Chizi and translated by Luo Dawei, *Socialism* (1903) by Nishikawa Koujiro and translated by Chou Baigao, *The Socialist Party* (1903), Socialism in General (Sakushinsei, 1903) by Shimada Saburo, and translated by Zhao Bizhen, are best illustrated by the publication of Chinese translations of such books.

The introduction of anarchism was a spillover of Russian populists' pursuit of Proudhon and Bakuninism. The earliest report on the activities of European anarchists was published on the *Multinational Communique*, covering the assassination of Sadi Carnot, the French president, by an Italian anarchist named Sauto J. Casecio. In 1900, Liang Qichao published an article accusing anarchism of its violent nature, and started calling European anarchism "wu zheng fu" in Chinese, literally meaning "the absence of government." Such a name was obviously borrowed from Japan's academic community.

Shusui Kotoku, a student of Chomin Nakae, who was a radical ideologist in Japan's freedom and civil rights movement, was one of the organizers of socialist movements in the early stage. Kotoku translated Errico Malatesta's *Anarchy* from Italian into Japanese. And then in 1903, Zhang Ji, a Chinese student in Japan, translated it from Japanese into Chinese, and had it published in Shanghai. Since then, "wu zheng fu" (anarchism) had been popularized in all sorts of writings. At that time, anarchism was regarded as a school of socialism and was used alongside Marxism. In 1903, the 8th issue of Tokyo's Zhejiang Tide was published under the signature Big-I, "The Theory of a New Society," which introduced communism by dividing the socialist trend that had taken Europe and the United States by storm into communism and extreme democratism.

This school was founded by the Frenchman Baboeuf, and its later development by the Jew Karl Marx can be seen in the labor of various countries today.

The principle says: land and capital, the production of capital. If the landlord, is the capitalist, why do we need it? Land, capital from the land, capital owners, and still exist. If the property is based on the first occupation, will be to the late starter no cone of land, if the property is based on labor, will be to the late starter no labor land. As machinery flourishes and wages are depreciated, the laborers will never be landlords or capitalists, so the system of private ownership must be abolished and returned to state ownership.

Introduction of Extreme Democracy:

It was founded by the Frenchman Proudhon, and the Russian Michael Bakounine, Max Schimel, its representative, now the Russian Nihilist Party, its image.

Its principle is that: where the society is said to be private and self and the nature of the difference between human beings, is Fantu.

The text explains that the ideals of anarchism are freedom and autonomy, "said: away from socialism in terms of freedom, the monopoly of privilege, injustice; away from freedom in terms of socialism, slavery, barbarism." The goal is: "Say: Today's enemy is not the landlords, not the capitalists, but the government officials. To give up the body of an aspirant, to run around in society, to perform iron blood means, is also a vocation."

The Big-I's liberal interpretation of anarchism and Marxism in accordance with the bourgeois view of democratic revolution, modifying the principles of socialism, did not lead to the spread of socialist ideas in China; only Sun Yat-sen and his followers were the ones who took socialism as the slogan of their revolutionary call and fought for it for the rest of their lives.

Some members of the League believed in anarchist ideology. Zhang Ji recalled that when he was studying in Tokyo, in addition to attending classes, he went to the library to read the books translated by Jiang Zhaomin during the Japanese Restoration period, such as the French Revolution and the Civil Pact, which made his "revolutionary ideas grow day by day."[16] Zhang Ji had already translated Mala Lustad's *Anarchy* and compiled his own book, *Socialism*. From then on, anarchism appeared in Chinese intellectual circles as a school of socialism. In 1907 Liu Shipei and He Zhen founded the bimonthly *Tianyi Bao* in Tokyo to propagate

[16] Tan Ruqian, ed. *A Comprehensive Catalogue of Chinese Translations of Japanese Books*, 59 pp.

anarchist ideas in the name of socialism, and it was officially published on 10th June. On 31st August, Zhang Ji and Liu Shipei organized the Socialist Seminar for the students who stayed in Japan and declared that they would follow the anarchists and take *Tianyi Bao* as their organ. Zhang Ji had a close relationship with Kou De Qiushui, who was very famous for his activities against imperialism and the war of aggression, and he gave a lecture on "*Liberal Socialism*" at the seminar, a summary of which was published in the 6th issue of Tienyi Bao, which was discontinued at the beginning of 1908. Li Shizeng and Zhang Jingjiang went to Paris to set up the *Weekly New Century* and established the New Century Society to advocate anarchism. With the help of anarchism, they criticized China's feudal principles and ethical concepts, exposed the evils of private property, and opposed the powerful government that bound the free will of individuals, which played a certain positive role under the specific historical conditions of China.

The *Minbao* (People's Newspaper), which was succeeded by the magazine "*Chinas of the Twentieth Century*" edited by Soong Chiao-yen, was launched in the winter of 1905 and began to propagate socialism with great fanfare, which was actually influenced by the ideology of the Russian Social Democratic Party (RSDLP). At that time, the Chinese revolutionary Song Jiaoren and the Russian revolutionary Pilstki held a meeting on March 10, 1906 and listened to the political ideas of the Russian revolutionary party, thinking that the political and social revolutions must be carried out at the same time and that "it is really the one who desires to improve both the political and the social aspects."[17] This political idea was endorsed by Sun Yat-sen, who broke through the framework of the bourgeois democratic revolution in political theory, thinking that political and social revolutions could be carried out at the same time, killing two birds with one stone, and "achieving the same result in one stroke." 1906 *Minbao*, No. 4, translated and published Kuchimoto Yukong's "*Types of Social Revolutionary Movements in Europe and the United States and Their Commentary,*" and No. 5 published Galloping Zhai (骋斋) as the author of "*Social Revolutionary Movements in Europe and the United States.*" No. 4 of *Minbao* published Song Jiaoren's *A Brief History of the Congress of the Universal Socialist Party*, and No. 5 published Zhu Jixin's "*On Socialism as a Parallel to Political Revolution,*" pointing out that Marx's

[17] Song Jiaoren, *My History*, Vol. 3, 19.

doctrines were "gradually being put into practice, and the world is calling it scientific socialism." This was the first time that the doctrine of scientific socialism was put forward. No. 7 contains Liao Zhongkai's translation of *Outline of the History of Socialism*. In translating Japanese literature, Zhu Jixin and Liao Zhongkai translated the bourgeoisie as the "rich" and the proletariat as the "poor," and discussed the contradictions and struggles between the two.

Shusui Kotoku, as the first theorist who introduced and studied Marxism in Japan, played a guiding role in China's revolutionaries. His book *Monster of the Twentieth Century: Imperialism* started circulating in China in 1900, but was banned by the Qing court in 1901. The book was translated into Chinese by Zhao Bizhen, a Chinese student in Japan, in 1902, analyzing and criticizing imperialism from the socialist perspective for the first time. It discussed the social origin of imperialism, as well as the political prospects for socialist revolution, which was quite different from another book called *Imperialism*. The latter was published in 1895, and it mainly focused on the expansionary policies of Germany and Russia. In 1903, Zhao translated and published another book on socialism, systematically introducing Marxism to China for the first time. The book consisted of 4 chapters. The second chapter introduced German socialism, with its focus on Marxism and Das Kapital, and highlighted the difference between Marxism and anarchism, which was that anarchism would resort to violence to annihilate governmental organization while Marxism would let nature take its course. The fourth chapter introduced the status quo of socialist parties in Western countries.[18]

Another theoretical work by Kōtoku Akisui, *Guang Changtong* (廣長舌), originally titled *Socialist Guang Changtong* (社会主义廣長舌), was translated into Chinese by the National Congshu Sha (国民丛書社), and published by the same society in 1902. *The Essence of Socialism*, translated by Shu Soul Yao (耆生), was published by the Socialist Seminar of the Chinese Students' Hall in 1906; there is also a translation by Chuang Sheng, published by Kyuwenkaku in Tokyo in 1907. Koude Akimizu sought to use the writings of Marx and Engels to discuss Marx's theory of socialism, an important treatise in the history of early socialist thought, and one that was particularly influential on the study of Marxism by students in Japan. Zhu Jixin, a political commentator of the League of

18 Rong Mengyuan, "*The Introduction of Marxism in Chinese Books and Magazines before the Xinhai Revolution*," *New Construction*, March 1952 issue.

Nations, published "*A Short Biography of the German Social Revolutionaries*" in No. 2 of the *Minbao*, in which he introduced for the first time the biographies of Marx, Engels, and Lassalle, and gave a very brief introduction to the *Communist Manifesto* (1848). Extracts of the *Communist Manifesto* were not published in the *Weekly Review* until 1919, when the May Fourth New Culture Movement emerged. The first translation of the *Communist Manifesto* into Chinese was the partial translation of the first chapter, which was directly translated by Sakai Toshihiko of Japan from the English text, and then translated into Chinese by Minming, with a preface to the *Communist Manifesto* written by Liu Shipei and published in the four volumes of *Tianyi Newspaper*, Volumes 16, 17, 18, and 19, which were not circulated at that time.[19] This is the earliest historical testimony to the propagation of Marxism by students in Japan.

### *Studying Abroad and the Trend of Western Studies*

The Qing court, proceeding from supporting students in the tide of studying in Japan, took a further step. In 1901 and 1904, policies were unveiled to encourage students to study in Britain, France, Germany, Belgium, and the United States. The U.S., fully aware of its disadvantages in competing with Japan in attracting Chinese students, such as higher cost and longer distance, came to the realization that the manipulation of China's cultural transformation and talent cultivation through the improvement of missionary school facilities and the promotion of new-type schools would only be possible if they succeeded in exerting influence on the youngsters. Arthur H. Smith, an American missionary, suggested that funds drawn from the cash indemnity by the Qing government since 1902, based on the Treaty of 1901, be refunded and invested into education, in an attempt to win over youngsters and increase American influence in China. A memorandum was submitted to President Roosevelt from Edmund J. James, the president of the University of Illinois in 1906, stating that countries would benefit spiritually, intellectually, and economically from their investment in the education of China's young generation during the drastic social transformation that was about to occur in China. The memorandum shed light on the fact that Chinese students studying in Europe, after returning to China, would persuade

[19] Lin Daizhao and Pan Guohua, eds, *Marxism in China* (Volume 1), Tsinghua University Press, 1983, 11 pp.

the Chinese government to follow the example of Europe instead of America, recommend teachers and engineers from Britain, France, and Germany, purchase goods from these countries, and grant them privileges in different industries. Mr. James advised that the trend of Chinese students studying abroad be steered toward and expanded in America, so as to control China through controlling its future leaders.

The U.S. government and Congress were persuaded. A bill was passed to refund the indemnity—which the Qing government was paying to the 14 foreign powers involved in the invasion of Beijing to crush the Boxer Uprising in 1900—from 1908 and until the expiry of the said indemnity in 1940. The money would be used, partly, to set up Qinghua (Tsing Hwa) in Beijing as a preparatory college, and partly, to subsidize the Chinese graduates from the preparatory college to continue their studies in American universities. This was an ingenious idea to train Chinese students in American institutes with Chinese money—in the hope that these Americanized Chinese returned students would eventually introduce American ideas to China's political, economic, cultural, and educational life. In September 1909, an agency, specializing in selecting and preparing Chinese students to study in the U.S., was established in Beijing. Three batches of students were sent to America in 1909, 1910, and 1911 respectively, with a total number of 180. Among them were Mei Yiqi, who later became the president of Qinghua University, and many more celebrated scholars including Bing Zhi, Hu Shi, Zhao Yuanren, and Zhu Kezhen. Qinghua School was established as a preparatory school based on a section under the mentioned agency. It later became China's Qinghua University and maintained a close link with several American universities. Since 1929, those who wished to study in the U.S. had been openly selected across China instead of being exclusively selected from Qinghua. From 1911 to 1929, the refund fully funded 1279 Chinese students, and the U.S. also subsidized 475 students with money drawn from the indemnity.

After the United States, Britain, France, Belgium, and other Western countries also decided to return the balance of the Geng money to the Chinese government for railway construction and cultural education. Chinese and American governments to manage the return of Jung money, decided by the Chinese and American people jointly formed the China Education and Culture Foundation (CECF), the board of directors of the Ministry of Foreign Affairs in Beijing on September 18, 1924 formally established, the first board of directors by the Chinese directors of the 10, the United States of America, composed of 5 people, are the relevant government departments, the education and culture of the authoritative personage. The institutions directly founded by the Foundation include the Chinese-American Association for the Advancement of Science and Technology in New York, the Scientific and Educational Advisory Committee (later changed to the Compilation and Translation Committee) in Nanjing, the National Beiping Library in Beijing, the Static Biological Research Institute, and the Social Research Institute. The most famous cultural programs funded by the Foundation include the Shanghai Oriental Cultural Programme Committee and the Beiping Chinese Construction Society.

To manage the return of the Geng money, China and France set up the Chinese Delegation of the Sino-French Education Fund Committee in 1925, which used the money to subsidize the Chinese College of the University of Paris and the University of Lyon and to send students to study in France. In China, the funds were used to support the Chinese-French University in Beiping, the Dafa High School, and many schools and academic institutions established by the French Catholic Church in China, the most famous of which were the Aurora University and the Xujiahui Observatory in Shanghai.

The first enthusiastic trend of studying abroad after the 1911 Revolution started with the establishment of non-governmental organizations offering work-study programs to students across the country. Applicants surged from dozens to thousands. Between 1912 and 1913, four batches of personnel went to France to study under the work-study program, including 100 students and 58 workers. However, the outbreak of the First World War interrupted the trend. Many Chinese workers were recruited by the French government to support the country in the war. Some students were transferred to Britain and France's southwestern region. 1916 saw the establishment of the Sino-French Society of Education, which improved cultural exchange between the two countries for the following 20 years. After the war ended in November 1918, the society requested the French government to provide education opportunities to the Chinese workers mentioned above. Under the approval of the authority, approximately 1700 students were sent to France between 1919 and 1921. This trend came to a halt in 1922 due to the economic recession in France. Some students, without guarantees on living expenses and study costs, returned home in 1927, and some others later chose Germany over France for its lower cost of living. Only 364 students remained in France by the end of 1927, and over 300 of them were at their own expense.

The second enthusiastic trend of studying abroad started in 1920 and lasted until 1938. The first phase, 1920–1929, saw students studying in Japan at official expense topping those in other countries in both quantity and proportion and peaking at 1025, 61.7% in 1929. America came in second, with 272 Chinese students studying there, accounting for 16.4% of the whole population. However, the second phase, 1934–1936, saw a decline in the number of students studying in Japan. 1937 hit a record low with only 49 Chinese students in Japan. Since 1931, liberal arts students have outnumbered science students in Japan, and the same change took place in Europe and America during this period. During the same period, the same change occurred in the number of students going to Europe at public expense. In the United States, on the other hand, from 1933 onwards, students of science and engineering outnumbered those of arts and sciences, and there has been no fundamental change in the decade or so since then.[20]

After the outbreak of the War of Resistance against Japanese Aggression, only those with foreign scholarships or subsidies were allowed to study abroad. The subjects mainly focused on military science, engineering, science, and medicine. This limitation was lifted in 1942. 327 students passed the entrance examination in 1943 and all of them were sent to the United States self-funded. 1934 students passed the examination in 1946, with 148 of them at official expense. Since 1944, the U.S. government has appropriated special funds from the C.E.F., Lend-lease Act, to train Chinese personnel in the fields of engineering technology, medical treatment and health care, and finance and economics. The number of Chinese overseas students and personnel trained in the U.S. reached 10,729, which set a new record in the history of Chinese students studying abroad.

[20] Shen Fuwei, *Western Culture and China* (1793–2000), 626–633.

## 9.3 Seeking Progress Under the Banner of Democracy and Science

### *Reform of the Academic Structure and the Recovery of Educational Sovereignty in the Early Republic of China*

In the early years of the Republic of China, Cai Yuanpei presided over the Education Department of the Nanjing Provisional Government and succeeded the Education Director of the Beijing Government. Based on his experience of studying and inspecting in Germany from 1907 to 1911, he fundamentally reformed the educational system issued by the Qing government. He replaced the five purposes which were loyalty to the monarch, respect for Confucius, respect for the public, martial arts, and practicality with military education, practical education, moral education, worldview education, and aesthetic education. He ordered the conversion of traditional academies into modern schools, the abolition of respecting Confucius and reading scriptures in primary schools, and the issuance of new textbooks that conform to the purposes of the Republic. The provision that primary schools can be co-educational greatly increased the opportunities for women to receive education. Secondary schools as regular education were no longer divided into Art and Science. This reformation swept thousands of years of feudal education into the trash. The school system announced in 1912, from enrollment at the age of 6 to graduation from university at the age of 24, had an 11-year academic system for primary and secondary schools, roughly referring to the French academic system. Universities have a 3-year preparatory program and a 4-year undergraduate program, totaling 7 years. Industrial education (equivalent to current secondary vocational schools) and teacher education were both valued, and teacher education was modeled after the Japanese educational system. Especially in industrial education, Type A industrial schools (4 years) and 5-year specialized schools are equivalent to higher specialized education. From the age of 11, when you enter a type B industrial school, to the age of 22, when you graduate from a specialized school, which is equivalent to a university college. The educational system from Class A industrial schools to specialized schools has pioneered the paradigm of vocational and professional education in China for an entire century.

Due to China's long-term failure to unify, the education sector was also divided between the North and the South. It was not until December

1921 that the education communities of the South and the North jointly formed the China Education Improvement Society, proposing specific plans for the implementation of the separation of religion and education for the first time. The basis of this plan is the advocation that Cai Yuanpei, when he was president of Peking University from 1917 to 1920, proposed to follow the ideas of the French positivism philosopher Auguste Comte (1798–1857) and his successors. Cai Yuanpei advocated replacing the education policy of a large number of missionary schools in China that used Christian doctrine as the education purpose with the idea that Comte paid attention to the scientific spirit. The subsequent movement to reclaim educational sovereignty began on August 1, 1917, with a speech titled "Replacing Religion with Aesthetic Education" at the Beijing Shenzhou Society. By March 1922, Cai Yuanpei had published "*Opinions on Church Education*" in Volume 4, Issue 4 of *New Education*, clearly advocating that education must be separated from religion, that all types of schools should exclude religious courses, and that students should not be forced to participate in religious ceremonies or join religious believers. He put it bluntly: "There is no need for theology in universities; there are no courses to promote doctrines in any school, and prayer ceremonies are not held; people who engage in missionary work do not need to participate in the education industry." He firmly demanded that the education industry learn from France, cut off the claws of foreign churches extending into China's educational gardens, and transfer the education sovereignty that cultivates talents and relates to the future destiny of the country to the management of the Chinese people.

After the establishment of the China Education Improvement Society, John Dewey (1859–1952), an American educator who advocated for life education, popular education, the purpose of education being to cultivate people who adapt to the environment, and opposed any external educational purposes, was invited to China for inspection from May 1919 to June 1921. During this period, he gave lectures in Beijing, Taiyuan, Shanghai, Nanjing, Changsha, Guangdong, Fujian, Liaoning, and other places, promoting the reform of China's educational system, using experimental education instead of respecting Confucius and reading scriptures in the early Republic of China. In November 1922, the Ministry of Education officially announced the school system reform plan and implemented the third school system reform, known as the Renxu school system, announcing the original meaning of education is to achieve the seven principles of "adapting to the needs of social evolution, leveraging the spirit of popular education for individual development, paying attention to national economic strength, paying attention to life education, making education easy to popularize and leaving room for expansion in various places." The new educational system introduced the 6-3-3 academic system from the United States, with children enrolled at the age of 6 and graduating from universities (graduate schools) at the age of 26. The reform has simplified and clarified the complex school system, laying the foundation for the shaping of the Chinese school system in the twentieth century.

The new school system and school management regulations pushed the recovery of the right to education to a climax. In 1926, young people and students from Beijing, Shanghai, Tianjin, Wuhan, Guangzhou, Hangzhou, Jinan, Chengdu, Fuzhou, and other places demanded the hand-over of the right to run Christian schools to the Chinese authorities, and the establishment of a non-Christian alliance to compete with the Chinese Christian Education Association under the control of the Christian church. In October 1926, the Nanjing National Government issued the *Regulations for Private Schools*, which determined to include missionary schools in private schools, and stipulated that missionary schools must register with the local government. Under the circumstances, the Chinese Christian Education Association had to decide in 1927 to return the school to the Chinese people for management and instructed local church schools to register with the government. In 1926, there were 15,000 missionary schools (including 6000 Christian schools and more than 9000 Catholic schools) with a total of 800,000 students (including 300,000 students from Christian schools and 500,000 students from Catholic schools) gradually being incorporated into China's national education system and subject to administrative supervision by the competent authorities. National schools were established as three major types which were national (or public), provincial, and private, and were not fully incorporated into the education system under the unified supervision of regulatory authorities until 1950.

As the highest level of missionary schools, missionary universities were gradually transformed into private schools after 1927 and incorporated into China's national education system. Missionary universities were gradually established in the last 20 years of the nineteenth century as Western studies courses were gradually promoted and offered by academies sponsored by various churches. Tengchow College and Christian College in China were once registered as universities in the United States, but after evaluation, they were only equivalent to high schools in the United States and only offered pre-university courses. In 1901, only to start the university preparatory course of the Suzhou Soochow University Hall was the first to hang out the sign of the University Hall, Wuchang Wenhua University Hall (1903), Lingnan Academy (1904), and North China Women's Concordia University (1905) also followed since the steps, but most of them did not formally open the university program. To upgrade their earlier colleges to universities is still many years later.

Of the 14 Christian church universities established by the United States in China, Changsha Xiangya Medical College was the first to break away from the church and become Chinese-owned. The school officially opened in December 1914 as Xiangya Medical College, with a preparatory course (two years), followed by a five-year undergraduate program. 1924 saw the expiry of the contract, and the Chinese took over, renaming it in 1929 as the Private Xiangya Medical College, and in 1940 as the National Xiangya Medical College. The elimination rate of students was more than 3/4.

In 1934, there were 13 universities in China with 5718 students, 3 medical schools, and 7 seminaries with a total of 6088 students.

St John's University, the earliest of the 13 universities established, was formerly known as Peiya College and Doon College, which were opened by the Anglican Church in Shanghai in September 1879, and was officially named St John's University in 1905; it was registered in the United States in the District of Columbia in 1906, and was set up with four disciplines of arts, sciences, medicine, and theology; the first batch of graduates in 1907 consisted of six students. Later, the university expanded its engineering and agricultural colleges, with a total of 17 departments in six colleges. The motto of the university was "Light and Truth," which fully embodied Christian education, and to cultivate interaction with Westerners and seek employment in foreign companies, the university fully implemented English-medium teaching and daily life, and the graduates' English proficiency topped the list in China. Attached to Tongren Hospital is the internship hospital of the Medical College. The library has a collection of 150,000 books, similar to that of Yanjing University in Beijing. The number of graduates who went on to graduate school in the United States ranged from a few dozens to several hundred, with only a small percentage returning to China. Among the graduates were Gu Weijun and Yan Huiqing, political figures in the early years of the Republic of China, and Song Ziwen, a financial magnate in the later years of the Republic of China.

Soochow University, located in Tianzhizhuang, Suzhou, was originally the site of the Boshi College, which was opened by the Board of Supervisors in 1884. The university was first opened as a preparatory school, and Zhang Taiyan was invited to give Chinese-language lessons; in June 1901, it was registered in Tennessee, USA, and only offered university courses in 1904; in 1908, the English name of Soochow University was established; in 1912, the Chinese and Western College in Shanghai was incorporated into Soochow University; in 1915, the Methodology Department was opened on the site of the former Chinese and Western College in Shanghai, and in 1927, it was upgraded to the School of Law, and in 1935, it was granted a charter by the Judicial Yuan, and was the best academic program of Soochow University. In 1935, it received a charter from the Judicial Yuan and is the most outstanding discipline of Soochow University.

Huazhong University was founded in 1903 by the Anglican Church in the U.S.A. The American Episcopal Church set up the Wenhua University Hall in Wuchang, which was later merged with the Wuchang Bowen College and the Hankou Bokuei College, and then formed into the private Wuchang Huazhong University in 1924, with the subjects of arts and sciences as well as librarianship, and then set up the Huazhong University of Hubei with three colleges of arts, sciences, and education, which was incorporated in 1931.

Zhijiang College of Arts and Sciences, expanded in 1907 from the Presbyterian Yuying College to form Zhijiang University Hall, moved to the vicinity of Liuho Pagoda outside Hangzhou in 1911 to form Zhijiang University, registered in the U.S. District of Columbia in 1920, and was founded as Zhijiang College of Arts and Sciences in 1931, with twelve faculties of arts, commerce, and engineering.

Lingnan University, the predecessor of the 1888 Presbyterian Church of the Gezhi College, moved to Macau in 1900 for four years, moved back to Guangzhou in 1904, the university department of only six students, known as the Lingnan School. 1916, after the expansion of the name Lingnan University, in January 1927 it became a private Lingnan University. In 1938, it set up 20 departments in 6 colleges of arts and sciences, agriculture, industry, commerce, medicine, and theology.

The University of Shanghai was founded in Shanghai by two Baptist churches, the Southern and Northern Baptist Churches of the U.S.A.

The Baptist University Church was formally established in 1908, with two divisions, a university, and a seminary, and the University of Shanghai was formally established in Yangshupu, Hudong, in February 1914, 1927 Dr. Zhanen Liu was elected as the first Chinese president of the university. In 1929, the English name of the university was changed to the University of Shanghai, with three colleges of arts, science, and commerce.

Jinling University was founded in 1910 by American Baptists, Methodists, Presbyterians, and Christians, registered in New York State in 1911, filed with the Beijing government in 1925, one of the earliest church universities, and then filed with the Nanjing government in September 1928, with 21 faculties of arts, sciences, and agriculture.

Qilu University, by the American Presbyterian Church, and the British Baptist Church, mainly a total of five American missionary societies, and five British missionary societies year by year through the opening of the parallel school, was formally established in Jinan in 1917. There were three colleges: Arts, Medicine, and Theology. In 1929, there were three colleges of arts, science, and medicine, and the medical school continued to expand as a major focus.

West China Association University, 1910, the United States of America Methodist, Baptist, and Anglican Church and other five churches jointly opened in Chengdu Huaxiba, 1913, the addition of medical, 1914, dental, 1934, registered in New York State, since then, dentistry, medicine, made rapid progress.

South China Women's College of Arts and Sciences, the United States Methodist Church in Fuzhou in 1914 officially established the South China Women's University, set up a 2-year specialty, and in 1917 opened a 4-year university program. 1933 case called South China Women's College of Arts and Sciences, with arts and sciences 6 departments.

Fujian Concordia University, a union of 3 missions, including the American Congregational Church and 3 missions, including the Anglican Church, established Fujian Concordia College in Fuzhou in 1915 and registered in New York State in 1918. Registered in 1931, there are 11 departments in 3 colleges of arts, science, and agriculture.

Jinling Women's College of Arts and Sciences was opened in Nanjing in 1915 by the American Baptist Church, the Board of Supervisors, and five other churches 1915, Jinling Women's University, which became a joint school of seven churches in the U.S.A., one church in England (the London Society), and Smith's Woman's University of Massachusetts, U.S.A. It was registered in the State of New York in 1918, and the first graduates of the school in 1919, with only five graduates, obtained the degree of recognition from the State University of New York.1930 In December, the case was filed and the name was changed to Jinling Women's College of Arts and Sciences. The departments of music and physical education were the largest in the country.

Yanjing University was the latest church university to be established, a joint venture of the Presbyterian Church (USA), the Methodist Church, the Methodist Women's Church, the Congregationalist Church, and the Church of London (UK), and was founded in Beijing in 1916. Men's and women's campuses. John Leighton Stuart (1876–1962), the pastor of the Southern Presbyterian Church in the United States, was a missionary to China from 1904 until January 1919, when he transferred from the Jinling Theological Seminary to Yenching University, where he solicited funds from the Princeton Consortium and Rockefeller Consortium, with support from the New York Department of Trusteeship, and received a total of $4 million from the United States from 1921 to 1937, with a total of $4 million in donations from The total donations from 1921 to 1937 were more than 4 million U.S. dollars.

In 1921–1937, the total donations from the United States were more than 4 million U.S. dollars. He moved the campus from the armor factory Huiwen University to Haidian in the northwest suburb of Beijing, with more than 80 large and small buildings designed by Murphy in the style of Chinese palaces. 1927 moved to the new site, the theological seminary became independent, and the whole university had a total of 16 departments in three colleges of arts, science, and law, with journalism and foreign languages being the most famous ones. After the filing of the case in 1929, Stuart took control of the board of trustees in his capacity as the president of the university and made Wu Leichuan, who had been a Hanlin of the former Qing Dynasty, the Chinese president. After the outbreak of the Pacific War, Stanton was imprisoned by the Japanese and did not leave Yanjing until 1946 when he took up his post as U.S. Ambassador to China.

The Catholic Church also ran four universities in China, the American ones being the Fu Jen University in Beiping (1925) and the Aurora Women's College of Arts and Sciences in Shanghai (1937). The French Jesuits ran the Aurora University (1903) and the Tianjin College of Commerce and Industry (1923).

All of these church universities filed their cases with the government one after another after 1925. Only St John's University, which was the earliest to open and had the largest number of students, insisted on filing a case until 1947 when the president was replaced. The rector, Bu Fangji, became Don Quixote who died with the cultural fortresses established by the Church in China.

### *Identification of Traditional Chinese and Western Medicine*

Although Western medicine was gradually implemented in coastal and riverside areas of China in the nineteenth century, the progress of Western medical education was slow, and the work of cultivating Chinese doctors who practiced Western medicine had little effect. Until the outbreak of World War I, there were still less than a thousand Chinese doctors who practiced Western medicine, making it like a drop in the ocean compared to the medical personnel of traditional Chinese medicine. Western medicine, as a new type of medicine, was difficult to fit into the imperial examination system. It cannot attract intellectuals who are eager for official positions and must also compete with traditional Chinese medicine in various branches. A minister of the Prime Minister's Office at that time then claimed, "In fact, I personally do not believe in Western medicine."[21] Due to the authorities not allowing students in the physiology class to intern at church hospitals, they only had medical textbook knowledge and had no hope of becoming specialized Western medicine talents. Although diseases that cannot be treated or delayed by traditional Chinese medicine can be cured by continuously improving Western medicine, traditional Chinese medicine has a vast team and broad audience, deeply rooted in the psychology and physiology of the people, hindering the promotion of Western medicine. The controversy between traditional Chinese medicine and Western medicine, sparked by smallpox and cowpox, had not yet caused a stir among the authorities in the nineteenth century, indicating that the promotion of Western medicine still required more anticipation.

Before cowpox was introduced to China from the West, China obtained the human pox inoculation method from the Dan family in Taiping County, Ningguo Prefecture, during the Longqing period of the Ming Dynasty (1567–1572) (Yu Maokun: *Poxidae jinjingfu jijie* 1727),

[21] W.A.P. Martin, *A Cycle of Oathay, Fleming*, New York, 1897, pp. 320–322.

and the *Zhang's Medical Tong* of 1695 listed the pox planting techniques as the pox clothes method and the nose seedling method, and the government-issued *Jinjian of the Medical Clan* of 1742 recorded the human pox inoculation technique as well. By the beginning of the eighteenth century, pox planting had become widespread in both rural and urban areas. Although the imported cowpox had all kinds of convenience and hygiene, and the structure of the virus was changed by the body of the cow, it was no longer harmful to enter into the body. However, due to the difference in the technology of cowpox inoculation, which was equal to the difference in the preparation and preservation of drugs, the effect was also different, resulting in the dissemination of a certain degree of obstruction; not to mention that the Chinese and Western medical treatment methods and equipment were also different, the Western medical practitioners used the knife and stab to inoculate the cowpox, which was very different from Chinese medicine, which used pulp and nasal seedling to inoculate the cowpox, so the public was suspicious of cowpox, and watched it, and refused to many people. The most crucial thing is that the more cowpox was spread, the more most Chinese doctors were against it. In 1885, when Xu Yue-bian made a preface to Shen Shanfeng's *New Series of Cowpox*, he pointed out with a pinch of salt: "Cowpox is perfect, the most impediment to the medical treatment of cowpox; cowpox is not necessary to prolong the medical treatment and is not favorable to the early childhood department; there is no leftover poison in cowpox and is not favorable to the surgical department; cowpox is not a medicine for the joy of the drugstore, and is not without a small loss. Therefore, every time some people shoot for profit, they regard the good deed as the end of jealousy and secretly instigate confusion. To stab for the amazing words, to re-emerge for the words of obstruction." Thus expanding, the promotion of Western medicine, not only to see whether medical education has made great progress, but also depends on the Chinese medicine can gradually accept some of the treatment methods of Western medicine, the transformation of traditional medicine.

By the end of the nineteenth century, the achievements of Western medicine in clinical practice had finally led to the resistance of some completely disbelieving arguments toward Western medicine, while others, who felt the need to learn from the strengths of Western medicine, had already felt it imperative to reform traditional Chinese medicine and called for the integration of Chinese and Western medicine. In 1892, Zhu Peiwen, a native of Nanhai, Guangdong, was influenced by Western medicine and believed that Chinese and Western medicine could connect while coexisting with their differences. His evaluation of Chinese and Western medicine was "each with its own right and wrong." He had seen the importance of anatomy in advancing surgical procedures.

The person who further requested the integration of Chinese and Western medicine was Ding Fubao (1874–1952), a native of Wuxi who had taught at the Beijing University and the Translation School. He founded a medical book company in Shanghai to spread new medical knowledge, served as the Vice President of the Chinese Medical Association established in 1904, and published the "Medical Journal," making Shanghai truly a center of Western medicine. In May 1909, Ding Fubao was ordered to investigate medicine in Japan. He saw that Japan, which had always been popular with Eastern medicine, introduced Western medicine and made rapid progress in medicine. After returning to his home country, he raised his own funds and established the Chinese and Western Medicine Research Association in Changshouli, Parker Road, Shanghai, which received support from the authorities. Starting from 1910, the publication "*Journal of Traditional Chinese and Western Medicine*" was published. This publication was the first non-Christian Western medicine journal published by John Kerry in Guangzhou, following the publication of the Chinese medical monthly "*Western Medicine New Newspaper*" in 1884 and the *Chinese Medical Missionary Journal* in 1888. The members of the Association for the Study of Traditional Chinese and Western Medicine are spread throughout Henan, Hebei, Zhejiang, Anhui, Jiangsu, Jiangxi, Fujian, Sichuan, and various provinces in the northwest and southwest. At the beginning of its establishment, it compiled a complete set of medical books such as the *German Medical Series* and *Physiological and Health Textbooks*, providing traditional Chinese doctors with many specialized new medical knowledge and technologies for learning Western medicine.

At the same time, China's Western medicine education also made a breakthrough in technology. After the establishment of the Beijing Medical School in 1912, it specially submitted papers to the authorities, requesting to carry out autopsy practice which Dr. Wu Lien-teh had carried out during the epidemic of the pestis in northeast China from 1910 to 1911 in pathology education. The Beijing government, forced by public opinion, issued papers in November 1913 and April 1914, officially approving medical schools and hospitals to use autopsies for scientific experiments, and retaining certain parts of the body for medical demonstration purposes when necessary. Surgical medicine has taken a step toward practical application since then.

The integration of traditional Chinese and Western medicine requires the emergence of combinations and enhanced connections in the academic field to be able to step up to the formal level. The establishment of the Chinese Medical Association in Shanghai in 1916, initiated by returned international students, was a new starting point. In 1917, the Chinese Medical Association and the Christian Chinese Medical Association jointly held an annual meeting in Guangzhou. By 1932, the two associations were merged and the Chinese Medical Association was established. Only then did Western medicine in China unite into one group. The rapid growth of Chinese doctors in Western medicine indicates that Western medicine will inevitably become the mainstream of future medical development. The existing medical education bore the responsibility of cultivating talents and delivering high-level Western medicine to society. The representational case was the expansion of Peking Union Medical School in June 1916, which was funded by the Rockefeller Foundation of the United States and changed its name to Peking Union Medical College. It was the first medical university founded by non-church foundations in China. For the next 20 years, the Peking Union Medical College was not only the best medical school in the Far East but also a renowned medical school with first-class level in the world.

According to Rethinking Missions: *A Layman's Inquiry after One Hundred Years* in 1932, Chinese doctors increased steadily around 1925, compared to Western doctors with 55:45 in 1920, 57:43 in 1925, and 67:33 in 1930. In the last year of the inquiry, 275 Western doctors were serving in church hospitals, and the number of qualified Chinese doctors had increased to 400, while the number of Chinese nurses had increased to around 700. Most of these Chinese doctors and nurses graduated from newly established medical colleges and received strict medical training. The growth of this team determined that the structure of the Western medical team would ultimately undergo fundamental changes, with Chinese doctors replacing Western doctors. Since the establishment of a medical school at Shandong University in 1926, many national and provincial universities have established medical schools. From Central University (Nanjing), Peking University, Wuhan University, Zhejiang University, Sun Yat-sen University (Guangzhou), Shanxi University (Taiyuan), Yunnan University (Kunming), to the establishment of Lanzhou University and Northwest University in 1946, the construction of medical schools had been regarded as a basic discipline in education, and support had been provided to enable China's Western medicine to have successors and vigorous development.

Chinese traditional medicine must also accelerate the pace of scientific transformation in the new situation, transform traditional Chinese medicine into a new medicine with modern diagnostic and treatment facilities and talent cultivation channels, continue to promote some of the advantages and strengths of traditional Chinese medicine and incorporate the old method of traditional Chinese medicine education from master to apprentice into the school education system, to cultivate new traditional Chinese medicine with modern medical knowledge. In 1928, to catch up with the increasingly standardized education of Western medicine, the traditional Chinese medicine community began to establish 11 schools, including Shanghai Traditional Chinese Medicine Vocational School, China Medical College (Shanghai), Shanghai Traditional Chinese Medicine College, Suzhou Traditional Chinese Medicine Vocational School, Zhejiang Traditional Chinese Medicine Vocational School, Lanxi Traditional Chinese Medicine Vocational School, Guangdong Traditional Chinese Medicine Vocational School, Guangdong Guanghan Traditional Chinese Medicine Vocational School, Meixian Chonggu Medical School, Henan Traditional Chinese Medicine School, and Wuxi Traditional Chinese Medicine Teaching Institute. In July 1929, 11 schools gathered at the Chinese Medical College in Shanghai to standardize the curriculum of traditional Chinese medicine schools, determine 29 courses, and begin writing textbooks. However, the National Federation of Medical Organizations initiated by the Shanghai Traditional Chinese Medicine community was finally forced to be abolished in 1931, two years after its establishment. It showed that only traditional Chinese medicine transforming and established hospitals can adapt to the trend of future medical development.

### *Industrial Salvation of the Nation and the Scientific Movement*

In the course of history in the nineteenth century, which is known as the century of science, China was unable to throw off the hat of scientific backwardness due to its being under the feudal and autocratic rule of the Qing Dynasty, and it was always beaten and became the laggard of the times. Some Western brain scientists, in their neuroanatomical studies, even argued, based on one-sided specimens, that the IQ of the Chinese human brain was close to that of apes. This view was put forward by J. L. Shellshear, a British scholar at the University of Hong Kong, in the British *Journal of Anatomy* in 1926. Afterward, he read a paper entitled "*A Comparison of the Chinese Brain and the Australian Brain*" at the Second International Anthropological Conference in 1934, stressing again that the occipitolunar sulcus of the Chinese brain was often in a primitive state, and was therefore inferior to the Caucasian brain, and similar to the ape brain. This argument was immediately and forcefully refuted by the participating Chinese scholars Ouyang Zhu and Wu Dingliang. Chinese scholars continued to conduct in-depth research in the relevant fields. Lu Yudao observed slices of the cerebral cortex of Chinese and Westerners and did not find any significant difference between the two; Wang Youhong studied the sinking area (insula) of the Chinese brain according to 30 cases of human brains and found that Chinese people were not different from whites or blacks in the form, length, and depth of the sinking grooves, which proved that even if certain differences were found in the shape or internal structure of the human brain, they could only have a reference value for the study of the physique and the brain's structure. It proves that even if some differences are found in the shape or internal structure of the human brain, they can only be used as a reference value in the study of physical fitness and anthropology, and have nothing to do with the judgment of IQ. As a result, the conclusions made by Xiao Siya with three to seven Chinese human brain specimens were completely bankrupt. However, the social reality is that due to China's long-term lag in the modernization process, Western scientists with racial prejudice deliberately planned to find a nearly "true" explanation, which was similar to the fact that China was often accused by scientists from some countries of being "a country without science" at international academic conferences at that time, resulting in the attendance of the conference. This, together with the fact that China was often criticized by scientists from some countries as "a country without science" at international academic conferences at that time, caused Chinese scientists attending the conferences to be often faced with dilemmas, and almost formed a trend. This directly prompted a group of Chinese students from the early twentieth century to study science and engineering in the developed countries of the West, to explore ways of further improvement in their own academic fields from the perspective of their country's destiny and future. The results could not be divided into two: one was to return to China to serve the country and make science blossom in the motherland

and have new generations; the other was to stay in the host country after completion of their studies and, relying on the advantages of the environment, expect to rise rapidly in academics and achieve even greater results. The requirements for international students who went to Europe and the United States to study science in those times were very severe and harsh, especially those gardeners who were determined to return to their home countries as early as possible, and who had to go through hardships to sow the seeds of science, the burdens they shouldered were dozens of times heavier than those of their classmates and colleagues in the countries in which they studied abroad!

The 10 international students who studied at Cornell University in the United States in 1914 experienced that there was no way to establish a country without science, and what China lacked the most was science. Therefore, their first thought was to establish the monthly magazine Science to convey the latest scientific achievements to China. This publication aimed to promote science, promote industry, approve terminology, and disseminate knowledge, to elucidate the essence and utility of science. It was signed on the manifesto "*The Origin of Science Monthly*" by Ren Hongjun, Hu Mingfu, Zhao Yuanren, Zhou Ren, Bingzhi, Zhang Yuanshan, Guo Tanxian, Jin Bangzheng, and Yang Quan. They first revealed the banner of "saving the country through science" in the novel style "*Science Monthly*" in January 1915.

Compared with the first scientific periodical "*Yazhuan Magazine*" issued by Du Yazhuan in Shanghai in 1900, the monthly "*Science*" embodied a more modern atmosphere and a completely scientific spirit from the form to the content, which was completely modeled on the Western magazines, adopting the Western horizontal format and published regularly by the Shanghai Huanqiu Chinese Students' Association. As the first modernized and comprehensive scientific periodical in Chinese history, it brought to the Chinese intellectual world a scientific spirit and a report on the dynamics of modern life that was completely different from that of the old traditional or Christian church journals. While planning the publication of scientific journals, the American students also organized a scientific society, which was formally established in October 1915, to promote the spirit of science, liaising with comrades, researching academics, and achieving the development of science in China. Once served as the secretary of the provisional president Sun Yatsen, Ren Hongjun (1886–1961) was elected president of the Board of Directors and president of the Chinese Science Society, and he devoted his whole life to the Chinese Science Society. In 1916 September, the Chinese Science Society held its first annual meeting in Endover Higher Education. Studying for a master's degree at Harvard University and also serving as the editor of the monthly magazine *Science*, Yang Quan (also known as Xingfo) called for the combination of science and industry to achieve the purpose of industrial salvation, advocated the establishment of the country's statistics and sociology, and to solve the problem of social livelihood.

In October 1918, Ren Hongjun and Yang Quan returned to China. In 1926, Yang Quan published an article entitled *Science and Revolution*, which advocated that "the only way to save the country is to co-operate between science and revolution. In other words, revolutionaries must have scientific knowledge, scientists must have the spirit of revolution, and work together to study the social problems, as well as all the personal problems of life, so that China can be saved, and the world can be bright!" (*Yang Xingfo Wencun*, Shanghai Pingfan Bookstore 1929, 69, pp. ff.). To change the international ridicule of China's lack of science, Ren Hongjun traveled around, through the Chinese Education and Culture Foundation established by the United States to return the balance of the Geng money, establishment of the Compilation Committee, the National Library of Beiping, the Institute of Social Research (location: Beiping), the Institute of Biological Investigation of the Jing Sheng Biological Survey (location: Beiping). The Institute of Biological Research, established by the Chinese Society for Science in 1922, became an internationally renowned research institute under the auspices of Bingzhi and Hu Xiansu.

In April 1928, under the leadership of Cai Yuanpei, the Academia Sinica promoted China's academic revitalization with the purpose of "carrying out scientific research, guiding, contacting and rewarding national research undertakings, and seeking scientific progress and human brightness."[22] Under the guidance of the thought of "scientific research being the foundation of all undertakings," Cai Yuanpei believed that it is necessary not only to change the backward material science, but also to adopt new methods and new horizons in the fields of Geisteswissenschaften (such as psychology, aesthetics), social science (such as economics, sociology), and other Western departments that have used scientific methods to study and to remove old methods and old ideas. To catch up with

[22] *Draft Amendment to the Organisational Regulations of the Academia Sinica*, *National Government Gazette*, No. 48, May 1928.

the scientific trend, he believes that the most important thing is to make scientific methods widely adopted by the public. He frankly believes that the development of science is first and foremost.

The Academia Sinica was officially established in Nanjing on June 9, 1928. It initially established 8 research institutes (institutes of geology, astronomy, meteorology, physics, chemistry, engineering, history and language, and social sciences). In 1929, it added animal and plant research institutes and psychological research institutes, which were expanded to 10. Later, it added 13 research institutes of mathematics, medicine, pharmacology, physical anthropology, geography, archaeology, and ethnology, with a total of 23 research institutes. The Academia Sinica Council, which Cai Yuanpei planned to establish, became the highest academic evaluation organ in the country. Academicians elected by the Council were honorary posts for life. They were academic leaders in various disciplines. The Academia Sinica has managed to overcome its funding difficulties, sent people to participate in many international academic conferences, put forward papers of academic value, and exchanged academic journals with the international academic community, thus improving China's position in the international academic community.

Cai Yuanpei published "*Reading and Saving the Country*" on March 12, 1927, advocating reading to save the country, i.e. education to save the country, science to save the country, and in May 1932, he delivered a lecture on "*Science to Save the Country*" at Wuchang Chinese University, advocating the use of science to save the country from peril. Science to save the country became a social movement of the Chinese public after the Japanese invasion of Manchuria during the September 18th Incident, and it was against this background that the movement emerged. Nanjing scholars Chen Lifu, Gu Yuchuan, Zhang Qiyun, and others established the Chinese Association for Scientific Movement, which was formally proclaimed on November 4, 1932 in Nanjing with the goal of "identifying the public and society." The aim of their book declared that they would send scientific knowledge to the people, "to make it the common wisdom of the general public, and they hoped that this knowledge, after being disseminated to the people, could generate a strong force to extend the life span of the nation, which was already at the point of life and death, and to revive our declining Chinese culture." They set up the bimonthly "*Scientific China*," which was popular in both content and text and was officially launched on New Year's Day in 1933. This publication was expected to make the spirit of science flourish by popularizing scientific knowledge and improving the quality of the nation, just as the popular science publication "*Scientific American*" did in the United States. The Chinese Scientific Movement Association also held science fairs, gave radio lectures, and opened schools for the public, calling on the public to break down stereotypes, disseminate scientific knowledge, and advocate scientific methods, to make society move forward in the direction of scientization. Chapters were set up all over the country, and the publication of various popular science journals reached a climax between 1933 and May 1938, promoting the spirit of science. Chinese scientists also went through many years of trials and tribulations and grew stronger and stronger in the midst of the national difficulties, with the Chinese Association of Scientific Workers established in Chongqing on 1 July 1945, and the Chinese Association for the Promotion of Science established on December 21, 1946 in Nanjing.

On July 1, 1945, the Chinese Association of Scientific Workers was founded in Chongqing, and on December 21, 1946, the Chinese Association for the Promotion of Science was founded in Nanjing. The backbone of the organization was the Chinese Science Society, which had 3776 members in 1949.

### *Technological Transformation of Agriculture*

Modern agriculture for the nineteenth century after the dwindling of China's rural areas and agricultural production seems very urgent, and agriculture from the traditional smallholder economy to science, and modernization, must be from farming methods, field management, seed selection, and breeding for a comprehensive technological transformation. Mechanized farming the application of chemical fertilizers, and cultivation of good varieties, from the end of the nineteenth century, have started, but until the Xinhai Revolution, there is still no fundamental change. Although the end of the Qing Dynasty has been in various agricultural regions such as agricultural experimental farms and other technological transformations of agricultural demonstration farms, the mechanization of agriculture, effective water conservancy projects, and the general promotion of chemical fertilizers are to the twentieth century after the 50 s to have significant progress. Before the 1950s, the technological transformation of Chinese agriculture was mainly focused on the selection of seeds and improved varieties, and the introduction of good varieties of agricultural products, as the main object of reference is the United States of America, which is situated in a latitude and in a land and climate similar to that of China. U.S. Department of Agriculture in the late nineteenth century to make greater progress in horticulture, began to systematically seek excellent varieties from all over the world to introduce, the United States of America's open northern prairies and the arid southwestern region more need to introduce from abroad drought and cold, disease and saline resistant crops, produced in China, an epidemic-resistant walnut, large persimmons without a nucleus, soybean origin in the Northeast, the wild pears, chestnut trees, drought-resistant Elm trees, as well as various types of millet, sorghum, wheat produced in northern China, Shaanxi, Gansu production of barley, are in the early twentieth century, more than a decade, by the U.S. Department of Agriculture sent to China's agronomic experts transplanted throughout the U.S., after improvement, into a good variety of adaptable to the new environment; some of these varieties, and from the United States back to China.

The United States of America's good soybean, wheat, and sorghum are introduced in China, a large area to promote the United States' production of land cotton, corn, and potatoes, but also China's efforts to promote, and efforts to cultivate new species planted throughout the country of cash crops and food crops.

Soybean: In 1903 in Jining East Seven Mile Castle in Shandong Agricultural Experimental Farm introduced the United States soybean. U.S. Department of Agriculture in India, Japan, and China's northeastern soybean to more than a thousand tests, in 1907 selected 600 kinds of good seeds. 1907 autumn, Chen Zhenxian, a graduate of the U.S. Agricultural returned to China, as the director of the Fengtian Agricultural Experimentation Farm in Shenyang, in the two years of the introduction of 18 varieties of beans, and gradually cultivated into excellent varieties.

Wheat: in the autumn of 1897, Shouxian, Anhui Sun Lixuan through his American friends in his hometown introduced the United States wheat, Chinese wheat, and American wheat artificial cross, selecting the best varieties. 1914 Jinling University established the Department of Agriculture, the United States of America Pei Yili (Joseph Bailie) as the head of the Department, the introduction of wheat from the United States, with a single wheat cultivation of the self-crossing line of new varieties, seven or eight years later, bred the "Jinling University 26." After seven or eight years, a new variety called "Jinling University No. 26" was produced, which was popularized in China from 1924 onwards. Jinling University bred more than a dozen new species to "Jinling University 2905" which is the best, in 1935 after the breeding and promotion. After World War II, the College of Agriculture of Peking University introduced early wheat from Kansas in 1946, named "Nongda No. 1," and promoted it; in 1949, Songhua River introduced Minnesota No. 2761 wheat from the U.S., renamed "Songhua River No. 2," and promoted it. American cotton: China first planted Asian cotton.

U.S. cotton: China planted Asian cotton and African cotton in the early days, and in the nineteenth century, China began to transplant upland cotton with big cotton bolls, long fiber, and high yield. The Zhili Agricultural Experiment Farm in Baoding, Hebei introduced the U.S. barley and U.S. cotton in 1902. U.S. cotton can be harvested more than 200 pounds per acre, which is more than 80 pounds of the local cotton. The late Qing Dynasty promoted the biggest achievements of American cotton in Shandong. Jining Qilipu introduced American cotton in 1903. In 1904, Shandong counties were fully planted with American cotton. Dongchangfu successfully tested planting, decided to enlarge the province's cotton areas, and figured out the U.S. cotton be planted earlier than the local cotton. With this method of planting, Shandong summed up the "early sowing, diligent irrigation" experience. Since then, cotton has gradually become an important commodity for export. During the Republic of China period, with the textile industry demand rising, the cotton planting areas in Hebei, Shandong, Shanxi, Shaanxi, Anhui, and Henan expanded. Even the areas like Jiangxi, Zhejiang, and Hubei which traditionally planted mulberry trees began to sow cotton seeds on a large scale. The Ministry of Agriculture and Forestry set up three cotton experimental farms in North China (Zhengding, Hebei), East China (Nantong, Jiangsu), and Central China (Wuchang, Hubei) in 1915 to introduce U.S. cotton. Mu Ouchu founded cotton planting experimental farms in Shanghai in 1914. Shandong has become an important base for the large-scale promotion of U.S. cotton since 1918, buying a large scale of U.S. cotton seeds every year.

At the invitation of the Shanghai Chinese Yarn Mills Association, Guo Tanxian, who received his master's degree at Cornell University, started to introduce U.S. cotton seeds to Nanjing, Tangshang, and Shanghai in 1919. He distributed 10 tons of Trice and Lonestar from the United States to farmers of Shaanxi and Henan. In 1922, the Chinese Yarn Mills Association set up the Cotton Improvement and Promotion Committee which was presided by Guo. Guo began to domesticate U.S. cotton seeds and improve Chinese cotton seeds at the Southeast University's experimental farm. In 1928, he cultivated a new cotton species, which was named "G. Guo" after his death. The Cotton Improvement and Promotion Committee also imitated the U.S. cotton farming tools and made five-tooth tiller, Chinese cotton planter, and American cotton planter.

After the establishment of the Central Agricultural Experiment Institute in 1931, Feng Zefang in 1935 selected the Stonville 4 for the Yellow River Basin and Delfos 531 for the Yangtze River Basin to promote new varieties. Since then, the American cotton has been fully promoted in the north and south of China. The output of Deltapine imported in the 1940s from the United States is higher than the Trice and Acala. In the 1960s, 15 species were introduced to China from the United States, domesticated, and subsequently promoted to become the main varieties of cotton.

Corn, rice, and wheat are the world's three major food crops. In 1910, geneticists studied the double hybrid corn yield method and opened the selection of self-inherited lines, and the preparation of double-cross corn and double-cross can increase yields. 1906 Fengtian Agricultural Experimentation Field, the earliest from the United States to introduce 14 varieties of maize experiment. 1930 American teacher Mu Rafale (R. T. Moyer) from the Midwest introduced the Golden Queen Silver Queen. In 1930, American teacher R. T. Moyer introduced 12 good maize varieties such as Golden Queen and Silver Queen from the Midwest, and planted them on the farm of Mingxian School in Taigu County, Shanxi Province, and compared with local varieties of Taigu Yellow and Pingding White, Golden Queen yielded an average of 273.5 kg per mu, with a maximum of 353.5 kg per mu. 1939, the whole province of Shanxi was rapidly promoted. Golden Queen in the next half a century as hybrid breeding materials and demonstration of good seed, after 1925 Chinese agronomists to carry out hybrid breeding of maize and the promotion of good varieties, plays a very important role in promoting. The University of Minnesota corn geneticist Hayes (H. K. Hayes) had come to China in 1936 to lecture and teach crop breeding techniques, and later 42 corn hybrids and more than 50 self-inherited lines through the Central Agricultural Experiment Institute distributed to Chengdu, Sichuan, Liuzhou, Guangxi, Guizhou, Guiyang, Yunnan, Kunming, test planting evaluation, have obtained yield increases. During the War of Resistance against Japan, the western region of the rapid expansion of maize sowing area. Sichuan,

Guizhou, Yunnan, Xiangxi, Shaanxi, Shanxi, and other mountainous areas mostly corn as the main grain, the whole plains of Hebei are also a vast cotton area, and a variety of corn is the main food crop. Li Jingxiong cultivated in Beijing in the single No. 2, mu yield of more than 400 kilograms, since 1977 in more than 20 provinces (autonomous regions) in the country to promote. During 1986–1990, the area of maize extension planting reached more than 65 million mu.

Potato: In 1906, Fengtian Agricultural Experimental Farm had already introduced red potatoes from New York State the United States, with large tubers and high-yielding gardens, which were also widely sown in Liaoning and Jilin. Canadian missionaries brought the red-eye nest to Sichuan, French missionaries to Sichuan the big white applesauce and yellow applesauce, and Russian missionaries to Jilin is hemp potato. 1934–1936, the evaluation of the promotion of the British and American potato varieties of seven million, gold, katatine, etc. In the 1940s, as a highly adaptable high-yield crop, enough to get through the grain shortage, in the country to promote the back of the great Shanxi, Shaanxi, Sichuan, Guizhou, and other provinces in particular publicity, promotion spared no effort. As a salvage crop suitable for cultivation in both the south and the north, the area under potato cultivation reached more than 23 million mu in 1950, and in 1986 China became the second largest potato producer after the Soviet Union, and the potato was also included in the list of the world's top five crops, together with rice, wheat, maize, and sorghum.

### *The Rise of the Modern City*

What best reflects the characteristics of the times is the change in the appearance of cities, and the process of modernization is also revealed by the modernization of cities. Traditional society for the administrative centers at all levels and commercial cities of the single pattern of urban distribution, due to the opium wars after the gradual expansion of the ports of commerce, began to shift to foreign trade as the main commercial and financial activities of the center, the new machine industry, mining and the rise of railways, ships, but also spurred along the rivers and rivers and land and water transport hub to the new city steps. The modern city, as a link to modern social life, first grew and prospered in the coastal and riverine areas.

Hong Kong, since January 1841, when it was forcibly occupied by British troops, and August 1842, when it was formally ceded to

Britain, has become quite Westernized in less than 20 years. However, by 1937, when the Chinese War of Resistance against Japan broke out, the population of Hong Kong had not yet exceeded the one million mark.

Among the ports opened along the coast and rivers, such powers as Britain, France, the United States, Russia, Germany, and Japan successively acquired the leased boundaries and wantonly expanded the boundaries step by step. Shanghai and Tianjin, which were originally the residences of the Ministers of the Southern Ocean and the Ministers of the Northern Ocean of the Qing Dynasty, had a three-country leased boundary in Shanghai and an eight-country leased boundary in Tianjin, which was gradually developed into a modern metropolis with the leased boundaries as the center. Some cities and ports were under the control of a single power for a long time, such as Germany in Qingdao, Russia in Harbin, Lushun and Dalian, and France in Guangzhou Bay, and the construction plans and cultural colors of these cities reflected the urban style of these countries, and they became the original centers of exoticism in modern Chinese cities.

Shanghai, since 1848, has been the largest seaport for foreign trade in the country. From November 1845, there was a Concession exclusively for the British to live and do business, followed by the American Concession and the French Concession, and then in September 1863 the British and the Americans merged the Concessions into the Public Concession, which was located on the west bank of the Huangpu River, spanning the north and south banks of the Suzhou River. By 1915, the area of the Public Concession had expanded to more than 30 times the size of the old county, about 36 square kilometers. The French Concession was located to the south of the Public Concession, to the north of Zhaojiabang, and to the west of the old county town, with an area of about 10 square kilometers. In the French Concession, the foreign powers established the Ministry of Industry as the administrative organ, enacted laws, stationed troops, moored warships, and set up an armed organization, the Universal Merchants' Association. Inside the Concession, there was a concentration of foreign banks, banks, shops, factories, warehouses, wharves, Christian and Catholic churches, as well as church schools, parks, recreational facilities, and coliseums.

Shanghai Concession east of the Bund on the west bank of the Huangpu River, there is the British Consulate and many large banks, hotels, newspapers, and European-style architecture introduced from abroad from Gothic, Baroque, neo-classical, Renaissance, British modern, pre-modern, there are all kinds of buildings, known as the Universal Architecture Complex. Western-style architectural style and building materials in the municipal construction of the Shanghai Concession play a dominant role, but also inevitably on the local original architectural style and folk housing to form a new style of li-long housing, a fusion of Chinese and Western cultures compatible with the Shanghai school of architecture, the use of green brick flat masonry in the building materials, cement polished porch, windows, in the traditional folk in the south of the country, the main hall, the intermediate compartments, the patio on the formation of the traditional style of the Shikumen Lilong (the old-style long housing). Old-style Lilong Residence). Later, after the First World War, there appeared the new style of li-long houses with Western-style undulating triangular roofs and gas windows on the front façade, surrounded by open space, which was a modification of the more mature architectural style of the Jiangnan water town houses in the Yangtze River Delta in the increasingly modern city. more European-style garden lilongs were built in the 1930s in patches in the western part of Shanghai where there was more space. This type of architecture led the way for a generation in the modernization of Chinese cities.

Shanghai in the municipal engineering facilities and public utilities, are at the forefront of the country, especially in the coastal areas. 1865 November, the country's first gas plant in Shanghai put into operation for gas supply; 1882, the earliest domestic power plant was built in Shanghai to provide electricity, away from the invention of the electric light is still only three years; 1883 June, the earliest national water supply plant was built in Shanghai water supply; 1870, Shanghai and Hong Kong began to erect submarine cables; 1880 Shanghai and Hong Kong started In 1870, Shanghai and Hong Kong began to set up submarine cables; in 1880, China's first land telegraph line from Tianjin through Qingjiangpu, Zhenjiang to Shanghai; in 1877, Bell invented the telephone six months after the first telephone line in Shanghai, Shanghai in 1882, the installation of telephone booths, became the second year after London after the passage of the telephone in the city. The public facilities closely linked to the invention of modern science and technology were first developed in Shanghai and then spread to other cities along the coast. In the first half of the twentieth century, Shanghai had developed into the largest industrial and commercial city in China, and an international financial and trade center in the Far East. The rapid development of industry and trade drove the population growth, the population reached

1 million in 1890, increased to 3 million in 1930, and ranked sixth among the world's metropolises. 1942, Shanghai's population reached 4 million, which increased to 6 million in 1945. 1915, the public and French Concession totaled 46 square kilometers, to 1945, Shanghai's entire built-up area of 80 square kilometers, the population increased to 3 times the number of the same period, but the number of people. By 1945, the entire built-up area of Shanghai was only 80 square kilometers, but the population had more than tripled in the same period. The city's 3 million people lived in 2000 hectares of neighborhoods, with a population density of over 1000 people per hectare, accounting for half of the city's population. The Huangpu, South, and Hongkou districts west of Fujian Middle Road and east of Xizang Middle Road had densities of 3000 per hectare, and the pre-1950 neighborhoods between Chengdu North Road and Xinzha Road had densities of up to 7000 per hectare, with many shanty towns clustered in these districts, a stark contrast between rich and poor in a modern, thriving metropolis. From the 1930s to the 1980s, Shanghai, with less than one percent of the country's population, created enormous wealth, and for quite some time it accounted for one-half of the country's industrial output, one-quarter of its trade, and one-sixth of its finances.

Among the coastal cities, Qingdao is a harbor city with a German architectural style and a famous summer resort. 1897, Germany sent troops to forcefully occupy the city under the pretext of the Caozhou Case, and in the following year, it signed a 99-year lease treaty with the Qing government, building Qingdao into a military harbor and a seaport for plundering Chinese raw materials and dumping German goods. 1900 the city plan was compiled, and in 1901, the Jiaozhi Railway was built, mining along the route and the Jiaozhi Railway. In 1905, the Jiaoji Railway was opened to traffic and the Dagang Wharf was built; in 1911, the shipping prison grew to one million tons, and in 1913, the railway freight site also reached more than 940,000 tons. At the outbreak of the First World War, Qingdao was occupied by the Japanese and was not returned to China until 1922, with the Japanese still enjoying all kinds of privileges. 1937 was again occupied by the Japanese, and by 1945 it was taken over by the Chinese government with the cooperation of the American army. German planning had.

The German zone was on the southern coast with a beautiful environment, while the Chinese zone was in the north with residential areas mixed with industry and commerce. Later, Japan established industrial zones in Sifang and Cangkou, and the workers' houses were distributed in the low-lying areas along the railway line, which made the conditions poor. The German District has wide roads, greenery with shrubs and flower beds, and most of the residences are German garden houses, and there are also a few low-rise apartment buildings consisting of 4-room and 6-room units. There are many small Japanese-style houses on Jehol Road. Electricity was originally available in the city, with cast iron poles along the roads within the German Quarter, some with underground cables. Some of the buildings face the road, and the Governor's Office and the Police Station make for a very elegant cityscape. The city has the characteristics of a modern city under the control of the Great Powers as an industrial and commercial city, but at the same time, it has been planned to fully reflect the function of leisure and summer residence.

In the first half of the twentieth century, along with the rise of many coastal and riverine cities, many traditional-style cities were also slowly transformed into more innovative cities, such as Jinan, Shenyang, Chongqing, Wuhu, Jiujiang, Suzhou, Hangzhou, Fuzhou, and Sha Shi. Most of the small and medium-sized cities are still under the original small peasant economy and maintain their traditional appearance. The most typical one is Beijing (called Beiping in the Republic of China), which has been the center of imperial autocracy for nearly a thousand years in the Liao, Jin, Yuan, Ming, and Qing Dynasties, with the inner city being the palace residences of the emperors and nobles and the official government offices, and the outer city being the residential area with a concentration of commerce, handicrafts, and transport and simple housing. 1900 saw the opening of the Jingfeng Road, the opening of the Beijing-Hanzhou and the Beijing-Zhangzhou Railway in 1906, and the completion of the Circle Railway, which resulted in the main streets and the traffic both inside and outside of the city becoming The main streets and the traffic in and out of the city became intersected in a plane. From the beginning of the twentieth century, the Qing government began to build some Western-style buildings, and stone slag pavement began to appear on the original dirt roads in 1904, and asphalt pavement was only available in the inner city in 1915. Running water was not provided until 1910. Electricity supply was initially limited to the Summer Palace and Zhongnanhai and Beihai and only took off with the construction of the Shijingshan Power Plant in 1919. Trams only appeared in 1924. The most significant change is that after the fall of the Qing Dynasty, the palace wall inside the Daqing Gate (China Gate) was demolished in 1913, opening up the Tianjie (Tian'anmen Street), and east–west arterial roads appeared in Jingshan to the north of the Forbidden City. In 1913, the Wenhua and Wuying Halls of the Forbidden City were opened to the public, and in 1925, the entire Palace was converted into the Palace Museum, while other royal forbidden gardens were opened one after another. 1914, the Temple of the Jikji Temple and the Temple of the First Peasants were successively converted into parks, and the Temple of Heaven, the Beihai Sea, and the Summer Palace were subsequently opened to the public as gardens as well. At this point, the entire function of this old city for the royal family and the powerful only began to take a big step toward urban renewal, but compared with Shanghai, Tianjin, the city in quite a long time to leave people's impression, can only be a long time to the memory of the old times.

## 9.4 The Chinese Art Boom in Western Countries in the 1930s

The Chinese fashion on both sides of the Atlantic Ocean in the eighteenth century was inspired by China's tea culture. In Europe, the pursuit of Chinese ceramic products, home decoration art, and garden layouts has become a fashion from the upper-class group to the urban residents. In the following century, the situation changed. After the 1870s, European great powers such as Britain, Russia, France, and Germany sent personnel from central Asia to enter the western border of China under the pretext of geographical exploration. As a result, many ancient sites, caves, and ruins preserved in the desert of western China attracted the attention of these inspectors, who were often escorted by armed personnel or were themselves professional soldiers. They searched for records in local chronicles and travel notes since the Ming and Qing dynasties through Chinese translators, exploring these Buddhist grottoes and temples. Grotto art, bamboo and wooden slips, and innumerable relics, which had no prominent position in the cultural history of China, and which had been buried underground for a long time were brought to light. From the perspective of cultural history and art archaeology, they show the history of past exchanges between China and Western Asia, which opens a very important page in the history of human civilization.

It is not the explorers and geographers but the Western sinologists and orientalists who devote themselves to reviewing these excavations that make people correctly understand this previously unknown Chinese art and civilization. Their research results were successively published during the two World Wars, and Chinese artists began to go abroad and engage in exchanges, leading to the rise of the Chinese art craze in Europe and North America in the 1930s.

### *The Chinese Art Craze Triggered by Grotto Art*

Emmanuel Edouard Chavannes of France (1865–1918) was a sinologist who first studied Chinese stone carving art and later became famous for his series of outstanding sinological works. In 1889, he went to Beijing to serve as a staff member of the French Embassy in China. In 1893, he served as a professor of Chinese at the French Academy. In the same year, he published *La Sculpture sur Pierreen China au temps des deux dynamies han*, which opened up the research gate on cave art. In 1907, he inspected the tombs of the Zhou and Tang dynasties in Shaanxi, the Yungang Grottoes, the Longmen Grottoes, and the Gongxian Grottoes, recorded the stone carvings in Jinan and the Goguryeo Tombs in Tonggou. He took 500 photos and printed them into two volumes, along with one volume of text. The overall title was *Mission Archaeologique dans la Chine Septentrionale*, which was published in Paris from 1913 to 1915. The earliest images of the three major grottoes in Yungang, Longmen, and Gongxian are therefore preserved, and some physical objects no longer exist. Buddhist art accounts for about half of the images. The Chinese documents obtained by Stein in Xinjiang were later handed over to Chavannes for interpretation. Chavannes published *Les Documents Chinois* in 1913, which laid the foundation for the study of bamboo slips in the Silk Road studies.

Within Xinjiang, numerous grotto temples are distributed along the ancient Silk Road. The largest in scale, the Kezil Grottoes, are located 60 kilometers west of Baicheng County. A pencil drawing of a woman on the west wall of Cave 213 and a Russian inscription dated January 15, 1879, indicated that Russians had already visited the site for exploration. It is speculated that the visitor might have been the archeologist Albert von Regel. Since the twentieth century, expeditions from Germany, Japan, Russia, France, and the United Kingdom have repeatedly visited Xinjiang to explore grotto temples, unearth cultural relics, strip murals, and transport Buddha statues. The German Berlin Folklore Museum organized four expeditions to Xinjiang between 1902 and 1914. In 1902, the Prussian Royal Turpan Expedition was dispatched to Turpan for archeological surveys, co-directed by Albert Grünwedel (1856–1935) and Georg Huth. They excavated the ancient site of Gaochang in Turpan from 1902 to 1903 and shipped 46 boxes of cultural relics to Berlin. The expedition report, Bericht über Archäologische Arbeiten in Idikutschari und Umgebung im Winter 1902–1903, was published in Munich in 1906, revealing to the world the Gandhara-style Buddhist art preserved in the ancient city of Gaochang. The murals in the temple featured Indian, Yuezhi, ancient Turkic, Uighur, Tibetan, and Chinese styles. The report pioneered the study of ancient city ruins.

From September 1904 to December 1905, the second expedition led by Albert von Le Coq (1860–1930) investigated the site in the Turpan and Hami regions. The inspection report *Chotscho: Facsimile-Wiedergaben der Wichtigeren Funde der Ersten Königlich Preussischen Expedition nach Turfan in Ost-Turkistan* was not published until 1913. There are illustrations of the murals in the Bezeklik Caves. Le Cog believes that the painting of this area, along with the Kuche and Bamiyan murals, belongs to the Central Asian painting style. During the third inspection, which was conducted by Grunwedel and Le Cog from the end of 1905 to April 1907, a large number of murals were stripped and statues were transported from Kuqa, Baicheng, Yanqi, and Turpan. From January 1913 to February 1914, Le Cog led the Fourth Prussian Royal Expedition to smuggle cultural relics from Tumushuke and Kizil cave temples. The murals taken from 36 caves in the Kizil Cave-Temple alone amount to 328.07 square meters, which have been cut into 252 pieces and preserved at the Indian Department of the Ethnological Museum of Berlin. Later, half of them were destroyed in the flames of World War II.

The investigation and research works of Grunwedel, Le Cog, Pelliot, and Stein on Xinjiang and Dunhuang grottoes revealed the colorful culture of ancient multi-ethnic integration in western China to the world before World War I. The most remarkable surprise was undoubtedly that on May 21, 1907, on the second expedition to Central Asia, Stein arrived at the Mogao Grottoes of Dunhuang to meet with a Taoist Wang Yuanlu, and opened the scripture cave. He defrauded 7000 volumes of ancient scriptures in 24 boxes, and 500 paintings (silk paintings, cloth paintings), embroideries, and works of art in 5 boxes, and shipped them to the British Museum in London. Later, Stein provided detailed descriptions of the *Ruins of Desert Cathay* and *Serindia*. The French orientalist Paul Pelliot (1878–1945) entered Xinjiang in 1906 and arrived at Dunhuang in February 1908. He stayed there for three months, numbered, surveyed, photographed, and copied the inscriptions of Mogao Grottoes, stole one-third of the remaining scriptures, and for the first time stripped the entire front murals in Cave 263 (the original P.117bis Cave). Sergei Odenberg of Russia (1863–1934) also followed the French and Germans' chemical methods to peel off murals. In 1914–1915, he searched for documents in Dunhuang and emptied the northern wall of Cave 263 of Mogao Grottoes. Pelliot secretly packed the looted cultural relics into 23 boxes and transported them to the Louvre and the Guimet Museum in Paris, with only one box destined for Beijing. From 1914 to 1918, Paris Geuthner published 8 volumes of masterpieces titled *Mission Pelliot en Asie Centrale*, gradually leading the study of cave art and multiculturalism in western China to a new level.

After the end of WWI, Langdon Warner (1881–1955), the Curator of Oriental Art at Harvard's Fogg Museum in the United States, and Horace H.F. Jayne, who later became the director of the University of Pennsylvania Museum of Archeology and Anthropology, went to Dunhuang in 1923. They took more than 20 murals from Dunhuang. Warner looted painted sculptures, ink, stone inscriptions, and documents along the way when he returned from Dunhuang to Jingzhou. In 1927, *The Long Old Road in China* included 31 illustrations depicting the process of extracting murals was published. The success of this matter led Warner once again to collude with Jayne and seven other people from the Metropolitan Museum of Art in New York to organize a second Northwest China expedition in 1925. They carried a large cart of cloth and prepared to peel off a cave mural. Eventually, they faced opposition from the locals and had to leave. This incident sounded an alarm for the Chinese authorities, and from then on, the protection of historical sites and artifacts gradually shifted from vigilance to legal protection in various parts of China. However, under the temptation of interests, it was difficult to eliminate the illegal sale and smuggling of cultural relics. Therefore, over the past few decades, Chinese cultural relics have been increasingly scattered abroad. It is reasonable that the center of Dunhuang and Silk Road research has shifted from China's headquarters to places such as London, Paris, and Berlin.

Many of the works by Lecoq, Stein, and Böschwer's studies of the caves of western China were not published until after the end of the European War, and most of these books were published in multiple volumes with fine plates. Lecoq's *Die Buddhistische Spätanike in Mittelasien* (Late Antique Buddhist Art of Central Asia), printed by the Lehmeyer Verlag from 1922 to 1933, is divided into five volumes (reprinted by Graz Verlag in 1973), and his *Bilderatlas zur Kunst und Kulturgeschichte Mittelasiens* (Illustrated Catalogue of the History of Art in Central Asia) was published by the same publisher in 1925. The French art archaeologist Victor Segalen, together with two of his assistants, examined Qin and Han tombs and Buddhist ruins in Shaanxi, Sichuan, and Nanjing, China, in his report, *Mission Archéologique en Chine* (1914 and 1917) Atlas. It was also published by the Gertner bookshop in Paris in 1923–1924. Burghardt edited *Les Grottes de Touen-houang* (Paris, Geuthner), a six-volume catalog of the Dunhuang caves, which was published from 1914 until 1924. At this time Stein also published his detailed report on his second expedition to Central Asia, Xinjiang Archaeological Records (Chinese translation as Archaeological Atlas of the Western Regions, published by Guangxi Normal University Press) in 1921. The book was divided into five volumes, volumes 1–3 were text with 345 maps, volume 4 contained 175 plates, and volume 5 was a boxed set of 95 maps. 1922 saw the publication in London of *The Thousand Buddhas* (Quaritch), which helped to promote the Thousand Buddhas Cave at Dunhuang as a treasure trove of art in Europe. The Thousand Buddhas Cave at Dunhuang was published in London. From then on, Europeans not only knew about the Achanta in India but also the Thousand Buddha Caves in Dunhuang.

Located on the "Roof of the World," Tibet was once a mysterious mountainous country governed by monks and lamas. Since 1903, when E. Younghusband led the British army into Tibet and occupied Lhasa in 1904, British scouts have been traveling to and from the Tibetan borderlands and inland hinterland, and many precious relics of Tibetan Buddhism have been searched for and transported abroad, making the study of Tibet a momentary craze in Western countries.

American Orientalist Berthold Laufer (1874–1934), originally from Germany, began publishing research papers on Tibetology in German in 1896, received his Ph.D. from the Institute of Oriental Languages in 1897, and came to New York, USA in 1898, shortly after transferring to the East Asian Department of the Fairfield Museum of Nature in Chicago. In 1908–1910, he took part in Mrs. Braxton's expedition to mainland China and Tibet three times, published many treatises on Chinese folklore, cultural anthropology, and archaeology of fine arts, and was the earliest one to collect Tibetan cultural relics for American museums after Zuo Keyi (W. W. Rockhill), and the earliest one to study the history of Chinese fine arts of American scholars of the twentieth century. *Chinese Gravestone Sculptures of the Han Period*, published in 1911 and distributed in New York, London, and Paris, is a book dedicated to Sha.

The book was published in New York, London, and Paris, and was dedicated to Chavannes as a book printed at the author's own expense. A.H. Francke collated Tibetan history and antiquities and published *A History of Western Tibet* (London, 1907) and *Antiquities of Indian Tibet* (1914–1926), the latter printed at the expense of the Anglo-Indian government. The latter was printed at the expense of the Anglo-Indian government and was a survey of antiquities from Xinjiang and Tibet. The French Bacot and Hackin, who specialized in Central Asian Buddhist art, had already published J. Bacot et J. Hackin, *L'Art tibétain*, Paris, Geuthner 1911) before the European War, enriching the international scholarly understanding of the civilization of western China.

In the post-war period of the 1920s, with the emergence of a large number of antiquities from underground excavations and the publication of research reports, the study of Chinese art in the West also showed an upward trend. the publication of the eight-volume *Bösch and Central Asian Expedition* and the five-volume *Archaeological Records of Xinjiang* by Stein gave impetus to the study of the archaeology of Chinese fine art, paved the way for the arrival of the climax, and triggered the publication of a variety of incisive theses in the following years. The publication of the five-volume Xinjiang Archaeological Records Volume 3 of the Xinjiang Archaeological Records contains ten thematic studies, including Raphae Petrucci's and Laurence Binyon's A Brief Discussion of Buddhist Painting in the Thousand Buddha Caves of Dunhuang, which is a much more comprehensive study than the 1905 book by the British sinologist Zhai Lishi (翟理士), a scholar of the Chinese language.

The 10 thematic works, including *An Introduction to the History of Chinese Pictorial Art*, published in Shanghai in 1905 by the British sinologist Herbert Allen Giles, are much more in-depth than those published in 1905 by the British sinologist Herbert Allen Giles, no matter from the style to the technique, due to the expansion of the material. Following Petrussi's publication of *Chinese Painters* in New York in 1920, A. Salmony's publication of *Chinesische Landschafts* in Berlin, and O. Fischer's publication of *Chinesische Landschafts* in Munich (1921). L. Ashton published *Introduction to the Study of Chinese Sculpture* (London, 1924), and A. Waley wrote *Introduction to the Study of Chinese Painting*

(N. Y.). Chinese Painting, N. Y. 1923), E. Boerschmann wrote *Chinesische Architektur*, 2Vols., Berlin, Wasmuth 1925), H. D'Ardenne de Tizac He published *Animals in Chinese Art* (London, 1923), *L'art Chinois Classique* (Paris, 1926), R. L. Hobson published *Chinese Art* (London, 1927) with 100 color plates, and R. L. Hobson published Chinese Art (London, 1927). Art, London, 1927), and R. L. Hobson published Chinese Art, London, 1927, with 100 color illustrations, which gave Western readers many excellent readings on Chinese art. Italian orientalist Giuseppe Tucci (1894–1984), a professor at the University of Rome since 1933, made many investigations and visits to Tibet and robbed a large number of cultural relics, and his Indo-Tibet (Indo-Tibetica), which was published by the Academia dei Leri in Rome from 1932 to 1941, was a book on the *Indo-Tibet (Indo-Tibetica)*, published by the Academia Lyle in Rome between 1932 and 1941, is an indispensable monumental work on the archaeology of fine arts in central and western Tibet, and authoritative work on the study of fine arts in Tibet, and he has published many papers on Tibetology, and has written *Tibetan Scrolls* (Rome 1949, 3 vols.) in English.

The master of Chinese art research is Osvald Siren (1879–1966), a Swedish historian of Eastern art. In 1925, he published a four-volume book titled *Chinese Sculpture, from Vth to the XIVth Century,* which includes many Chinese inscriptions and a clear study of cave temples and Buddhist statues of the most important periods of Chinese sculpture art history from the Western Jin Dynasty to the Yuan Dynasty. It includes 623 images and over 900 photos, some of which are missing, making these images even more precious. Afterward, there were three volumes of *The Imperial Palaces of Peking*, two volumes of *Chinese Paintings in American collections*, and four volumes of *Histoire des Arts Anciens de la Chine.* The two-volume edition of *Histoire de la Peinture Chinoise* comprehensively described the entire process of Chinese painting from ancient times to the Yuan Dynasty. In addition, he also published *The Chinese on the Art of Painting* in Beijing in 1936.

### *The Western World Praises China's Three Major Crafts*

Jade carving, bronze ware, and ceramics are the three world-renowned crafts in ancient China, which have long been noticed by Western countries.

Jade was first used in China, where it was regarded as a symbol of virtue, and has a history of 7000 years in the Yangtze River Valley. While not much ancient jade has been handed down, more and more objects have been unearthed from the ground since the twentieth century, which has led to significant progress in the study of jade. Bishop published *Investigation and Studies in Jade*, N.Y. in 1906. Stephen W. Bushell, in 1904, issued by Weiman's Bookshop, London, Chinese Fine Arts, 2 vols. This book, which first introduced Chinese fine arts from the multiple perspectives of paintings, carvings, architecture, and arts and crafts collected in foreign museums, had already begun to pay attention to the study of the history of Chinese jades, and a revised edition was published soon after in 1909, with a Chinese translation. Based on Wu Dacheng's four-volume *Ancient Jade Chart Examination*, Laufer cited many new heirlooms and wrote a 370-page tome, *Jade, A Study in Chinese Archaeology and Religion*, (The Mrs. T. B. Blackstone Expedition, Chicago 1912), with 68 beautifully illustrated plates, gave a new approach and perspective to the study of ancient jade. Later, Bergschwer continued to delve deeper, based on the collection of Lu Qinzhai, who dealt in antiquities in Paris and published *Jades Archaïques de Chine*, Collection Loo, Paris, Van Oest, 1925. Laufer, having studied the collection donated by Bachar to the Museum of Natural History in Chicago, also published Jades Archazques de Chine (1927), printed in Chicago at Bachar's private expense, a 52-page book with 36 plates. But Laufer, like Bishop, did not believe that Zhou-Han ancient jade came from Ho-Tep, Xinjiang. They do not believe that the "Jade of Kunshan" in the pre-Qin texts is Ho-Tepan jade, preferring to believe Stoliczka's argument in an 1874 article in the British Journal of Geology, vol. 13, that Ho-Tepan jade did not exist in the Central Plains until Zhangqian passed through the West (F. Stoliczka, Note Regarding the Occurrence of Jade in the Karakoram 1874). Occurrence of Jade in the Karakash Valley on the Southern Borders of Turkestan, *Quarterly. Geol. Soc.*, London 1874, Vol. 13, pp. 568–570). A large number of Ho-Tsang jade carvings unearthed in the Yinxu excavations in the 1930s and the discovery of the Liangzhu culture on the banks of the Qiantang River in Zhejiang Province completely refreshed the history of jade in China, and the large number of penannular jade pieces, jade jewelry, jade kui, and jade cong, as well as the discovery of jade silkworms and jade gorgets, have revealed the ancient craftsmanship of jade carving and the use of jade for jade carving to the world. The discovery of a large number of jade penannulars, jade bibs, jade gui, and the discovery of jade silkworms and jade go have shown the world the craftsmanship of jade carving in ancient times and its use in rituals.

Shang and Zhou bronze ware and inscriptions on it were a great craft during the peak of the ancient Chinese foundry industry, with achievements surpassing similar crafts in Europe and the Mediterranean world. The Shang and Zhou bronze vessels have advanced casting techniques, rich patterns, and numerous masterpieces of giant structures. From ritual vessels to daily utensils and musical instruments, they are abundant in quantity and exquisite in craftsmanship. With the increase of underground artifacts and handed-down artifacts, they were recognized by the world's academic and archaeological communities in the twentieth century. The publication of A. J. Koop (*Early Chinese bronzes*, London 1924), E. A. Voretzsch (*Altchinesische Bronzen*, Berlin 1924), and W. P. Yetts (*Chinese Bronzes*, London 1925) added knowledge of Chinese bronze age arts and crafts to Western countries. Later, M. L. Rostovtzeff's *Inlaid Bronze of the Han Dynasty* and *The Animal Style in South Russia and China* were even more refined, comparing Chinese bronze casting art with the bronze craftsmanship of the grassland ethnic groups in western Asia, earning high praise from the academic community. Yetts also compiled the *Catalogue of the Chinese and Corean Bronzes, Sculpture, Jade, Jewelry, and Miscellaneous Objects* for Eumorfopoulos during 1929–1932. Later, Dong Zuobin and Li Ji excavated Anyang, and a large number of bronze vessels were unearthed in Shaanxi, Anhui, and other places, which impressed the world and provided indisputable physical evidence for exploring the origin of Chinese civilization. In his book *The Birth of China: A Study of the Formative Period of Chinese Civilization*, H. G. Creel admired the bronze craftsmanship of the Shang and Zhou dynasties, while he couldn't help but lament that it was impossible to replicate these artifacts.

Ceramics among the three major Chinese crafts are particularly popular. China is the home country of porcelain, with a long history of pottery-making techniques, and the art of painted pottery is similar to that of Babylon. Ceramics had already appeared in the cultural layer of Yangshao in Henan Province by 3500 BC. Swedish geologist J. G. Anderson made outstanding contributions to the origin and evolution of Chinese art. In 1921, he discovered painted pottery with red patterns on a red or gray at the Neolithic site in Yangshao Village, Mianchi County, Henan Province. These painted pottery fragments were once studied by

Hobson, a specialist in ceramics at the British Museum, and were believed to belong to the period of 2000–1500 BC. In 1923, Andersson published *"An Early Chinese Culture"* in volume 1, no.5 of *Bulletin of the Geol. Survey of China*, opening up a discussion on the Western perspective of Chinese painted pottery culture. In 1934, Andersen published *Children of the Yellow Earth: Studies in Prehistoric China* (London), he continued to argue about the distribution of painted pottery culture in western China and sparked controversy over the origin of Chinese ethnicity. For a while, there was a mixture of believers and skeptics, and it was difficult to distinguish between their views.

The study of the origin of porcelain and the relationship between porcelain and glazed pottery was still in the primitive stage at that time. R. L. Hobson had already published *Chinese Pottery and Porcelain*, as early as 1915, and later *The Wares of the Ming Dynasty* in 1923, and *The Later Ceramic Wares of China*, in 1925. R. L. Hobson and A. L. Hetherington co-authored *The Art of Chinese Pottery from the Han to the Ming* in 1923. C. Hentze's book *Chinese Tomb Figures* is dedicated to studying Chinese funeral customs and pottery figurines. Laufer is extremely proficient in the study of Chinese ceramics, and has published *Chinese Pottery of the Han Dynasty* (1909), *Chinese Clay Figures* (1914), and *The Beginnings of Porcelain in China* (1917). There are 339 pages of text, 65 plates, and 55 illustrations in *Chinese Pottery of the Han Dynasty*, and 64 plates, and 55 illustrations in *Chinese Clay Figures*. However, the origin of porcelain is relatively complex, and consensus was not yet reached in the first half of the twentieth century. It is now widely recognized that high-temperature fired porcelain only existed after the first century AD.

The art magazines published in Berlin, Paris, London, and Shanghai have driven the trend of Chinese art. In 1912, Berlin founded *Ostasiatische Zeitschrift*. Since 1924, Paris has also founded the *Revue des Arts Asiatiques*. London regularly publishes the *Transactions of the Oriental Ceramic Society*. In the East, J. C. Ferguson established the *China Journal* in Shanghai, which was founded in 1923 and discontinued in 1938. In addition to publishing various articles on Chinese art, archaeology, and history in the magazine, the Commercial Press also published the massive *Survey of Chinese Art* in 1939.

### *Chinese Art Exhibition and European Art World*

One of the world events of Chinese art before the Second World War was the "International Exhibition of Chinese Art," which was officially opened on November 28, 1935 at the Royal Society of Arts, Burlington House, London, and closed on March 7, 1936 after 100 days of the exhibition. More than 420,000 people visited the exhibition. It was second only to the Italian Art Exhibition held in London. The exhibition set 17 countries and more than 50 private collections of ancient Chinese cultural relics, exhibits are divided into bronze, porcelain, painting, and calligraphy other four parts, a total of more than 1000 pieces, directly from China participate in the exhibition of fine art accounted for the majority. The exhibition displayed many precious relics, to the world in metallurgy, silk weaving, papermaking, engraving and printing, ceramics, stone carving, painting, architecture, and other fields of creation, and invention, there are the Shang Zhou bronze, Loulan Changle brocade, Tang topography Tang framed the "*hot spring inscription*," the "*datang publication of the fallacy complement Que cut rhyme*," Song Huizong Zhao evaluation paper "*pond autumn color*" hand scrolls, and other rare treasures. At the same time, there are many exhibits from foreign collectors, such as the National Museum of Germany the exhibition of the Spring and Autumn King of Wu bronze (42 cm high, diameter 70 cm), Yungang, Longmen, Tianlongshan Buddha, Hebei Yixian colorful sculpture of Lohan sitting statue, Tang Zhaoling six steeds in the Zaolu Purple, the Song Dynasty, the Department of Jade made by the Xiu Nei jade building. Many cultural relics have been lost overseas, especially in the cave temples in the murals, statues, and Dunhuang paper scrolls, such as the foreigners and Chinese antique dealers smuggled out China's cultural relics, the number of members of the huge, high-quality, heard and seen by the people of the country are disturbed, deplored. European countries and Chinese students in Europe have come from all over the world to watch, and the major newspapers and magazines have reported in the articles, said the exhibits of the best. Visitors were all for China's deep, colorful artistic attainments, breathtaking, so on both sides of the Atlantic set off a wave of China fever. Shanghai Commercial Press in 1936 printed with *Chinese and English instructions to participate in the London International Exhibition of Chinese art products* illustrated in four volumes, only to become a permanent memorial of this event.

Many of the precious relics displayed at the International Exhibition of Chinese Art in London were dispersed throughout Europe and the United Kingdom, including the Britannic Museum (British Museum) and the Victoria and Albert Museum in London, the Musée Guimet in Paris, the Museum of Ethnology in Berlin, the Museum of Chinese Culture and Art established in 1939 by the Hermitage Museum in St. Petersburg, the Far Eastern Museum of Antiquities in Stockholm, the Metropolitan Museum of Art, Boston Museum of Fine Arts, Harvard University, the Museum of Fine Arts, the Museum of Fine Arts of Boston and the Museum of Contemporary Art and Design of Harvard University. The United States has the Metropolitan Museum of Art in New York, the Museum of Fine Arts in Boston, the Fogg Art Museum at Harvard University, the Freer Gallery of Art in Washington, D.C., and other museums, which have a large collection of Chinese cultural relics, many of which are fine works of art, masterpieces.

In the 1930s, under the pressure of external and internal disasters, the Chinese students who went abroad to study and study, especially Chinese painters, were deeply impressed by the artistic charm of Chinese art in Europe and sought the support of Western countries in international affairs and cultural strings of China's necessity to go out of the country, single-handedly into the center of Western art, such as Paris, Berlin, and other places to hold exhibitions, to personally preach the attainments of Chinese art and teach the Chinese art of painting. He took it upon himself to preach about the attainments of Chinese art and to teach Chinese painting. This attempt was initially provided by the 1930 Belgian centenary of the founding of the Universal Exhibition, when Xu Beihong got the help of the Belgian Embassy, held a personal painting exhibition in Brussels, is the first Chinese painters in foreign countries into the exhibition, the Queen of Belgium came to visit, won a good reputation. After that, the Beijing University of China and France in Lyon on behalf of Liu Dabei and Xu Beihong's works and Belgium's centenary of the Universal Exhibition of Paintings, transported to the University of Lyon held an exhibition of paintings, within three days attracted 4000 visitors, Chinese painting attracted the French public. At this time in Germany, there is also a wave of heat in China. 1931 Shanghai art school headmaster Liu Haisu went to Europe during the investigation, in Frankfurt, Germany, the Chinese Academy lectured on Chinese painting and held it in the local museum of Modern Chinese painting exhibition. June exhibition moved to Heidelberg, a sensation of the whole of Germany so that those of the

Oriental art of the German people who are confined to the imitation of Japanese art in Europe, for the first time will be able to closely observe the essence of the Oriental art. In June, the exhibition moved to Heidelberg and became a sensation throughout Germany. Thereafter, this appreciation of Chinese art became a hunger in France and Germany, and the public increasingly demanded more and more opportunities to get close to it.

From 1933 to 1934, Xu Beihong organized a series of exhibitions of Chinese paintings in Europe, holding seven successful exhibitions in five European countries: France, Belgium, Germany, Italy, and Russia, and setting up four showrooms of modern Chinese art in major museums and universities. The exhibition of Chinese art was held on May 10, 1933 at the Beaumont Museum in Paris, specializing in contemporary foreign art, with the support of the French Minister of Education, Jean-Pierre Témontzi, and with a foreword by the great poet Van Lochte, a member of the Académie des Maîtres de France. The exhibition was divided into two parts, ancient and modern, with a total of 70 ancient paintings, of which the Louvre Museum loaned 5, the Guimet Museum loaned by Dunhuang robbed the treasures, from the European collectors borrowed from the exhibition of more than 40 ancient paintings, the domestic provision of the Song and Yuan paintings more than 10. There are about 200 modern paintings. The total number of 270 up and down, is an unprecedented scale of the exhibition. Belgium Stoclé collection of the Tang dynasty's "*crazy drunkenness*," Yuan Dynasty Zhao Mengfu's "*double horse*," and France Rivière collection of the early Yuan masterpiece "*Monks*," are rare works, the exhibition shocked the Chinese and foreign painting world. Due to public demand, the exhibition was extended to 45 days, with more than 10,000 visitors. France's National Museum of Modern Foreign Art decided to buy the exhibition works 12, to set up another special room, these works are Fang Yaoyu's "*Bird*," Zheng Manqing's "*ink anemone*," Wang Yiting's "*Dharma*," Gao Qifeng's "*sailboat*," Qi Baishi's "*palm trees*," Chen Shuren's "*banana*," Wang Yaden's "*summer*," Zhang Daqian's "*Lotus*," by Heng Yi's "*Orchid Stone*," Zhang Iguang's "*Kingfisher*," Zhang Shu's "*Peach Blossom*," and Xu Beihong's "*Ancient Cypress*." Since then, modern Chinese paintings have been permanently exhibited abroad, and contemporary Chinese paintings have begun to occupy a place in Paris, the center of world art. The success of the exhibition exceeded that of the Japanese art exhibition held in the museum in 1929, and more than 300 reviews were published in newspapers and magazines, which unanimously appreciated the achievements of Chinese paintings.

Paris exhibition shook the whole of Europe, countries invited to exhibit. In 1933 June, Xu Beihong second to Brussels, Belgium, held a solo exhibition. In December of the same year in Italy, Milan Palace held a never before Chinese painting exhibition. The exhibition period was from December 19, 1933 to January 19, 1934, Italian celebrities were listed as members, and film popularity of the country. In 1933 November Berlin Fine Arts Association also opened a personal exhibition of Xu Beihong, in February 1934, Italian exhibits moved to Germany, Frankfurt, Germany, National Museum of Fine Arts, and the Chinese Institute of the President Lu Ya-wen, Vice President Hua Hua. President Lu Yawen, vice president of the Chinese Ding Wenyuan's strong support. Then Xu Beihong with the desire to exchange art with the Soviet Union, at the invitation of the Soviet Union Foreign Culture Association, arrived in Moscow by sea via Odesa. On May 6, 1934 in Moscow State Historical Museum held an exhibition of modern Chinese paintings. Visitors are very enthusiastic, unlike Western European countries, the audience, especially workers, and farmers. Many workers wrote letters asking Xu Beihong to convey their respect to the Chinese artists. Xu Beihong in the exhibition during the impromptu brush painting horse, by the love of horses into a fetish of Marshal Bu Jongni asked to go. Xu Beihong in the Moscow Academy of Fine Arts gave a "Chinese Ancient and Modern Art Changes" lecture. In July, the exhibition moved to the Petersburg Ermitage Museum to continue to display. Xu Beihong and the Soviet Union oil painters, print-makers, sculptors, cartoonists Gong Chelovsky, Grabar, Leilov, Konenkov, Fasuelsky, etc., and the poet Arosev (President of the Soviet Union foreign cultural associations), writers Riding, Pirinek, had been teaching in Peking more than a year of the script of the "*Roar, China*," such as Tretyakov exchanges. 1935, in January, Moscow, the Sino-Soviet art exchange ceremony was held. In January 1935, a ceremony was held in Moscow for the exchange of artworks between China and the Soviet Union, at which Arosev pointed out that there had been many exhibitions of foreign art in the Soviet Union, and that the exhibition of Chinese paintings had been the most successful, and that the exhibition of paintings "will become the cornerstone of cultural co-operation between the Soviet Union and China." This year, the Sino-Soviet Cultural Association was formally established in Nanjing and regularly published the magazine

"*Sino-Soviet Culture*," which carried out a series of influential activities for the cultural exchanges between the two countries.

At the same time, Britain, Czechoslovakia, and Poland were also holding exhibitions of Chinese paintings, and in February 1935, the exhibition of modern Chinese art opened in London's Burlington House, which was well attended and refreshing to British ears. Liu Haisu was invited by the Chinese Society of London to report on "*Trends in Modern Chinese Art*" on 26 February, with Xiong Shiyi as the English interpreter; in March, the Czech Oriental Institute and the National Museum of Fine Arts, with the assistance of the Chinese Embassy, organized an exhibition of Chinese paintings, displaying Liu Haisu's works in the exhibitions of Berlin and Geneva, and the Chinese artworks collected in Prague to take part in the display, which was later moved to the Polish capital, Warsaw, for display. Later it was moved to Warsaw, the capital of Poland. The success of the exhibition led the Chinese authorities to decide to organize the largest international exhibition of its kind, which led to the "International Exhibition of Chinese Art" in London. This led to the launch of the "International Exhibition of Chinese Art" in London.

### *Mei Lanfang's Peking Opera Was a Hit in the U.S. and the U.S.S.R.*

In addition to the plastic arts, the performing arts play an important role in the visual arts and often leave a deep impression on the audience through the contagion of dynamic beauty. In Chinese stage art, Peking Opera, which combines music, dance, martial arts, rap, performance, art, and costumes, has a large audience in China and is called "Chinese opera" by foreigners. It is a theater with multicultural factors and deserves to be a window into China's performing arts. At the end of the Qing Dynasty, when Chinese classical dance was on the verge of extinction, Mei Lanfang (1894–1961) rose to discover the cultural heritage, innovate Peking Opera, establish the Mei School of Drama, and give Peking Opera a new splendor. Mei Lanfang learned the art at the age of 8 and made his debut at the age of 11 as a Qing Yi. Mei Lanfang, who was in his prime, because Chinese music had fallen into disrepute and ancient dances had been lost, wrote, sang, and acted in many new operas, forming new ancient costumes and new songs, and his sword dances, raincoat dances, and feather dances were particularly successful in both the ancient and modern worlds, making him the best of all time.

In 1919 and 1924 Mei Lanfang successively led the troupe to perform in Japan, the Tokyo ladies and gentlemen of the streets to see Mei Lanfang's play, and the Japanese dance community rallied to follow suit, and the name of the plum dance spread throughout Japan. Later, Mei Lanfang took his troupe to Hong Kong to perform, and the local people in Europe and America were overwhelmed by the performance, which caused a sensation even more than that in Japan. Beijing's envoys in China, such as U.S. Minister Klein coming to China, expressed to President Hoover, could inviting Mei Lanfang to the United States, through his art to show the charm of Chinese culture, would be able to enhance the hearts of the American people adore China. Mei Lanfang, who was famous in the East, also intended to study Western theater in Western civilized countries, so that Chinese traditional art could be developed and improved by the times.

Mei Lanfang founded the Chinese Academy of Theatre Arts in Peking at that time, to reform Chinese opera, and Mei Lanfang went abroad to study in the name of the Academy's representative. At that time, Qi Shan, Professor of Theatre at Peking University, and Mei Lanfang co-edited *The Origin and Flow of Chinese Theatre Organisations* (English version) with Mei Lanfang's biography and scripts, and Mei's *Examination of the Plays Prepared for Performance in the U.S.A.* (English version), which was used as a popularity pamphlet. The troupe compiled line scores, face scores, instrument scores, and stage diagrams in English and Chinese, which could be publicly displayed to enable Americans to understand Chinese theater arts.

On January 18, 1930, the Mei Lanfang Peking Opera Troupe, consisting of 24 members, embarked on the Canadian Empress from Shanghai to the United States for an inspection tour and to perform Peking Opera. After arriving in Seattle, the troupe took a train to New York on February 8th. On February 14th, Mei Lanfang went to Washington D.C. for the inaugural performance, where he showcased three acts: "Chang'e Running to the Moon", "Duel on the Green Stone Mountain", and "A Thousand Gold for a Smile". Many government officials and dignitaries, including the Vice President, attended the performance. Starting from February 18th, the troupe officially opened at the Broadway Theater on 49th Street in New York, with the first performances of "Fen River Bay" and "The Palace Lady's Tiger-Killing" being a great success. Mei Lanfang took more than ten curtain calls at the audience's request, and the tickets for the two-week run were quickly sold out. Departing from New York, the troupe traveled through Chicago, San Francisco, Los Angeles, San Diego, and Honolulu, performing for a total of 72 days. Wherever they went, they were warmly welcomed. Mei Lanfang's performances allowed the American public to instantly find in Peking Opera the distinct elements of opera, dance, and drama that are separate in the West. The art of Peking Opera performance provided a vivid blueprint for the American film industry, which was transitioning from silent to sound films and exploring how to skillfully integrate speech, singing, dancing, and martial arts into a cohesive whole. Hollywood greatly benefited from this, and soon after, the rise of living newspapers in the United States was also inspired by the expressive movements of Mei's performances. On January 18, 1930, 24 members of the Mei Theatre Troupe traveled from Shanghai on the Queen of Canada mailboat to the United States to study and perform Peking Opera. After arriving in Seattle, the troupe took the train to New York on February 8, and on February 14, Mei Lanfang went to Washington, D.C., to give the first reception performance, with the following programs.

With rich sets, colorful costumes, beautiful singing, and mesmerizing dances, the American public was caught up in an intoxicating frenzy for three months. The Peking Operas with strong Chinese colors, such as "The Swinging of the Lush Flowers," "The Drunkenness of the Concubine," "The God of Luoshen," "The Fisherman Kills His Family," and "Farewell My Concubine," etc., unexpectedly won the admiration of the American public, who in the past had only known Opera, Ballet, and Drama. The American public, who used to know only opera, ballet, and drama, admired and appreciated them. Mei Lanfang became a big hit in the United States in 1930. Hollywood stars, directors, producers, and theater critics were all eager to meet Mei. Columbia University, the University of Chicago, and the University of San Francisco held symposiums to explore the mysteries of Peking Opera and invited Mei Lanfang to lecture on the art of Peking Opera to show the academic community's admiration for traditional Chinese art. Pomona College in Los Angeles and the University of Southern California competed to award Mei Lanfang the degree of Doctor of Literature in recognition of his work in bridging East and West.

The Chinese Peking Opera, represented by the Mei Opera, has now stepped out of the country and is officially included in the world's theater arts. Mei Lanfang and his creation of the Mei School of Art has become a hot topic in American public opinion, the community also appeared Mei

Lanfang flowers, and Mei Lanfang song, showing that the Americans on the Chinese Peking Opera frenzy is no less than the Italian opera star Eleonora Duse (Eleonora Duse) traveled to the U.S. performances.

In 1933, the British literary giant George Bernard Shaw arrived in Shanghai on his voyage around the world and also met Mei Lanfang, expressed his hope that the Mei Opera Troupe would come to the UK to perform. This wish was not fulfilled, but two years later, the invitation from the Soviet Union's Foreign Cultural Association to the Mei Opera Troupe to perform in Moscow and St. Petersburg finally became a reality. February 22, 1935, the troupe, a group of 22 people, with six plays and dances from *Xi Shi* and *Ma Gu Xian Shou*, boarded the Soviet Union's special Northern liner from Shanghai to Hai Shen Temple, and then transferred to Siberian Railway, arriving in Moscow on 22 March. The arts, cultural organizations, and the media held a reception performance, packed 1500 people in the Concert Hall music pool, to watch Mei Lanfang perform two plays "*Stabbing the Tiger*" and "*Fenhewan*". 23 evening of the official public performance, before the performance of the Ambassador to the Soviet Union Yan Huiqing, made a "spirit of the Chinese theatre" speech. After the meeting, Mei Lanfang performed the plays of the previous day, and all the ambassadors were present to watch them. 6 days later, the troupe went to St. Petersburg to continue the performance, and 16,000 audiences watched the performance in 8 days. 13 April, the troupe gave a farewell performance in the Moscow Bolshoi Theatre, and performed the three plays of "*The Night Runner,*" "*The Fishers and the Killers,*" and "*Rainbow Pass,*" with Mei Lanfang starring in the latter two plays. Mei drama also stirred up a sensation in Moscow, Stalin, and Gorky also watched the farewell performance, after the end of the curtain as many as 18 times, the opening of the Bolshoi Theatre stage performances never before the highest record. Mei troupe was warmly received by the Moscow Theatre Association, the First Art Theatre Director Stanislavsky and Meyerhold, Nemirovich Danchenko, and film director Eisenstein were members of the reception committee, and Mei Lanfang had cordial exchanges. *Rainbow Pass* was made into a film by Eisenstein.Please check the sentence "23 evening of the official public performance, before the performance..." for clarity.On February 22, 1935, a troupe of 22 people, carrying six plays and dances from "Xi Shi" and "Magu's Birthday Present", departed from Shanghai to Vladivostok. They arrived in Moscow on March 22. A reception performance for the art community, cultural groups, and media was held in a

150-seat concert hall, where Mei Lanfang performed the plays "Tiger-Killing" and "Fenhe Bay". On the evening of the 23rd, the official public performance took place, preceded by a speech on "The Spirit of Chinese Drama"by the Chinese Ambassador to the Soviet Union, Yan Huqing. Mei Lanfang performed the same plays as the day before. Diplomats from various countries attended the performance. Six days later, the troupe continued their performances in St. Petersburg, where 16,000 people watched over eight days. On April 13, the troupe gave a farewell performance at the Moscow Grand Theatre, performing three plays: "Night Run", "A Fisherman's Struggle" and "Rainbow Pass", with Mei Lanfang starring in the latter two. Joseph Vissarionovich Stalin and Maxim Gorky watched the farewell performance. After the show, there were as many as 18 curtain calls, setting an unprecedented record for the stage performances at the Grand Theatre. The Mei Troupe was warmly received by the Moscow Theatre Association. Konstantin Stanislavsky,the head of the Moscow Art Theatre, Vsevolod Meyerhold, Vladimir Ivanovich Nemirovich-Danchenko, and film director Sergei Eisenstein all had a cordial exchange with Mei Lanfang. The Rainbow Pass was filmed by Eisenstein.

Mei's plays also unexpectedly fascinated Bertolt Brecht (1898–1956), a German dramatist and poet living in Moscow. Brecht, who believed in Marxism and loved Chinese literature and opera, found in Mei Lanfang's Peking Opera the benchmark of his theory of dramatic expression. He pursued the "non-Aristotelian" theory of narrative theater in his theory of theater, and in contrast to the emotional approach of European theater, he focused on rational complaints, advocating a certain distance between the actors, the characters, and the audience, and a rational and scientific understanding of the realities of life as expressed in the play, and Mei Lanfang's use of ideological techniques to refine and artistically reproduce the life of the play. Mei Lanfang's use of ideological techniques to refine and artistically reproduce life on stage coincided with his admiration for the creation of such a play, which had already been raised to a high level of rationality, and he particularly appreciated "Fishing and Killing Families." He attended a symposium of dramatists in Moscow and thought that Mei Lanfang's performances were "as if there had never been a fourth wall" (D. Willett, *Brecht on the Theatre*, New York, Hill & Wang, 1964, 91). This led Brecht to realize that the "interstitial effect" would prevent the audience from entering into the play and making rational judgments about the play. This famous "interstitial effect" first appeared in an English translation of a miscellany in his *The Life and Literature* (London, 1936), which was later included in the book *Interstitial Effects in Chinese Theatre Art*, where it was written in pencil on the manuscript: "This miscellany is an observation after seeing a performance by the Mei Lanfang troupe in Moscow in the spring of 1935" (ibid., 97).

These exchanges between Chinese and European dramatists had far-reaching significance through Brecht's reform of Western theater, which was dissolved from the performing arts of Chinese Peking Opera. This fulfilled the prophecy of the American writer Thomas Dalbaut in his essay on *Mei Lanfang*: "Though Mr. Kipling had said that East was East and West was West and that these contracted pairs had never met, in Mei Lanfang they now competed." Mei Lanfang also hoped at a reception in Moscow: "Soon there will be a new art, a blend of Chinese and Western art in one furnace." Mei Lanfang was the first to introduce Chinese theater to the Western world.

The first to introduce Chinese theater to the Western world was Chen Jitong, a diplomat in Paris at the end of the Qing Dynasty. His book *Le Theatre des Chinois* (Theatre of China), written in French and published in Paris in 1886, was the first to recount the history of Chinese theater performances, with selections and translations of Chinese opera pieces. This book, of course, hardly made any impression on the French about Chinese theater. Later on, Zuker published *The Chinese Theatre* (1925), Arlington published *The Art of Chinese Theatre* (1930), Arlington and Harold Acton co-authored *Famous Chinese Plays* (1937), and Cecilia Zong published *The Wonders of Chinese Theatre Art* (1937), which gave a new understanding of the formation of the Peking Opera in China and the innovation of the contemporary performing arts, and the narrator in the Peking Opera was particularly backward and narrow in view to the art of Mei Pai, which was the first Chinese opera performed in Paris. The narration in Peking Opera was also particularly appreciated by the Western theater industry, and the integration of music, dance, and rap in the Mei School of Chinese Opera was regarded by the theater industry on both sides of the ocean as the pinnacle of the performing arts. Although the London public did not have the opportunity to listen to the Mei Opera performances in person, there was the play "*Wang Baochuan*" written and directed by Xiong Shi—in English—that they could watch, and in 1936 they were actually able to stay in the British Isles on a foreign stage for a long time, so was the charm not from the Mei School of theater? At the same time, there is also the afterglow of the London Chinese Art Exhibition, which has led to the enthusiasm of the European public for Chinese costumes and Chinese art still being maintained at one stage.

## 9.5 The Ongoing Cultural Convergence

### *Taking the Pulse of the Economic Globalization Process*

Since the establishment of the World Trade Organization Agreement on Tariffs and Trade (GATT) in 1946, the strategy of globalization has been the main objective of economically developed countries in exporting capital and technology abroad, and significant progress has been made with the conclusion of the Uruguay Round of negotiations. The economic relations between developed and developing countries have changed accordingly, with a gradual shift from the earlier international division of labor, which prevented the developing countries from industrializing, to a horizontal division of labor, with the developed countries at the center and the developing countries as their subsidiaries, and a vertical division of labor between sectors and industries. Since the end of the 1970s, developing countries have vigorously reformed their economic systems, giving full play to the dynamism of the market economy and facilitating the process of globalization. Foreign direct investment (FDI) absorbed by developing countries rose from an average of $26 billion a year in the second half of the 1980s to $97 billion in 1995, accounting for 30% of global FDI. Private investment has replaced the earlier foreign aid and foreign debt and has gradually become the main source of financing for world economic development. The World Bank's report Global Economic Prospects and the Developing Countries, published in 1996, pointed out that in the 10 years 1985–1994, the share of foreign direct investment in the gross domestic product of developing countries had doubled compared to the previous 10 years. The increase in FDI has contributed to the development of manufacturing in developing countries and the expansion of the division of labor at the international level. At the same time, international trade had grown faster than production as a result of expanding international investment and trade in manufactured goods. According to the World Bank, the growth rate of the volume of world trade as a proportion of the GDP of all countries in the decade 1985–1994 had been three times that of the previous decade. One-third of world trade in exports and another third in trade is carried out between the parent companies of multinational enterprises and their subsidiaries located abroad. In other words, the main factor dominating the development of world import and export trade is trade within transnational corporations. This new type of international division of labor has not only strengthened the world's productive capacity, but also expanded market capacity, and is forming the integration of the world economy and promoting the process of globalization of the world economy.

China, as the largest developing country, began to implement reform and opening up in the 1980s, shifted from a planned economy to a market economy, correctly dealt with the world system, grasped the development opportunities, and, like most of the East Asian countries, created an economic miracle of sustained and rapid growth, and by 1996 had risen from 11th place in the 1980s to 6th place in terms of comprehensive national power. According to the World Bank's Annual Atlas of the World Economy released on December 30, 1995, the fastest-growing Gross National Product (GNP) per capita in the world over the past 10 years was in Asia, with Thailand, South Korea, China, and Singapore. 1996 saw China's Gross Domestic Product (GDP) growth rate of 9.6%, or US$816.8 billion. By the end of 1997, China's utilization of foreign direct investment (FDI) exceeded US$200 billion, making it the second largest importer of FDI after the United States. For some time until 2005, China's GDP maintained an annual growth rate of over 8%, moving it from a low-income country to a lower-middle-income country, and in 2010 China's GDP surpassed Japan's, making it the world's second largest economy after that of the United States. In 2010, China's GDP surpassed Japan's and became the world's second largest economy after the United States. Many people in the West are therefore looking forward to the possibility of China becoming the world's largest economy in the next 20 or 30 years.

The changes in the international division of labor brought about by the wave of economic globalization have triggered worldwide competition between the front and the back. The transfer of industries from developed countries to developing countries has facilitated the modernization of industries in developing countries. For developing countries, there is also a demand to achieve this goal as soon as possible to change the division of labor pattern that is unfavorable to them. At the same time, some of the more advanced developing countries also seek to transfer industries that have lost their competitive advantage within their borders to relatively backward countries or regions, where the competitiveness of the latter's upgraded industries against the former will in turn force the former to upgrade further and go on to catch up with the developed countries. In East Asia, the transfer of industry and technology is particularly typical. Following the revitalization of Japan, the first Asian "Four Little Dragons" (Singapore, Hong Kong, China, Taiwan, and South Korea) in the 1970s economic take-off, followed by China, Thailand, Malaysia, and Indonesia's sustained high-speed growth. In 1986, East Asia's developing countries' access to foreign capital gradually exceeded that of Latin American countries, becoming the main input of foreign capital. After 1986, foreign investment in developing countries in East Asia gradually surpassed that in Latin America and became the main input region for foreign investment. China in the 1980s adjusted its industrial structure, beginning from a labor-intensive to a technology-intensive knowledge-intensive, capital-intensive industry transition, the rapid growth of economic scale, has made the East Asian economy achieve marketization, and integration process an indispensable member.

In 1973, with mankind for the first time 10,000 components mounted on 1 square centimeter integrated circuits, a step from the electronic age into the information age, mankind began to enter an era of high economic growth marked by scientific and technological progress from the industrialized societies, the United Nations research institutes in 1990, this economy is called the knowledge-based economy, the Asia–Pacific Economic Cooperation (APEC) in 1996, more explicitly for the "Knowledge-based economy" is an economy with high-tech industry as its pillar. The high-tech industry, with high-tech as its most important resource, is not a simple innovation of traditional industrial technology but mainly includes information science and technology, life science and technology, marine science and technology, high-tech environmental protection, and the accompanying management science. The knowledge-based economy is a means of achieving sustainable economic development without further depletion of natural resources, relying on the integration of the world economy and the input of intangible assets. The new international division of labor has given rise to changes in the world's industrial structure and trading system. With the establishment of information technology, especially the Internet, time and space distances have been greatly reduced, and the operation of enterprises and the development of markets have shifted from a national to a global scale so that the economy of any country has to be oriented toward the world and be integrated into the orbit of global economic development. The

World GATT has been replaced by the World Trade Organisation (WTO) since January 1, 1995, and China has been admitted as a full member since November 11, 2001 after a long period of negotiation. China must, for a considerable period in the future, change the purpose of technology introduction from production and use and "import substitution" to digestion and innovation and participation in international cooperation; change the mode of introduction from the mere import of production lines to the introduction of soft technologies and necessary key equipment; and change the object of introduction from "product-oriented" technologies to "product-orientated" technologies. The mode of introduction has changed from merely importing production lines to paying more attention to the introduction of soft technologies and necessary key equipment; the object of introduction has changed from mainly "product-oriented" technologies to industrial basic technologies, major technologies, and high technologies. It took the United States half a century to achieve the transition to a knowledge-based economy, with the knowledge-based economy accounting for only 9% of the total economy in the 1950s and reaching 65% in 2000; in contrast, China's knowledge-based economy still accounts for only 26% of the total economy, and there is still a long way to go in the future.

### *Chinese Culture and Cultural Convergence*

Conflicts caused by cultural differences are indeed common in history, but cultural differences do not necessarily lead to conflicts. There is another cultural phenomenon that is parallel to cultural differences, which is convergence. Compared to differences, convergence plays a more mainstream role.

The rise of the East Asian economies, especially the "Four Little Dragons" (Singapore, Hong Kong, China, Taiwan, China and South Korea), has created a miracle that has attracted world attention since the 1970s when they led the world's economic growth train with an average annual economic growth rate of more than 8%. In 1988, the declaration issued by the G7 Summit stressed the importance of the dialogue between the seven industrial powers and the "Four Little Dragons," and began to use the "Newly Industrialized Economies" (NIES) to call the "Four Little Dragons" in Asia. The "Four Little Dragons" transformed from being on par with the lower to middle-income countries of Latin America and impoverished regions of Africa in the 1950s to leaping into the ranks of the world's moderately developed countries by the end of the 1980s. Their trajectory from poverty to prosperity is clearly visible, serving as a model for latecomer nations and regions to rapidly join the ranks of the originally early-developed modern countries within a short period. The economic take-off of Japan and the "Four Little Dragons" in Asia have caused overseas academics to re-examine the social functions of traditional culture in favor of the current economic development of these countries or regions that were earlier part of the Confucian cultural circle.

Herman Kahn (1922–1983), director of the Hudson Institute in the United States, compared the Confucian ethics practiced in East Asian societies with the early Protestant ethics in the West and realized that Confucian ethics differed from Western Protestantism in highlighting the individual and personal gain, and advocated individual loyalty, dedication, and responsibility to the group, and valued the cooperation and coordination of interrelationships, and referred to modern Confucianism as the new Confucian culture. Kahn refers to Japan and the "Four Little Dragons" as Neo-Confucian societies, arguing that the key to their great achievements is "cultural factors conducive to development, excellent management of the economy, an international environment conducive to growth, technological conditions and foreign aid, as well as hard work and dedication." By nurturing dedicated and enterprising individuals with a sense of responsibility and expertise, whose loyalty to society and organizations led them to take pleasure in helping others and thus place little emphasis on personal gain, Neo-Confucianism has demonstrated its vitality in promoting modernization, higher rates of economic growth and social progress. In his 1979 book *World Economic Development: 1979 and Beyond*, Kahn argued that China, Japan, and the Four Little Dragons of Asia, as well as the ethnic Chinese groups of Thailand, Malaysia, Indonesia, and the Philippines, had an outstanding potential for promoting economic development and introducing modern technology. In the words of Leon Vandermeersch of France in *A New Chinese Cultural Circle* (1986), Confucianism has died out and will not be resurrected, but its spirit is still preserved in the system of Chinese characters, which serves as a spiritual temple, just as Christianity has declined in Western Europe, but the spirit of evangelicalism is still there. This view is very much in line with the idea of establishing a moral religion and practicing moral supremacy as advocated by Mou Zongsan, a Neo-Confucian scholar from Taiwan, in the 1970s. Although this view may not be entirely consistent with the reality of building a modern society in East Asia, it is at least consistent with the view of thousands of years of history. Although this view may not be entirely consistent with the actual construction of a modern society in East Asia, at least Chinese culture, with its thousands of years of tradition, does have certain advantages in uniting different cultures and constantly renewing and extending its own vitality; thus, the East Asian region has been able to keep abreast of the times and make remarkable achievements in catching up with the wave of globalization.

Among the 30 million overseas Chinese worldwide, 24 million in Southeast Asia played a pivotal role in the "growth triangle" economies during the mid-1980s, as international trade became increasingly interconnected. The changing industrial structures and trade systems in this new era were particularly evident in several developing Southeast Asian countries. These nations managed to overcome their vastly different cultural backgrounds, varying stages of economic development, and distinct social and economic systems to organize localized economic cooperation zones, utilizing economic complementarity among several relevant countries to form these "growth triangles." The "growth triangles" included only a small portion of the land area of the participating countries.

The emergence of these "growth triangles" was driven by the Southeast Asian Chinese business community to align with the trends of globalization, leveraging capital accumulation, industry partnerships, and enhancing market competitiveness. The "Singapore-Malaysia-Indonesia (Riau Islands)" growth triangle was the first to appear, taking advantage of the inexpensive land and labor in the Riau Islands and the low-cost semi-skilled labor in Johor, Malaysia. Finished products from Johor and the Riau Islands were marketed through Singapore's service sector with well-developed infrastructure. Subsequently, various other growth triangles promoted by Chinese enterprises emerged. These included the "North Growth Triangle" consisting of Phuket in southern Thailand, the northern Malaysian states (Penang, Kedah, Perak, and Perlis), and Aceh in Sumatra, Indonesia; the "Greater Malaysia Growth Triangle" encompassing Penang in Malaysia, southern Thailand, and Medan in Indonesia; the "East Growth Triangle" comprised of Sulawesi and East and West Kalimantan in Indonesia, Sarawak in Malaysia, Mindanao in the Philippines, and Brunei; and the "Malay-Indochina Growth Triangle" consisting of the Malaysian East Coast states of Terengganu and Pahang and the Indochinese countries of Vietnam, Cambodia, and Laos. There is also the "Philippine-Singapore-Malaysia Growth Triangle." These growth triangles have vigorously driven regional economic cooperation among the ten ASEAN nations, forming the bedrock of cultural convergence among Southeast Asian countries.

A more mature and robust growth triangle emerged between China's Hong Kong, Taiwan, and the Pearl River Delta in South China due to their economic cooperation. Regions such as Shenzhen in Guangdong Province, the Special Economic Zones of Zhuhai, and many areas in the Pearl River Delta experienced significant transformations in their economic landscape since the 1980s, attracting investment from Hong Kong and Taiwan. This led to the development of export-oriented economies, resulting in substantial changes in the economic outlook of these regions. People's living standards improved rapidly, urban development flourished, and numerous development zones sprang up along the transportation routes. The urban–rural gap further narrowed. As of the most recent rankings, the top five cities in terms of comprehensive strength include Shanghai, Beijing, Shenzhen, Guangzhou, and Tianjin. Chinese enterprises gained new vitality and dynamism through reforms and transformations, with 115 Chinese companies making it to the Fortune Global 500 list in 2017.

With the opening of mainland China, the number of students studying abroad and pursuing further education abroad has continued to expand, becoming a strategic task to cultivate advanced specialized talents and raise the nation's technological level. On January 31, 1979, the leaders of China and the United States signed a series of agreements at the White House to strengthen cultural cooperation, including comprehensive agreements on scientific and technological cooperation, cultural exchanges, and student exchanges. From 1978 to 1992, China sent over 190,000 students to nearly 100 countries and regions around the world, making China the country with the most university students studying abroad. Japan, which ranked second, had only 39,000 students studying abroad. Morocco, ranking third, had 36,000 students studying abroad. As of the end of 2000, the cumulative number of Chinese students studying abroad exceeded 320,000, with 110,000 returning to work in China. Starting from 2001, due to policy adjustments, the number of students studying abroad has been around 100,000 each year. Nearly half of the students studying abroad go to the United States. In 2013, the number of Chinese students studying in the United States reached 190,000, while the United States sent 20,000 students to China. In 2015, the number of Chinese students studying in the United States reached 270,000, accounting for one-third of the total number of foreign students studying in the United States. The exchange between China and the world plays a leading role in China's foreign cultural exchanges.

To provide a favorable entrepreneurial environment for returnees, China has established student entrepreneurship parks in more than 20 cities, where businesses founded by returnees themselves have exceeded a thousand. By the end of 2013, overseas Chinese had returned to China to start businesses, and the total number of returnees reached 300,000. These returnees are known as the "sea turtle" group, and their role in revitalizing China is growing.

The expansion of international exchanges is also evident in the significant growth in the number of foreign students coming to China. From 1998 to 2002, over a five-year period, the cumulative number of students coming to China to study reached 240,000, a 1.5-fold increase compared to the previous five years. By the end of 2008, the number of international students studying in China reached 490,000. Foreign students' majors expanded from traditional subjects such as Chinese language and traditional Chinese medicine to applied disciplines like science, technology, economics, management, and law. The number of students pursuing degree education increased year by year, with significant growth in those pursuing master's and doctoral degrees. Currently, there is a global trend of learning the Chinese (Mandarin) language. Chinese language and courses related to Chinese culture and history are offered in 109 countries worldwide. China, as a nation worth further exploration and research, has not only become the subject of study for some countries' Sinological institutions but has also introduced Chinese courses at the secondary school level in some countries to facilitate comprehensive learning and mastery of the Chinese language, thus expanding international exchanges. The number of people worldwide learning Chinese has exceeded 40 million. The world's understanding of China is also progressing. With more people proficient in Chinese, the Chinese government's initiative to establish 500 Confucius Institutes worldwide by the end of 2004 was realized by 2015, providing a convenient pathway for China to reach out to the world and the world to come closer to China. There are now Confucius Institutes in 109 countries. The world is moving toward finding a common language among its 6000 languages: peace, progress, and prosperity.

The manufacturer's authorised representative in the EU is Springer Nature Customer Service Centre GmbH, Europaplatz 3, 69115 Heidelberg, Germany. If you have any concerns regarding our products, please contact ProductSafety@springernature.com

Printed and bound by CPI Group (UK) Ltd, Croydon, CR0 4YY
30/09/2026
02225286-0010